Advances in Solar Technologies

Advances in Solar Technologies

Topic Editors

Jayanta Deb Mondol
Annamaria Buonomano
Biplab Das

MDPI

Basel • Beijing • Wuhan • Barcelona • Belgrade • Novi Sad • Cluj • Manchester

Topic Editors

Jayanta Deb Mondol
Belfast School of Architecture and the Built Environment
Centre for Sustainable Technologies
Ulster University
Belfast
UK

Annamaria Buonomano
Department of Industrial Engineering
University of Naples Federico II
Naples
Italy

Biplab Das
Department of Mechanical Engineering
National Institute of Technology Silchar
Assam
India

Editorial Office
MDPI AG
Grosspeteranlage 5
4052 Basel, Switzerland

This is a reprint of the Topic, published open access by the journals *Energies* (ISSN 1996-1073), *Photonics* (ISSN 2304-6732) and *Technologies* (ISSN 2227-7080), freely accessible at: https://www.mdpi.com/topics/7D181ADKD1.

For citation purposes, cite each article independently as indicated on the article page online and as indicated below:

Lastname, A.A.; Lastname, B.B. Article Title. *Journal Name* **Year**, *Volume Number*, Page Range.

ISBN 978-3-7258-3465-5 (Hbk)
ISBN 978-3-7258-3466-2 (PDF)
https://doi.org/10.3390/books978-3-7258-3466-2

Contents

About the Editors

Jayanta Deb Mondol

Jayanta Deb Mondol is a Professor of Solar Energy Technologies and Research Director at the Belfast School of Architecture and the Built Environment, Ulster University. He previously served as Course Director for the Architectural Engineering and Energy and Building Services Engineering programs at Ulster University. He holds a BSc (Hons) and MSc in Physics from Visva-Bharati University, India, and an MTech in Energy Science and Technology from Jadavpur University, India. He earned his Ph.D. from Ulster University in 2004. Professor Mondol is a member of the Energy Institute, the Institute of Physics, and the International Solar Energy Society and a Fellow of the Higher Education Academy. His research focuses on multifunctional solar facades, energy efficiency in buildings, and net-zero carbon buildings. An experienced research leader, he has successfully coordinated numerous multidisciplinary projects and secured over £18 million in research funding from EPSRC, Invest Northern Ireland, the Royal Society, the British Council, Innovate UK, and the EU. Jayanta has served as the Work Group leader for the EU COST Action project (TU1205) and is a working group member for the Bio-Energy action plan in Northern Ireland. He is also an active member of the Climate Emergency Committee for The Royal Society of Ulster Architects (RUSA). He serves on the Editorial Board of the Journal of Renewable Energy and the Sustainable Buildings Journal and has acted as a Guest Editor for multiple special issues. A regular peer reviewer for more than 30 international journals, he has published over 140 research papers and contributed to more than 25 research reports. His contributions have been recognized with prestigious awards at international conferences.

Annamaria Buonomano

Annamaria Buonomano is an Associate Professor of Applied Thermodynamics and Heat Transfer at the Department of Industrial Engineering of the University of Naples Federico II, Italy. She received her B.Sc. and M.Sc. in Engineering Management summa cum laude, in 2004 and 2006 from the University of Naples Federico II, and a Ph.D. in Energetics from the University of Palermo in 2010. She was also a visiting scholar at the Energy Performance of Buildings Group of the Lawrence Berkeley National Laboratory (Berkeley, USA) in 2009 and a researcher at the Ben Gurion National Solar Energy Center of the Jacob Blaustein Institutes for Desert Research of the University of Ben Gurion (Sde Boqer, Israel) in 2011, to conduct studies on hybrid ventilation and concentrating photovoltaic thermal systems, respectively. Since 2015, she has been a visiting scientist at Concordia University (Montreal, Canada), where she was appointed an Affiliate Assistant Professor in 2017 in the Department of Building, Civil, and Environmental Engineering. She has received about EUR 2.0 million worth of research grants from the European Commission and the Italian Ministry of University and Research, as public and private funding to develop comprehensive studies and simulation models for the energy efficiency of buildings and transportation systems, as well as advanced simulation models for energy communities based on renewable energy sources. She has authored more than 220 refereed papers, published in international journals and conference proceedings, receiving six best paper awards. She serves as Editor of the Elsevier Journals *Renewable Energy* and *Energy Reports* and is a member of the Editorial Board of the Elsevier Journal *Energy Conversion and Management*, as well as of the MDPI Journal *Energies*.

Biplab Das

Dr. Biplab Das is currently serving as an Associate Professor at the Department of Mechanical Engineering at the National Institute of Technology (NIT) Silchar, India. He completed his Ph.D. at NERIST, India, in 2014, followed by postdoctoral research at the University of Idaho, USA, under the prestigious Bhaskara Advanced Solar Energy (BASE) Fellowship, awarded by the Indo-US Science and Technology Forum (IUSSTF) and the Department of Science and Technology (DST), Government of India. Additionally, he has been awarded a DBT Associateship by the Department of Biotechnology, Government of India, to visit Ulster University, UK, as a visiting researcher. With over 16 years of experience in teaching and research, Dr. Das has published more than 150 papers in refereed international and national journals and conferences. He is currently involved in three sponsored projects. He has completed 10 sponsored projects funded by the SERB, DST, the Ministry of Power, and the Ministry of Climate Change, Government of India. Dr. Das has supervised ten Ph.D. students in the past, and he is currently supervising seven Ph.D. students.

energies

MDPI

Review

Transformative Impacts of AI and Wireless Communication in CSP Heliostat Control Systems

Quanwu Liu, Zengli Dai, Yuan Wei, Dongxiang Wang and Yu Xie *

CSP Research Centre, Solar Power Generation Technology Research Institute, SEPCO3 Electric Power Construction Co., Ltd., Qingdao 266100, China; liuquanwu@sepco3.com (Q.L.); dzl@sepco3.com (Z.D.); yuan.wei@sepco3.com (Y.W.); wangdongxiang@sepco3.com (D.W.)
* Correspondence: xieyucsp@sepco3.com

Abstract: In this review, the transformative impact of integrating artificial intelligence (AI) and wireless communication technologies into the heliostat control systems of concentrated solar power (CSP) plants are explored. Heliostat control systems are categorized based on wired and wireless implementations, and calibration methods are analyzed from traditional methods, auxiliary equipment, and AI in detail. The applications of artificial intelligence, machine learning, and deep learning techniques enhance the accuracy, control ability, and prediction performance of CSP heliostat control systems. At the same time, wireless communications play an important role in reducing costs, enhancing scalability, and enabling more flexible deployment. The synergistic impact of AI and wireless technologies improves the efficiency, reliability, and economic viability of heliostat systems, and shows great potential in global energy transition.

Keywords: heliostat control; artificial intelligence (AI); wireless communication; heliostat calibrate

Academic Editor: Massimo Dentice D'Accadia

Received: 23 January 2025
Revised: 11 February 2025
Accepted: 20 February 2025
Published: 22 February 2025

Citation: Liu, Q.; Dai, Z.; Wei, Y.; Wang, D.; Xie, Y. Transformative Impacts of AI and Wireless Communication in CSP Heliostat Control Systems. *Energies* **2025**, *18*, 1069. https://doi.org/10.3390/en18051069

1. Introduction

The escalating demand for sustainable energy has underscored the increasing significance of CSP as a feasible substitute for fossil fuels [1,2]. The performance of the heliostat field within CSP plants, especially those utilizing the solar power tower architecture, is of paramount importance, given its function of precisely reflecting sunlight onto a central receiver [3]. These systems encompass a multitude of individual mirrors, termed heliostats, that must exhibit high tracking and alignment accuracy to effectively capture and reflect solar radiation; this is contingent upon advanced control and calibration systems. The heliostat system, constituting the linchpin of solar thermal power generation technology, comprises a heliostat, a central receiver, a heat exchanger, a circulating working fluid, a turbine, a compressor, and an energy storage system. This system meticulously directs sunlight onto the central receiver via the heliostat, thereby achieving exceptional solar energy absorption and conversion efficiencies. The captured thermal energy is subsequently relayed to the circulating working fluid via the heat exchanger, propelling the thermal cycle. This energy is then harnessed by the turbine and compressor to produce mechanical work, which is subsequently converted into electrical energy by the generator. Additionally, the energy storage system plays a pivotal role by amassing surplus thermal energy for deployment during periods of sunlight scarcity. The efficiency of the heliostat field—defined as the ratio of energy absorbed by the receiver to the incident solar energy—serves as a paramount metric for evaluating system performance. This efficiency is contingent upon various factors, including reflection and geometric characteristics, as well as occlusion losses [4]. Achieving SunShot's efficiency goals for concentrated solar power requires advancements in molten salt (MS) technology. This includes increasing operating temperatures to 650–750 °C and exploring alternative salt chemistries like chlorides or carbonates. Future MS power tower designs (Figure 1) envision incorporating dual thermal energy storage (TES) sections and a flexible power module. Current systems use solar salt at 565 °C coupled with a steam Rankine cycle; the proposed systems would instead leverage higher-temperature salts and a supercritical CO_2 (sCO_2) Brayton power cycle for improved performance [5].

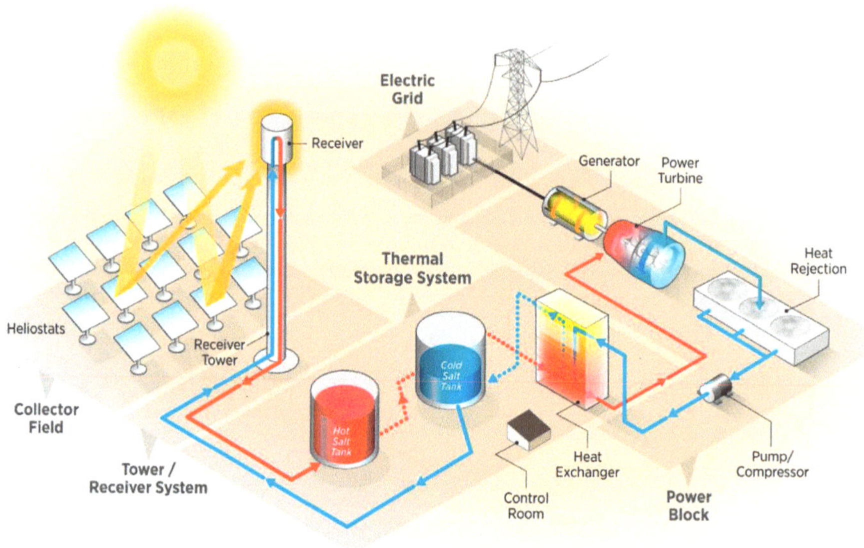

Figure 1. Molten salt power tower with direct storage of salt. Current and advanced salt designs are conceptually similar but future designs envision higher salt temperatures with a sCO_2 Brayton power cycle [5].

Conventional control mechanisms, predicated on wired networks and manual calibration methodologies, exhibit substantial limitations regarding cost, flexibility, and scalability, thereby necessitating a search for superior alternatives to augment the overall performance of CSP technology [6–9]. Artificial intelligence (AI) offers a promising approach to controlling solar heliostats by leveraging machine learning for enhanced adaptability. AI-driven control systems can learn from historical data to predict and compensate for environmental disturbances (e.g., wind, temperature) and device-specific variations, leading to improved tracking accuracy. Furthermore, AI algorithms can optimize control strategies to maximize energy production. The advancement of artificial intelligence (AI) and wireless communication technology provides promising alternative solutions for addressing these limitations. Artificial intelligence facilitates more complex and adaptive control methods, while wireless communication enables more cost-effective and adaptable implementations. The synergistic integration of these technologies has laid a novel trajectory toward more efficacious, robust, and economical heliostat systems [10–13].

This review endeavors to furnish a thorough analysis of the extant literature concerning heliostat control systems, control systems are divided into wired control systems and wireless control systems based on communication methods, and three main calibration methods for heliostats are introduced: traditional calibration, manual light source calibration, and AI-/data-driven calibration methods. There is a focus on the utilization of artificial intelligence for meticulous calibration as well as the application of wireless technologies in cost-effective and adaptable control. A detailed analysis of present control methodologies, calibration techniques, and wireless communication strategies is undertaken.

2. Heliostat Control System

A heliostat control system is crucial for tracking solar radiation and directing it to the central receiver. According to its communication method, it can be roughly divided into two categories: wired and wireless.

2.1. Wired Heliostat Control System

In traditional wired control systems, each heliostat requires a physical cable connection to the central control system. The high installation and maintenance costs and limited layout flexibility of wired systems pose challenges for the implementation of large and complex heliostat fields [6,9], as shown in Figure 2.

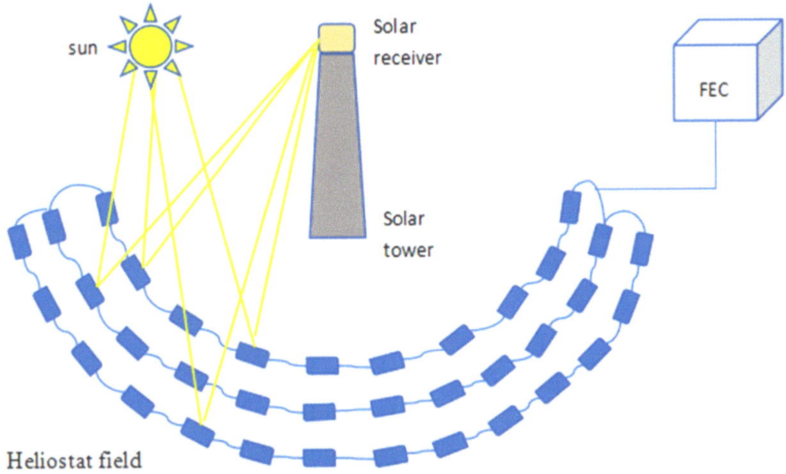

Figure 2. Schematic representation of a conventional wired heliostat control system in a solar power tower plant.

Furthermore, optical detection, mechanical installation, and other factors can influence the stability and tracking accuracy of heliostat control systems in CSP plants [8,14]. These factors can greatly limit the applicability of such control systems in certain application scenarios [15]. Therefore, future improvements in solar tracking technologies should focus on reducing costs, simplifying system design, and ensuring stability in various environmental conditions while enhancing efficiency. Control methods can be categorized into adaptive control, model predictive control, and master–slave control.

2.1.1. Adaptive Control System

Concentrated solar power (CSP) utilizes a heliostat array to focus sunlight onto a thermal receiver, transferring heat energy through a high-temperature medium to drive a generator for electricity production [16]. Traditional control methods often struggle to adapt to these variations due to the fluctuating nature of solar irradiance, weather conditions, and equipment performance. Adaptive control can deal with environmental changes, variations in system dynamics, and uncertainties, thereby enhancing the stability, reliability, and efficiency of system operation [17]. Therefore, the application of this control method in solar thermal power generation is of significant importance. However, adaptive control strategies can require significant computational resources and careful adjustments, which may present challenges in practical implementation.

The control system described by S. J. Freeman et al. incorporates a combination of inertial measurement units (IMUs), piezoelectric actuators, photodiodes, and a fast Fourier transform (FFT) analysis system to effect precise direction of sunlight via heliostats toward a central receiver [6]. The implementation of piezoelectric actuators and IMUs provides a scalable solution for large-scale solar thermal projects. Through experiments, it was found that the reflector angle error range of the system is between 0.52% and 0.65%. However, practical deployment mandates complex signal processing and control logic to address initial calibration, thereby substantially augmenting the attendant technical challenges. Further research is needed to address the challenges of real-time implementation and robustness in highly variable conditions.

Gamra et al. proposed an adaptive fuzzy logic control (AFLC) method, which is particularly suitable for nonlinear, uncertain, and adaptive systems, capable of providing robust performance under parameter variations. Unlike traditional control methods, fuzzy logic control (FLC) does not require a mathematical model of the controlled system; its rules can be expressed based on a series of logical statements [17]. This method dynamically adjusts the parameters of a proportional–integral (PI) controller in real time through fuzzy rules, ensuring the system's self-adaptive capability to parameter changes. Specifically, the inputs to the fuzzy supervisory controller include the error and its derivative, while the outputs are the gain parameters of the PI controller. In comparison to the traditional PI controller, the adaptive fuzzy logic controller-enhanced PI (AFLC-PI) exhibits superior performance in terms of response time. Specifically, the AFLC-PI surpasses the PI controller's response time of 5.55 s. Furthermore, it demonstrates a notable advantage in reducing overshoot by 6.66%. Additionally, the peak time of the AFLC-PI is also improved, being shortened by 2.96 s compared to the PI controller. Notably, both controllers achieve zero static error, indicating their effectiveness in maintaining steady-state accuracy. These findings highlight the advantages of the AFLC-PI in enhancing the dynamic performance of control systems.

Based on the review by Nsengiyumva et al., as shown in Table 1, each tracking strategy has its own applicable scenarios and limitations, and requires careful consideration of cost, accuracy, and environmental adaptability. Due to their high precision and composite characteristics, the combined systems have more advantages in applications requiring high precision [18].

Table 1. The advantages and disadvantages of tracking system category.

Tracking System Category	Description	Advantages	Disadvantages
Time- and Date-Based Tracking Systems	Uses pre-programmed algorithms based on time and location.	Generally lower cost, does not require complex sensors or control devices. 2. Can achieve precise daily and seasonal solar tracking via programming.	1. Unable to adapt to real-time weather or environmental changes, such as cloud cover. 2. Tracking errors may occur if programming errors or time calculation errors exist.

Table 1. *Cont.*

Tracking System Category	Description	Advantages	Disadvantages
Microprocessor- and Electronic Optical Sensor-Based Tracking Systems	Uses sensors that detect differences in sunlight to track the sun.	Ensures precise tracking of the sun through real-time detection of the sun's position by sensors. 2. Highly adaptable, able to adjust rapidly according to environmental changes.	1. High cost due to the need for more sophisticated electronic components. 2. System performance is susceptible to sensor failures.
Combination Trackers	Integrates sensors with date-/time-based algorithms.	Achieves a balance between cost and performance.	Requires complex integration.

2.1.2. Model Predictive Control (MPC)

Model predictive control (MPC) is an advanced control strategy that utilizes a dynamic model of the system to achieve precise control by optimizing future control inputs at each control instant [19,20]. In solar thermal power systems, MPC is widely employed to enhance operational efficiency, stability, and reliability. When external control parameters exhibit dynamic variations and uncertainties, MPC can maintain accurate tracking through prediction and error correction [21,22]. As shown in Table 2, each MPC system strategy has its own applicable scenarios and limitations.

Table 2. The advantages and disadvantages of MPC tracking system category.

MPC System Category	Description	Advantages	Disadvantages
Linear MPC	Uses linear models of the system. Often used in simpler scenarios.	1. Computationally feasible for real-time control; lower hardware costs. 2. Can be implemented in complex systems with several components and devices.	1. Requires a linear model, may not be accurate for highly nonlinear systems or wide operating ranges. 2. Performance depends on the accuracy of weather forecasts (solar irradiance, temperature, load demand).
Gray box MPC	Combines physics models with data-driven methods.	1. Improved accuracy in modeling complex dynamics (solar field efficiency, receiver heat losses, thermal storage dynamics, turbine efficiency). 2. Enhanced robustness to disturbances and uncertainties (solar irradiance fluctuations, ambient temperature variations, load demand variations, equipment degradation).	1. Can be more complex than simpler white box (physics-based only) or black box (data-driven only) models. 2. Still depends on the quality and quantity of available data.
Data-driven MPC	Uses data to build the system model, especially when a physics-based model is difficult to obtain.	Can adapt to changing system conditions and performance degradation over time by retraining the data-driven model with new data.	Susceptible to overfitting if the model is too complex or the training data are limited. Regularization and validation techniques are crucial to prevent overfitting.

Soo Too et al. designed a transient solar thermal model with a matrix controller that integrates a heliostat aiming control strategy, employing dynamic matrix control (DMC) for multi-input, multi-output systems to address transient issues in solar radiation [23].

This model addresses solar irradiance fluctuations by dynamically adjusting the aiming of heliostat in real time, thereby reducing system stress and material wear, while simplifying optical and thermal aspects to accelerate simulation speeds. By addressing various interactions among heliostats, the model achieves more precise control compared to standard open-loop methods. However, the model has certain limitations in handling extreme conditions, particularly under rapidly changing cloud cover, where some assumptions may affect its accuracy.

García et al. proposed a closed-loop aiming control strategy model based on DMC which can disperse and alter the target points of focused heliostats according to real-time direct normal irradiance (DNI) changes caused by cloud movement [24]. This model demonstrates significant advantages in handling dynamic responses by adjusting control to circumvent transient overshoots caused by rapid DNI changes due to cloud cover. Under basic comparison, the N-S cloud disturbance model uses a basic control strategy, resulting in a heat loss of approximately 0.65% due to southward-moving clouds, and an energy loss of 1.13% using this model. It has been validated for its adaptability and effectiveness in various cloud movement scenarios, but the implementation of a DMC strategy with a large number of tuning parameters is computationally complex.

Abreu et al. developed a predictive model for circumsolar normal irradiance (CSNI) at varying half-opening angles and explored its effects on CSP system performance specifically under clear sky conditions. The model was created and tested using data from five globally distributed sites, utilizing the libRadtran radiative transfer model (RTM) to simulate both DNI and CSNI datasets. A polynomial relationship was then formulated to predict CSNI, and its influence on energy capture and interception factors within CSP systems was subsequently examined. This approach directly addresses the discrepancy between measured irradiance and the actual irradiance received by solar energy systems. This difference, it was found, arises from the large half-opening angles of traditional pyranometers. Statistical analysis reveals that, for an opening angle of 0.8°, the model exhibits a range of relative Mean Bias Error (rMBE) values between −1.96% and −10.16% at the test site. In contrast, other models demonstrate a broader range of rMBE values, spanning from −3.69% to −49.26% [21]. This model is capable of predicting CSNI across numerous locations in diverse climate zones, even in the absence of high-quality atmospheric data. However, the effects of varying aerosols and atmospheric characteristics were found to substantially influence both model fitting and predictive accuracy. Furthermore, the somewhat limited scope of this study, utilizing data from only a small number of sites, could potentially limit the model's general applicability across various climatic regions.

Sarr et al. explored adaptive neuro-fuzzy inference system (ANFIS) and artificial neural network (ANN) predictive models for heliostat tracking errors in solar power tower systems. The models were constructed from experimental data, leveraging temporal parameters, with date and time, as inputs to estimate the reflected beam location on both the elevation and azimuth axes. When compared to both the artificial neural network (ANN) model and traditional geometric methods, the adaptive neuro-fuzzy inference system (ANFIS) model demonstrates an enhanced predictive capability. Specifically, the ANFIS model yields average Mean Squared Error (MSE) values of 0.028 and 0.037 on the respective axes, outperforming the ANN model (with MSE values of 0.06 and 0.134) and the geometric model (with MSE values of 0.16 and 0.208). Additionally, in terms of the coefficient of determination (R^2), the ANFIS model exhibits robust performance, achieving an R^2 value of 0.97, accompanied by root mean square error (RMSE) values of 0.167 for altitude and 0.192 for azimuth. In contrast, the ANN model yields lower R^2 values of 0.964 and 0.925, with corresponding RMSE values of 0.244 and 0.366, respectively. Furthermore, the traditional model demonstrates an even lower R^2 value of 0.8, coupled with RMSE values of 0.4 and

0.456 [25]. Despite these achievements, this work is not without limitations, including a reliance on data from limited periods and specific locations, which could hinder its widespread use in different geographic regions and variable seasonal conditions, and the use of temporal parameters as inputs might necessitate additional influencing variables, such as specific heliostat locations, to improve overall model application.

2.1.3. Master–Slave Control

Master–slave control is frequently employed for the coordinated control of multiple heliostats or solar receivers. These systems optimize the operation and performance of solar thermal power systems and simplify the control process by dividing the heliostat field into clusters controlled by one or more master heliostats, thus establishing a hierarchical relationship between master and slave control systems.

Xie et al. divided the entire heliostat field into several clusters, assigning a high-precision closed-loop control master heliostat to each cluster, with the remaining heliostats acting as followers. Within the context of the designed affine group, the primary solar reflector demonstrates exceptional tracking accuracy, approaching zero error under cluster control conditions. Furthermore, when employing the master–slave control approach, the pitch angle deviation of the master heliostat remains confined within ± 0.02 radians, while the azimuth angle deviation stays within ± 0.03 radians. Simulation studies have affirmed that, even in the presence of external disturbances, the calibration of the master heliostat maintains a high degree of precision, enabling rotations with minimal error. This strategy enhances the automation level of solar power towers while reducing control costs and computational complexity [26]. The proposed master–follower strategy reduces closed-loop control costs by minimizing sensor deployment, thereby improving heliostat field control accuracy and energy efficiency, and enhancing system scalability and disturbance rejection capabilities. However, the system still faces limitations due to potential cumulative communication errors and a lack of comprehensive empirical studies.

Röger et al. investigated the application of a master–slave control architecture in a linear moving bar system within an external receiver configuration for solar power generation, specifically focusing on the precise control of heliostats, which direct concentrated solar radiation onto a receiver. The control method facilitates real-time adjustment of heliostat aiming points to optimize flux distribution at the receiver and thus improve collection efficiency. The proposed approach leverages a unique main axis module design to enhance stability, reduce dynamic loads, and improve operational safety [27]. This method exhibits several advantages, including enhanced operational flexibility, improved system safety, demonstrated high measurement precision that remains stable under elevated temperatures, and superior scalability for larger receivers when compared to rotational systems through the implementation of a master–slave control mode for a linear bar system. However, limitations are present, including elevated technical complexity, which increases installation and maintenance costs, as well as higher demands on the technical expertise of personnel for system setup and operation.

Ahlbrink et al. introduced a hybrid master–slave control system for central receiver solar power plants, integrating Dymola/Modelica for detailed air loop modeling, STRAL for precise optical simulations, and MATLAB/Simulink Version R2008b for developing MPC. This architecture employs TCP/IP and TISC for distributed simulation, enabling specialized modeling within each software environment [28]. The system benefits from modularity, scalability, high-fidelity optical modeling, and predictive control capabilities. However, it faces challenges related to computational overhead, lack of experimental validation, and the development of complex dynamic control strategies. Further research is needed to validate its practical application.

2.2. Wireless Heliostat Control System

Traditional wired networks, due to costs associated with road excavation, cabling labor, and land usage, cannot satisfy low-cost requirements [29]. Wireless communication, as an alternative solution, can reduce cabling costs, decrease installation and maintenance complexity, and provide a more cost-effective solution [30]. Wireless communication provides an efficient and flexible means of communication for solar power generation and concentrated solar power systems [31]. Wireless communication technologies enable heliostats to transmit data in real time, allowing for status monitoring, control feedback, and error calibration without reliance on complex wired connection systems.

Modern wireless communication systems for heliostats commonly employ technologies such as Wi-Fi, Bluetooth, Zigbee, LoRaWAN, and 5G. The reliable data transmission and control capabilities provided by wireless communication not only enhance the automation level of heliostats but also optimize their real-time adjustment and precise tracking capabilities. Heliostat sensor data can be transmitted wirelessly in real time to the control center for immediate analysis and processing, thereby enabling precise calibration of the heliostats and ensuring optimal system performance. Figure 3 illustrates a schematic diagram of the central controller and heliostat wireless communication system in a tower power plant, where the local control system controls the movement of the heliostat through wireless nodes. The wireless heliostat field control system consists of a central control unit, a local controller, and a heartbeat and status message system that ensures continuous signal transmission.

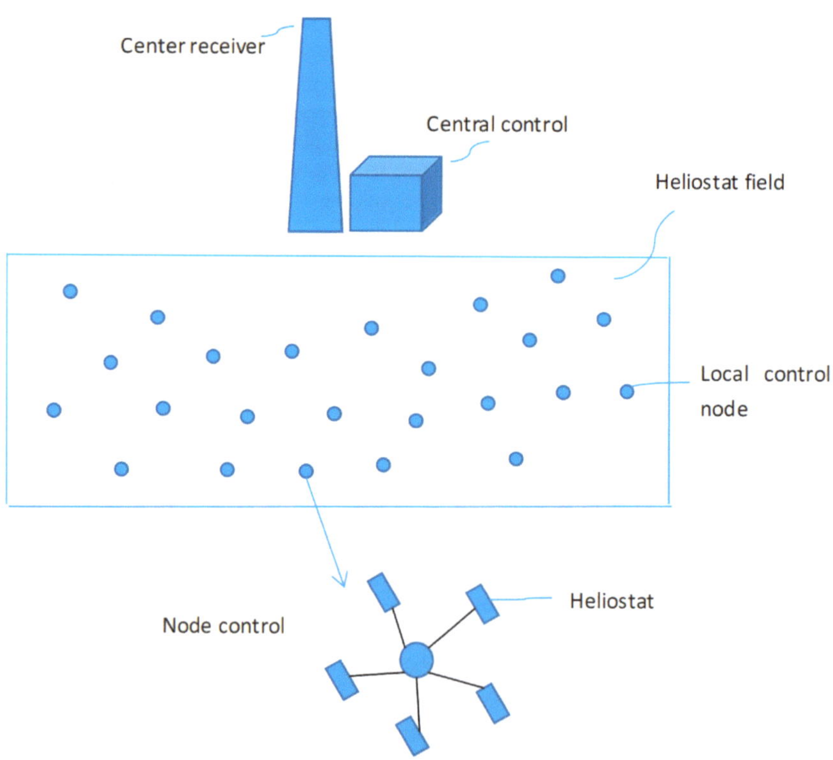

Figure 3. Schematic diagram of wireless system control nodes.

2.2.1. Low-Power Wireless Network

Technologies such as ZigBee, LoRa, Bluetooth, and 5G offer low-power solutions with varying degrees of bandwidth, transmission distance, and reliability. These technologies are suitable for low-data-rate applications in wide-area deployments, where low energy consumption is required [32]. Table 3 shows the commonly used wireless communication protocols for CSP heliostat fields. Wireless communication plays a crucial role in the operational reliability of CSP plants. These facilities often incorporate extensive heliostat fields, comprising hundreds to thousands of individually controlled mirrors, which demand precise and synchronized movement to accurately concentrate solar radiation onto a central receiver. Consequently, failures in the control system of even a single heliostat can significantly impact the overall efficiency of the solar energy capture process. Furthermore, the high concentration ratio achieved by heliostats necessitates robust safety mechanisms. Erroneous positioning or control signals can lead to excessive heat flux, potentially resulting in component overheating, equipment damage, or even posing a fire hazard.

Table 3. Comparison of wireless communication technologies for heliostat control.

Communications Technology	Working Frequency Band	Communication Distance
Wi-Fi 6/6E	2.4 GHz, 5 GHz/6 GHz	The maximum communication distance is 120 m
Zigbee	868.0–868.6 MHz (Europe), 902–928 MHz (North America), and 2400–2483.5 MHz (global)	Communication range is 10–100 m
LoRa (Long Range)	863–870/873 MHz (Europe), 902–928 MHz (North America)	Supports long-distance low-power transmission from 4.8 km (urban) to 16 km (rural)
Bluetooth	between 2.402 GHz and 2.480 GHz.	Communication range is 10–100 m
5G	24 GHz, 28 GHz, 37 GHz, 39 GHz, 47 GHz	Communication range is 1–10 km or more

Shariff et al. designed a system for monitoring and controlling power plant data using a PIC18F553 microcontroller, an LM35 temperature sensor, a solar irradiance pyranometer, and LEM voltage and current sensors. The system employs a point-to-point Zigbee wireless communication network to connect components distributed throughout the facility [33]. While Zigbee offers a low-power solution, its lower data transmission rate compared to other wireless technologies, such as Wi-Fi, could limit its applicability in situations requiring high data transfer volumes. In addition, Zigbee's signal penetration is less robust in complex indoor environments, even though outdoor transmission ranges can reach up to 1.5 km.

Paredes-Parra et al. have proposed an IoT solution based on LoRa technology, designed to monitor distributed solar power systems, facilitating real-time surveillance over a large area, including the acquisition of electrical and meteorological data, to ensure the stability and reliability of the systems. LoRa technology exhibited variations in performance based on distinct spreading factors (SFs), prompting the conducting of two experimental sets. Notably, the packet transmission rate achieved with SF12 reached a notable level of 91%, marking a substantial enhancement when compared to the 55% rate observed with SF11 [34]. This cost-effective wireless solution enables long-range communication, characterized by low power consumption and reduced maintenance requirements. The limited bandwidth, low data rates, and duty cycle restrictions pose limitations when applying LoRa to scenarios involving large data transmissions.

The Solar Thermal Research Group at Stellenbosch University has developed a modular HelioPod technology, which employs a range of hardware, including 802.11n standard

(Wi-Fi 4) wireless access points and Raspberry Pi devices, as well as software, such as the ZeroMQ message library, to measure performance metrics like communication latency and throughput under varying capacity, distance, and heliostat density conditions across different testing environments. During long-distance testing, specifically within a range of 110 m involving a single Local Control Unit (LCU), a substantial signal attenuation was observed, leading to elevated packet loss rates and a subsequent decline in node performance. Furthermore, Round-Trip Time (RTT) exhibited a positive correlation with network capacity, density, and range. Notably, the masking effect exacerbated the RTT latency, contributing an additional 13.6% increase [12]. This system validates the feasibility of employing low-power wireless communication systems in Concentrated Solar Thermal (CST) fields, particularly for wireless communication within modular heliostat fields.

2.2.2. Wireless Tracking Control System

Wireless tracking control systems utilize radio frequency communication to transmit data and control signals between heliostats and a central control unit, resulting in benefits including lower capital expenditures, improved operational flexibility, and diminished maintenance requirements [13,35,36].

HELIOCOMM facilitates optimization of the energy efficiency of heliostast, achieves real-time closed-loop control, and provides support for larger-scale heliostat arrays through the utilization of Wi-Fi 6 and multi-unit architectures, thus enhancing overall system performance. By implementing shared nodes as a strategy to enhance the efficiency of wireless communication, and through the adoption of a multicellular architecture for facilitating communication within the 5 GHz frequency band, it is possible to achieve approximately 42% reductions in power consumption and control-related costs [7]. The implementation of wireless control technologies continues to present challenges, particularly with regard to stability and reliability. Wireless communication is inherently susceptible to interference, and data security necessitates robust reinforcement. Furthermore, photovoltaic panels and batteries integrated into heliostats may experience compromised energy harvesting due to weather variations, potentially leading to communication disruptions.

The wireless control system architecture for multi-tower concentrated solar power (CSP) plants features wireless multifunctional nodes installed atop each receiver tower, forming the backbone network. These nodes communicate with solar reflectors and sensors and are connected to a central control server. This architecture offers high flexibility and scalability. The wireless devices include Wi-Fi, wireless Ethernet, and ISA-100.11a-compatible multifunctional nodes; wireless sensors for temperature, pressure, and differential pressure/flow measurements with an accuracy of ±0.075% for pressure and differential pressure/flow; and wireless adapters for converting wired signals to wireless. A meteorological station periodically collects environmental data via wireless transmission for optimizing plant operations [37]. The advantages of the wireless control system lie in significantly reducing cable requirements, thereby lowering construction costs, shortening construction cycles, and minimizing mechanical failures. Furthermore, the flexibility and real-time monitoring capabilities of the wireless system enhance the operational efficiency of the power plant. However, wireless systems also face challenges such as signal interference, security concerns, and maintenance complexity. Despite being equipped with internal firewalls and security management software, the security of the wireless network requires continuous attention.

The Solar Thermal Research Group at Stellenbosch University undertook an investigation of a wireless control system specifically designed for the Helio40 heliostat array. The system adopts a model-based, real-time error correction technique, which, through rotation, translation, and the tuning of adjustment coefficients, optimized daily tracking precision. Prior to calibration, the root mean square (RMS) error of the heliostat's normal vector was measured at 27 mrad. Subsequent application of the calibration coefficient led to a marked reduction in this error to 2.05 mrad, signifying an improvement in tracking performance by an order of magnitude. This outcome verifies the efficacy of the open-loop tracking and error correction method grounded in the model foundation. Concurrently, the Microchip MRF24J40 radio module was utilized for communication within the 2.45 GHz frequency band. This module possesses the capability to transmit 625 data packets per second and, theoretically, supports network control for up to 10,000 nodes [38]. This system demonstrates reductions in CSP costs and improvement in efficiency. However, the need for further refinements to enhance its practicality and dependability remains.

2.2.3. Wireless Network Topology

The design of wireless topologies in solar power systems plays a crucial role in enhancing the communication efficiency, reliability, and flexibility of the systems [39]. In large-scale solar power tower systems, wireless communication networks are utilized for real-time monitoring, data transmission, and the precise control of heliostats. Common wireless network topologies include star, mesh, and hybrid topologies, each offering unique advantages in terms of reliability, range, bandwidth, and latency, with their suitable application scenarios summarized in Table 4. The commonly used wireless network topology diagram for controlling the heliostat is shown in Figure 4.

Table 4. Network topology classification.

Topology Type	Application Scenarios
Mesh Topology	This network is more robust and versatile, but has a more complex network design, making it suitable for large areas with numerous obstacles.
Star Topology	Provides a simpler network topology with a central coordinator, in which each node is directly connected to the main station.
Hybrid Topology	By combining different network topologies and different communication standards, it is possible to create highly adaptable and robust systems, leveraging diverse technologies to achieve optimal network performance levels.

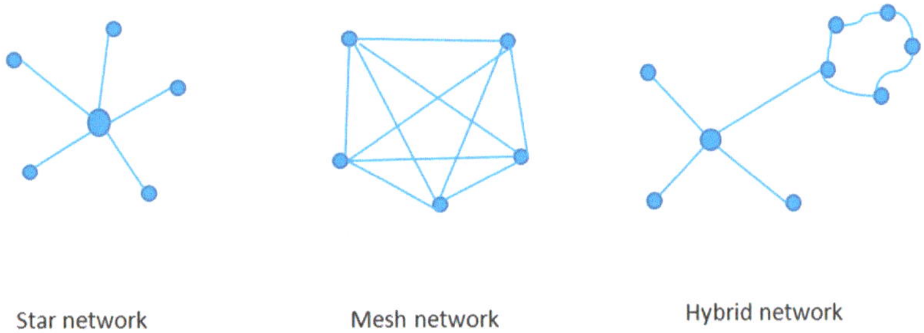

Figure 4. Illustrative diagram of common wireless network topologies for heliostat control (star, mesh, and hybrid).

The HELIOMESH project utilizes a wireless mesh network to control and monitor a solar power tower field. Each heliostat is equipped with a HelioNode module, which includes a microcontroller, an IEEE 802.15.4-2006-compliant radio transceiver, PV cells, and energy storage devices to maintain nighttime operation. The network adopts a virtual backbone structure and connection dominance set algorithm, supporting various scheduling strategies and redundant paths to enhance energy efficiency and fault tolerance. Each node in the network both transmits data and acts as a repeater, ensuring full coverage and high redundancy. Sixteen non-overlapping radio frequency channels are employed, supporting up to sixteen subnets to enhance bandwidth and reduce latency. A single broadcast can quickly cover an average of 98% of HelioNode nodes within 50 milliseconds, and the communication of the farthest node only requires up to five intermediate node relays [13]. The HELIOMESH project demonstrates an efficient, low-cost, and easy-to-maintain control scheme with enormous potential and scalability, providing feasibility for future large-scale solar power projects.

Liebenberg et al. proposed that a star network architecture offers a compelling topological solution for wireless communication networks in modular heliostat fields. This topological structure establishes connections from all end devices (EDs) to a central base station (CBS). Each device transmits solely its pertinent data and is able to transition to a low-power mode during periods of inactivity, thereby ensuring low communication latency. This structure employs a control interval of 1.5 s to attain a tracking error of no greater than 1 mrad [40]. The star topology presents a simplified approach to troubleshooting and device expansion method, thus facilitating operational ease. While the potential for a single point of failure at the central base station is a recognized consideration, its inherent simplicity, scalability, and reduced maintenance requirements render it a suitable option for heliostat applications.

The adoption of Wi-Fi 6 (IEEE 802.11ax) within a star network topology presents a highly suitable approach for the demanding network environments typical of CSP plants. Such an implementation facilitates sophisticated control of heliostats, a critical factor in maintaining precise solar tracking and achieving optimal energy efficiency [10].

Hillocom presents a multi-unit wireless control topology for heliostat fields within concentrating solar power (CSP) systems. The wireless control architecture divides a large heliostat field into multiple distinct zones, each encompassing several access points. Every access point is responsible for managing communication within its designated subset of heliostats. These access points are interconnected to a central control system by means of wired links [41]. The proposed architecture exhibits a high degree of scalability, enabling adaptation to varying field sizes that can encompass tens or hundreds of thousands of heliostats. It also offers reduced deployment costs compared to conventional wired systems. Furthermore, this architecture allows for the flexible and dynamic assignment of heliostats to access points, thus adapting to changing environmental and communication conditions while enhancing spectrum utilization through frequency reuse. The method also presents challenges associated with signal interference and obstruction, which are inherent to wireless communication. It additionally requires reliance on advanced communication technologies and protocols, increasing financial investments during both development and testing. Future work should prioritize efforts to address these signal interference problems in order to ensure the reliable performance of this system.

3. Calibration Method for Heliostats

A precise and efficient calibration system is a crucial component of heliostat control, ensuring optimal focusing of solar radiation onto the central receiver for maximum efficiency. The calibration method of a heliostat can be categorized based on various parameters, including measurement techniques and system light source types. This section will introduce three primary heliostat calibration methods used in concentrating solar power (CSP) plants: traditional calibration, artificial light source calibration, and AI-/data-driven calibration approaches. The applicability of these methods across different CSP system types will also be discussed.

3.1. Traditional Calibration Methods

3.1.1. Manual Calibration

In solar power systems, manual calibration involves adjusting the heliostat's orientation and alignment accuracy through human intervention to ensure the precise focusing of solar radiation onto the receiver [42]. Manual calibration typically employs tools such as ground targets, reflected images, or solar images for manual or semi-automated error correction [43]. While this method can provide high accuracy, it is characterized by lower efficiency and the need for regular execution, particularly when equipment performance degrades. With the increase in the size of the heliostat array, manual calibration faces challenges such as substantial workloads and time consumption, which has spurred the research and application of automated calibration techniques. Nevertheless, manual calibration still plays a crucial role in small-scale systems, particularly during initial commissioning and maintenance phases.

Elsayed et al. presented a mechanically driven single-axis solar tracking system, specifically customized through 3D cam profiles, that does not require electricity, electronic components, or specialized materials. The experimental findings indicate that the system exhibits the greatest variability in accuracy during the morning and evening hours of the day, with deviations spanning from $-3.36°$ to $1.65°$. Nevertheless, when considered over the entire year, the arithmetic mean of the absolute deviations remains below $0.5°$, while the standard deviation remains under $0.75°$ [44]. This approach, although requiring a lot of manual labor, is commonly employed in regions where advanced electronic systems are not feasible.

Table 5 summarizes various calibration methods for solar concentrator optical characterization, as discussed by Ren et al. [45], and provides a comparative analysis of their advantages and disadvantages.

Table 5. Manual calibration method.

Calibration Method	Method	Advantages	Disadvantages
Light Flux Mapping	This method experimentally maps the irradiance distribution on the receiver to characterize basic optical properties, directly capturing images of the focused solar radiation with a camera, and using computer simulations to adjust parameters to match experimental results, thereby inferring optical errors.	Intuitive and straightforward; direct observation of irradiance distribution via experimental evidence.	Lacks uniqueness, as different mirror geometries can generate similar irradiance distributions; computationally intensive.
Hartmann Test and Related Methods	This method detects local slope deviations by observing the deviations of light rays reflected from different parts of the mirror. It employs techniques such as lasers or screen projections to measure the deviations of reflected light rays on a point or line basis, with data processing via computer.	Capable of measuring local slope information with relatively high accuracy; high spatial resolution.	Can be complex for large mirrors or in situ measurements, requires costly equipment and significant time.
Photogrammetry	This method uses photographic techniques to capture images of a series of marked points on the measured surface from different directions, then calculates the three-dimensional coordinates of these points to reconstruct the mirror shape.	Commercially available software, mature technology, applicable to multiple types of solar concentrating mirrors, and relatively inexpensive equipment.	Long data acquisition and processing time, limited resolution, requires algorithms to reconstruct surface shape, potential for large errors.

Table 5. *Cont.*

Calibration Method	Method	Advantages	Disadvantages
Long-Range Observer Method	Evaluates the areas on the mirror where reflected light misses the receiver by observing the image of the reflected tube in the mirror from a long distance.	Simple and fast, intuitive in operation, applicable to outdoor and large equipment.	Requires long-range observation, often difficult to achieve due to space limitations in actual solar power plants, only suitable for trough collectors.
Fringe Reflection Method	This method evaluates local slope deviations using known fringe patterns and their deformation on the mirror surface, requiring computer processing of the image data.	High resolution and accuracy, low cost and time requirements, widely applicable.	Complex mathematical processing, susceptible to ambient light, requires strict calibration of equipment.

The methods of light flux mapping and long-range observer, while relatively simple lack specificity. Hartmann tests and photogrammetry offer higher accuracy but require expensive equipment and complex operation. In contrast, the fringe reflection method achieves a good balance of cost-effectiveness and applicability. Each method has specific applications and limitations, and the optimal choice depends on the particular needs and environmental conditions.

3.1.2. Camera-Based Approach

In solar power systems, camera-based calibration methods can enhance heliostat alignment accuracy, ensuring precise solar concentration onto the receiver [46]. High-resolution cameras capture images of the reflected heliostat beams, and image processing algorithms analyze the beam's reflection location to calculate deviations and subsequently correct them. Common camera calibration techniques involve using ground targets, solar images, or reflected images to detect misalignments in heliostats, and subsequently adjusting the heliostats' azimuth and elevation angles via automated or semi-automated control systems Advantages of camera-based calibration include high precision and automation; however, its performance relies on the camera's resolution and the accuracy of its installation location. Compared to traditional manual calibration, camera-based approaches offer enhanced efficiency and reduced human intervention, making them suitable for large-scale heliostat arrays. However, environmental factors (e.g., varying illumination and weather conditions) and equipment performance can influence the system's stability.

Koikari et al. proposed a site calibration-based heliostat calibration system to improve solar concentration efficiency and reduce energy losses. This method ensures a stable illuminated spot on the receiver by selecting the appropriate first rotation axis (g-axis) direction, thereby eliminating the rotation of the concentrated spot. Specifically, the g-axis is perpendicular to the horizontal plane containing the heliostat pivot point of the heliostat and the center of the receiver, ensuring the horizontal edge of the illuminated spot intersecting the receiver remains horizontal, thus preventing rotation. When aligning heliostats, the g-axis should be perpendicular to the plane containing the receiver and the tangent of the heliostat rows, minimizing obstructions and optimizing the concentration path [47]. This method has the advantages of eliminating spot rotation, ensuring uniform heat distribution, and improving photovoltaic conversion efficiency; row alignment reduces obstructions, enhancing energy collection efficiency; and the design process is simplified, eliminating complex calculations. However, this method assumes perfectly flat heliostats and precise tracking accuracy, potentially challenging in practical applications. Further validation is needed when considering factors such as terrain, wind, and temperature.

HelioControl represents a technology for accurately assessing the aim points of heliostats within operating solar central receiver systems (CRSs). This technology employs a 13-bit resolution analog-to-digital (A/D) conversion process alongside a high dynamic range camera capable of capturing images at a frame rate of 25 frames per second (FPS), enabling continuous monitoring of the remote area of the high-temperature receiver. By analyzing frequency signals introduced by the periodic motion of heliostats, a more refined analysis of the system is achieved. Subsequently, Discrete Fourier Transform (DFT) is subsequently utilized to identify the individual aim points of each heliostat. After testing, the detection error angle is less than 0.32 mrad at a viewing angle of 256 × 256 pixels, covering an area of 10 m × 10 m. This method facilitates real-time assessment of multiple heliostat aim points, obviating the need for additional targets. This consequently enables precise control over receiver irradiance distribution, and this approach is applicable to both new and existing installations [15]. HelioControl offers a promising avenue to improve both the cost-effectiveness and efficiency of solar power generation by mitigating light losses and optimizing heliostat layouts without augmenting the complexity of mechanical or drive systems; however, field testing is necessary for its industrial application.

Sattler et al. devised a technique for calibrating small single-faceted heliostats, based on the extraction of invariant moments. This technique uses a high-resolution CCD camera to capture images of reflected light spots, enabling the calibration of heliostat base tilt and wind-induced deformation errors [48]. As shown in Figure 5, the system design is simple and low cost compared to other complex calibration methods, and the use of invariant moments enhances robustness to image translation, rotation, and scale changes, achieving a recognition accuracy of 94.05%. However, current research is primarily focused on small single-faceted heliostats, and its applicability to larger systems remains unverified.

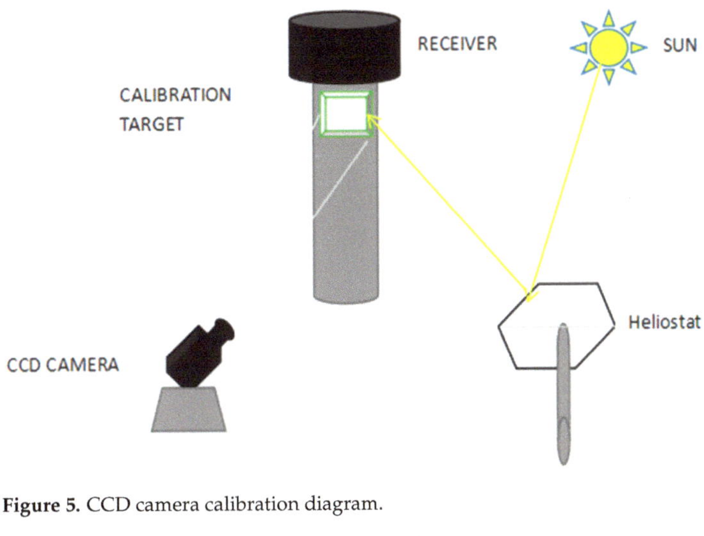

Figure 5. CCD camera calibration diagram.

3.1.3. Non-Intrusive Optical (NIO) Method

Non-intrusive optical calibration methods utilize optical detection techniques to evaluate and calibrate heliostat reflectivity and alignment accuracy without direct contact with the heliostats [49]. These methods rely on high-precision imaging systems, such as cameras or laser systems, coupled with image processing, to capture the projection locations of reflected light spots or solar beams. Analysis of spot shape, position, and symmetry guides adjustments to the heliostat's orientation, thereby maximizing beam focusing accuracy [50]. Advantages include reduced maintenance costs, streamlined operational procedures, enhanced tracking accuracy, and improved photovoltaic conversion efficiency [43]. However, these methods are sensitive to environmental factors such as light intensity and background noise and require substantial investment in specialized hardware and software for initial implementation. Non-intrusive optical calibration provides an effective approach to improving heliostat precision and reducing human intervention.

Mitchell et al. proposed a non-intrusive optical method that precisely calibrates heliostats and calculates errors through a series of steps. Initially, camera lens distortion is calibrated using Photomodeler software to determine the distortion coefficients. Subsequently, the camera's position and orientation are calculated using the least squares method through collinearity equations, with image processing techniques based on the Otsu method used to identify edges of the reflected images. Following this, the tower edge position relative to the heliostat is calculated, and mirror slope errors and tracking errors are derived based on reflection laws, with tower positions then adjusted to enhance accuracy. Upon testing, the results indicated that the uncertainty associated with the measurement of slope error on the heliostat surface was less than 0.22 mrad. The heliostat's tracking error exhibited an average measurement value of 1.71 mrad, accompanied by an uncertainty of 0.12 mrad. Furthermore, the average measured root mean square (RMS) value of the slope error on the heliostat surface was determined to be 1.22 mrad, with a corresponding uncertainty of 0.05 mrad. This method is used for measuring and calibrating optical errors in heliostats within large-scale power tower solar plants [51]. This method, which requires no additional equipment and utilizes structures such as the natural tower, derives multiple optical errors from reflected images. In addition, large-scale measurements can be completed in a short time using UAVs. However, additional marking or multiple shots are needed for measuring two-dimensional optical errors, and data analysis and storage processing demands are high.

Mitchell et al. introduced a novel non-intrusive optical (NIO) methodology for quantifying heliostat optical misalignments in large-scale solar power plants, combining theoretical imaging models with aerial imagery obtained via unmanned aerial vehicles (UAVs). The method encompasses several key stages: (1) using photomodeler software for camera calibration and lens distortion correction; (2) using the principle of collinearity and the known size of the heliostat to determine the camera position and direction; (3) reflected edge identification through grayscale conversion and segmentation of reflected tower edges; (4) calculation of ideal reflected tower edge positions based on established tower geometry; (5) calculating the surface slope error using Snell's Law and deriving the surface normals from camera positions and reflected points; (6) calculating the facet misalignment error through regional averaging of slope errors and comparison against a reference heliostat; (7) iterative correction of the tower position relative to heliostats based on the tracking error, followed by recalculation of optical errors; and (8) uncertainty analysis of results from a series of reflected images to determine measurement precision. During the experimental testing, a minimum of ten images were acquired from each mirror surface, each possessing a resolution of 300 × 300 pixels. The uncertainty in the camera position was limited to less than 0.048 m, while the uncertainty in the tower position was constrained to less than 0.350 m. This methodology enabled the effective control of measurement errors, maintaining them within a range of 0.25 mrad [52]. This method has the following advantages: it is non-intrusive, minimizing operational disruptions to the greatest extent possible; it demonstrates high efficiency, enabling rapid, large-scale inspections; and it exhibits high accuracy due to its insensitivity to the heliostat–tower distance. Despite these advantages, the method is inherently one-dimensional, and two-dimensional measurements necessitate additional markers or multi-angle geometric inferences. Moreover, the method's stability is contingent upon computationally intensive image processing and geometric calculations, as well as precise camera positioning, lens selection, and stable ambient illumination. Future work should prioritize the development of techniques that enable two-dimensional measurements and enhance robustness.

Peña-Lapuente et al. proposed a novel measurement system based on an array of photodetectors. The system utilizes non-tracking heliostats, allowing scanning of reflected beams as the sun moves. Heliostats are positioned near a vertical post. The detector array completes the beam scan within minutes and employs a dynamic gain adjustment to accurately measure radiant differences. The system comprises a sensor array and lens mounted on a post to enhance signal power density and improve the accuracy of remote measurements. Multiple systems can be deployed in parallel to rapidly measure multiple heliostats, thus reducing the overall measurement time. In the simulated environment, the system achieves an error margin of less than 0.5% for heliostats positioned up to 1200 m away, while successfully accomplishing a measurement of such heliostats within a timeframe of six minutes [53]. This approach enhances signal-to-noise ratio and measurement accuracy by minimizing nonlinear responses, target non-uniformities, and noise. Its adaptability and portability make it suitable for various commercial power plants, especially those that have already been put into operation.

3.2. Calibration Method for Auxiliary Equipment

3.2.1. Artificial Light Source

Artificial light calibration (ALC) is a technique used to calibrate heliostats by mimicking sunlight using carefully controlled artificial light sources. This technique relies on stable artificial light sources, such as halogen lamps or LED arrays, to create a well-defined spot of light. Then, the heliostat adjusts its reflective surface and tracking to align with this spot. One advantage of this approach is the ability to calibrate regardless of sunlight or weather conditions, which makes it useful for commissioning and maintaining large solar energy systems. The accuracy of this method depends on how stable the light source is, and how well its intensity is controlled. Additionally, differences between the artificial light and real sunlight can influence how precise the calibration is. Artificial light calibration is a useful way to achieve a flexible and controllable calibration, but further improvements are still necessary to make it both more precise and more adaptable.

Zavodny's team developed an artificial light calibration (ALC) system specifically for tower-based CSP plants. This system uses artificial light sources, such as LEDs, for calibration at night. As shown in Figure 6, the ALC method utilizes multiple short towers strategically positioned around the heliostat array, each tower equipped with a camera and LED light source. Calibration is accomplished by directing the reflected LED light from each heliostat into its assigned camera. During nighttime operation, this system utilizes real-time camera feedback to allow a central control unit to precisely adjust each heliostat's orientation [54]. The ALC system employs parallel data acquisition techniques, enabling each camera to concurrently gather data from more than 100 heliostats. Within a period of two weeks, it is capable of calibrating over 24,000 heliostats, achieving a precision of reflection error below 1.5 mrad. Furthermore, the system incorporates low-power (approximately 10 watts) LED light sources with extended longevity (up to 30,000 h), which are capable of emitting light of varying colors. This effectively mitigates stray light interferences, including background light and moonlight, thereby ensuring the reliability of the calibration process. This approach offers a practical solution to the challenges associated with heliostat alignment. eSolar's ALC system builds upon existing distributed camera systems. Instead of relying on sunlight, eSolar uses LED sources mounted on towers surrounding the field to calibrate heliostat pointing precision. By carefully controlling the relative positions of the light sources and cameras at night, the system refines the motion model for each heliostat by precisely directing the reflected light into the corresponding camera. This method provides a means to achieve more accurate heliostat alignment.

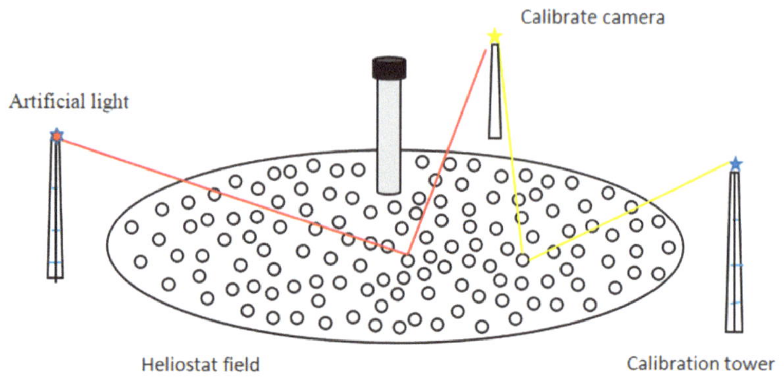

Figure 6. Field geometry for artificial light calibration depicting tower-based cameras and lights (the color of line and star represent the color of the light).

3.2.2. Installing a Camera on a Heliostat

Camera-on-heliostat calibration is a technique that adjusts heliostat orientation by real-time capture of the reflected light spot positions from the heliostats [55]. This method employs high-resolution cameras and image processing algorithms to precisely analyze the shape, position, and symmetry of the reflected light spot to calculate tracking errors and adjust the azimuth and elevation of the heliostat, ensuring accurate alignment with the receiver. Advantages include the ability to perform real-time and efficient precision calibration, continuously monitor the operational status of the heliostat, and reduce external disturbances. However, this technique depends on high-precision cameras and image processing systems and can be affected by factors such as surface contamination and weather conditions, which may degrade calibration accuracy. Therefore, while this method can significantly enhance the efficiency of solar thermal power systems, practical application still needs to overcome challenges such as environmental adaptability and system costs. Figure 7 is a schematic diagram of the camera-on-heliostat.

Burisch et al. proposed a method for high-precision estimation of heliostat motion parameters by installing low-cost cameras on each heliostat and observing artificial light source targets within the field, coupled with image processing techniques and geometric models [55]. The experimental results demonstrate that variations in pixel position errors, attributed to targets situated at diverse distances, elicit tracking inaccuracies spanning from 0.9 to 1.8 mrad. Specifically, a discrepancy of 5 cm in the target's positional alignment introduces an error of 0.2 mrad. Furthermore, when the heliostat exhibits an axis value error of 0.5 mrad, this contributes to a tracking error of 1.2 mrad. This method allows for parallel calibration, making it suitable for large-scale heliostat fields; however, it is impacted by environmental effects, noise interference, target installation requirements, and computational resource demands.

Burisch et al. achieved high-precision positioning by installing cameras on each heliostat and observing infrared targets distributed throughout the solar field. This methodology employs target images captured by a camera from diverse angles, in conjunction with a kinematic model, to estimate the geometric parameters of the heliostat. Based on these parameters, it predicts the heliostat's attitude at various axis angles. Following calibration, a high-quality camera is utilized to capture the position of the reflected light spot from the heliostat onto a Lambertian surface. The centroid of this light spot is then calculated to assess the reflection accuracy. Ultimately, by comparing the deviations between the actual detected angles and the theoretically computed angles, the root mean square (RMS) error of the beam accuracy is determined [56]. This method is primarily suited for smaller-scale heliostat fields and enables complete calibration of the entire heliostat field overnight. Infrared light sources and cameras with infrared filters are used to detect the target positions, and a motion model is established by adjusting the heliostat angle and recording the target position within the pixels; the motion model parameters are then iteratively optimized through multiple observations. The method utilizes a 5-megapixel CMOS camera for calibration purposes, conducting measurements at two specific angles: 45° and 10°. When assessing angle prediction accuracy, the maximum errors observed were 0.4 mrad and 0.3 mrad for these angles, respectively. The corresponding average errors amounted to 0.26 mrad and 0.22 mrad, with root mean square (RMS) errors of 0.29 mrad and 0.22 mrad. Furthermore, throughout the entire day's tracking process, the RMS error associated with the measured beam accuracy was approximately 0.6 mrad. This method is highly efficient, automated, and flexible. However, it faces challenges including limitations in light source application, error accumulation, equipment complexity, and environmental interference.

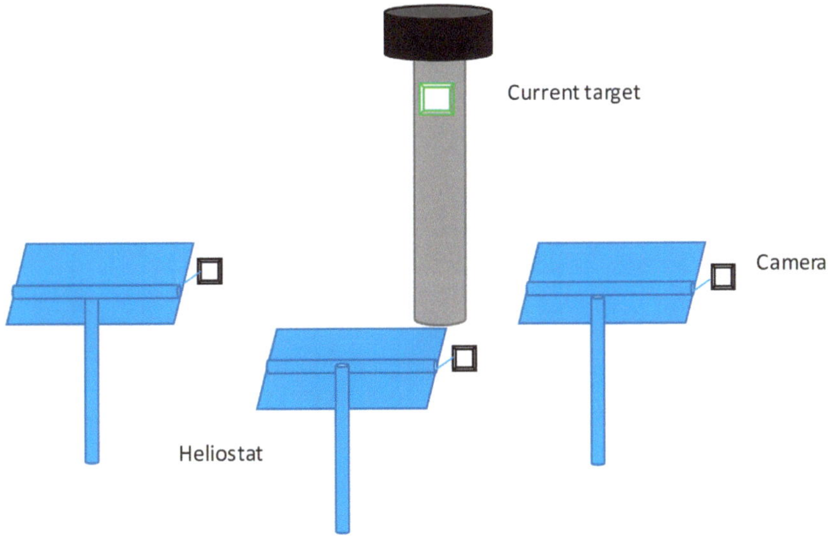

Figure 7. The camera is mounted on a heliostat.

Les et al. proposed a scalable heliostat calibration system based on low-cost cameras, known as SHORT (Scalable HeliOstat calibRation sysTem), which utilizes high-power infrared light sources to calibrate heliostats within a heliostats field [57]. Each camera associated with a heliostat is equipped with a corresponding infrared bandpass filter during periods of high solar radiation to aid in the detection of these targets. The system is highly scalable and automated, and capable of calibrating all heliostats within a single night, achieving a calibration accuracy of less than 0.5 mrad in heliostat positioning error. Camera aberration issues and camera detection accuracy can affect the stability of the system during periods of high daylight irradiance.

Morales-Sánchez et al. proposed a novel computer vision-based method for detecting reflections from heliostat panels in solar concentrating towers. This method employs semi-automated edge detection techniques to identify mirror reflection edges and detect distortions caused by mirror curvature, as shown in Figure 8. Detection is performed using cameras attached to adjacent heliostats, avoiding issues associated with long focal length lenses and camera position uncertainties [58]. The technical process includes, initially, lens distortion being corrected, followed by conversion to grayscale. A Gaussian blur is then applied to reduce noise prior to adaptive thresholding, which generates initial candidate edge pixels. These raw edges are subsequently refined using a Canny edge detector and morphological operations. Linear regression and median-based calculations are employed to fit lines to the optimized edge data. The technical process includes pre-processing steps to eliminate lens distortion, use of a Canny edge detector and morphological filtering to identify edges, the approximation of edge lines using either linear fitting or median filtering and detection of corner points by calculating the intersection of line segments. This method uses a high-resolution camera to simulate images captured at a distance of approximately 5 m. The distance between two consecutive corner points in the image is about 2120 pixels combined with the size of the metal plate (side length of 1000 mm), it is calculated that each pixel is about 0.47 mm, and the deviation calculated by the linear regression method is 3.63 mm. This method also provides an option for manual correction, suitable for situations where edge detection is difficult, thus providing a new tool for enhancing the alignment precision of heliostats in solar concentrating towers.

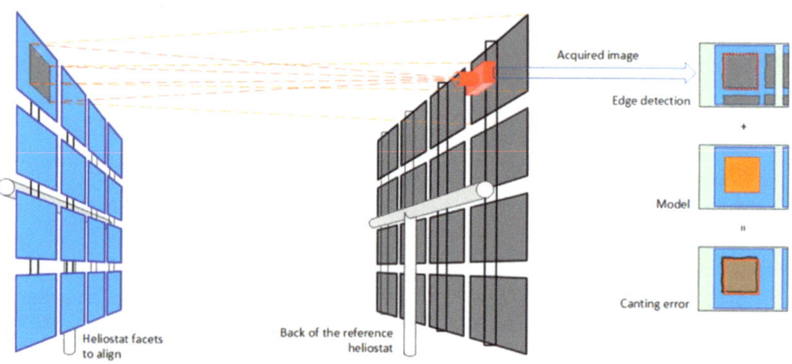

Figure 8. Canting error detection conceptual overview. On the left side, the picture shows two heliostats facing each other, with a camera on the reference one looking at the facet to align. On the right side, the alignment method relies on an accurate edge detection technique combined with a theoretical model to measure the canting errors [58].

3.2.3. Multi-Copter Calibration

The calibration methods of heliostats based on unmanned aerial vehicles (UAVs) utilize UAVs equipped with high-precision camera systems to conduct aerial inspections of heliostat arrays, accurately acquiring the positions of reflected light spots, and analyzing heliostat alignment errors via image processing, as shown in Figure 9 [59]. This method enables rapid and efficient coverage of large areas, making it suitable for large-scale solar power systems while avoiding the complexities associated with ground-based equipment installation and operation. The flexibility of UAVs allows for data acquisition from various angles, enhancing measurement accuracy. However, this method places high demands on hardware and computational capabilities, and weather conditions can affect flight stability and the precision of data acquisition. Nevertheless, the use of UAVs or multi-rotor aircraft equipped with cameras and lights for faster and more flexible calibration in large-scale applications demonstrates significant potential.

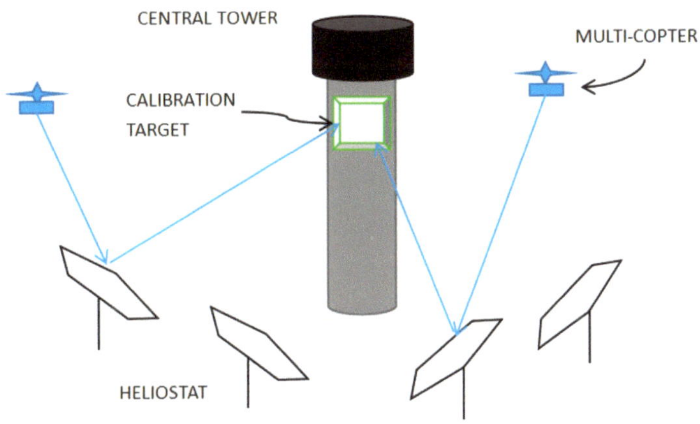

Figure 9. Multi-copter calibration.

Lombard et al. proposed a novel method for heliostat calibration using multi-rotor UAVs. In comparison to traditional Beam Characterization Systems (BCSs), this method achieves efficient calibration by controlling the UAV to fly to known locations and then capturing the targets reflected from the heliostats via camera. In the experiment, a total of twenty data points were gathered, yielding mean errors of 1.6 mrad for azimuth angles and 2.6 mrad for elevation angles, respectively [60]. Despite the challenges of precise positioning, automated control, safety, battery management, and environmental constraints, this method allows for simultaneous calibration of multiple heliostats, shortening the calibration cycle, and does not rely on a control tower. UAVs can also access locations inaccessible to the sun, thereby providing a more comprehensive dataset.

Milidonis et al. introduced a novel approach employing unmanned aerial vehicle (UAV)-assisted close-range photogrammetry for precise geometric characterization of heliostats in concentrated solar power plants [61]. Leveraging advancements in drone technology, this method achieves high-precision mirror reconstruction, boasting an average spatial resolution of 0.35 mm/px and encompassing over 27 million points. The measurement process is insensitive to mirror orientation, eliminating the need for complex ground-based instrument movement. While RTK technology mitigates the effects of magnetic interference, pre-measurement preparation involving mirror spraying and cleaning remains time-consuming. Furthermore, 3D reconstruction accuracy is compromised at the mirror edges, potentially influenced by discrete point artifacts. Calibration accuracy analysis, based on slope deviation, reveals a maximum slope deviation of 6.3 mrad, primarily localized at the mirror periphery. The root mean square deviation (RMSD) for 2D slope measurements is 1.4 mrad.

3.3. *Artificial Intelligence Calibration Method*

In solar power systems, AI, coupled with data-driven calibration methods, leverages machine learning and data analytics to optimize heliostat calibration [62]. By acquiring real-time data relating to reflected beam positions and ambient conditions, and by incorporating AI algorithms, such as deep learning and neural networks, the system autonomously identifies and corrects heliostat tracking errors, thus enabling efficient orientation adjustments and improved positioning accuracy. This method offers key advantages, including a high level of automation, real-time responsiveness, and robust adaptability, which reduces the need for manual intervention while also enabling the ongoing optimization of system performance within complex operating environments [63]. Although this approach is not without its challenges, including stringent data accuracy requirements, complex model training procedures, and enormous computational demands, AI and data-driven calibration methodologies offer innovative and effective solutions for enhancing energy collection efficiency and bolstering system reliability, especially in the context of large-scale solar power systems.

3.3.1. Deep Learning Methods

In solar power systems, image analysis calibration methods based on deep learning automatically extract key information from the images reflected by heliostats, using techniques such as deep neural networks (DNNs) or convolutional neural networks (CNNs), to accurately evaluate and correct heliostat orientation [64,65]. These methods rely on high-resolution imaging systems to capture reflected light spots or projected images, which are then processed by deep learning models to automatically identify the position, shape, and symmetry of the light spots, enabling real-time correction of tracking errors. Deep learning models possess a powerful self-optimization capability, enabling adaptation to complex environmental conditions such as variations in lighting and weather, thus enhancing both the automation level and precision of heliostat calibration.

Pargmann et al. compared the application of the traditional Levenberg–Marquardt (LM) algorithm and deep neural network methods in heliostat calibration. Traditional methods employ regression analysis and the Stone method, utilizing camera-captured focal point positions to determine geometric models and adjust deviations. In contrast, deep neural network methods integrate self-regularizing neural networks (SNNs) and transfer learning, trained using a dataset from the Jülich solar power tower [66]. This method significantly enhances calibration accuracy, achieving a threefold improvement over traditional methods, with a precision of 0.42 mrad. Simultaneously, through pretraining and transfer learning, the neural networks achieve high accuracy with smaller datasets (only 300 measurement points), while also correcting various nonlinear errors, including deformation and backlash. This method also presents a black box problem, as the computational processes of neural networks are difficult to interpret, lacking explicit and traceable control functions, which increases uncertainty in field applications. Furthermore, insufficient data coverage may lead to a decrease in accuracy under extreme operating conditions. Finally, the physical meaning of neural network models is somewhat ambiguous, as they cannot provide specific physical error parameters, potentially limiting their ability to correct particular types of errors.

Deep learning methods, using neural networks for calibration, can manage complex models influenced by multiple factors and address nonlinear errors that are difficult for traditional methods to model. In order to reduce data requirements, both unsupervised and supervised pretraining methods have been utilized. Unsupervised pretraining involves layer-by-layer training of the network's hidden layers using stacked autoencoders and is suitable for situations with limited data, as shown in Figure 10. Supervised pretraining, on the other hand, utilizes ray-tracing software to generate large amounts of virtual data, helping the model to learn relevant features and apply them to actual heliostat calibration tasks [67]. In the experiment, the calibration accuracy of the benchmark model was about 0.8 mrad per increment of error, and the optimal baseline of the pretrained neural network reduced the error to about three times that of the benchmark model. This method demonstrates the ability to significantly enhance accuracy with limited data and possesses strong adaptability and substantial room for improvement. However, the training process for this method relies on substantial data, parameter adjustments require experience, and if there are large differences between the pretrained model and the actual data, the model may converge to a local optimum.

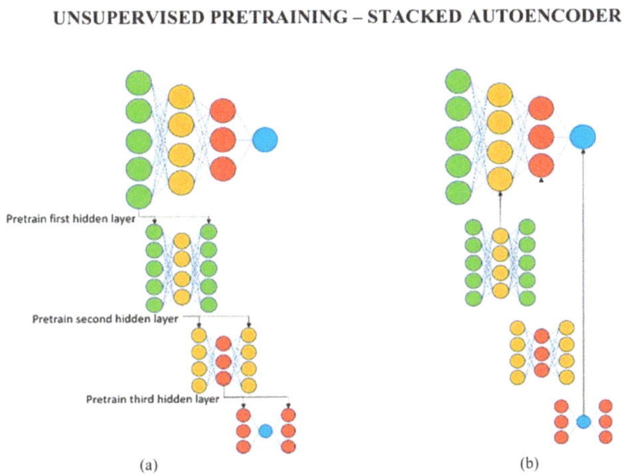

Figure 10. (**a**) First unsupervised pretraining step. In this "greedy layer-wise" pretraining, several autoencoders are created. The first takes the original training data as input and output layer and is trained until no enhancement is recognizable anymore. After this, for every layer of the original neural network, a new autoencoder is created and trained using the one step earlier trained hidden layer as the new input and output layers. (**b**) After all autoencoders are fully trained, the original neural network is initialized by the trained hidden layers of each. After this, the real training step takes place [67].

Bonilla et al. devised a novel closed-loop heliostat tracking calibration method based on computer vision and deep learning. This method determines the vector from the heliostat to the sun (VS) and the vector from the heliostat to the target (VT), subsequently aligning the heliostat's normal vector (VA) with the bisector of these two vectors. An alternative method involves mounting a camera at the heliostat's center point (O0) to provide a planar view of the scene (CP) (see Figure 11c). An artificial neural network (ANN) detects the center points of the sun (S0) and the target (T0); the midpoint between them (A00) is the desired alignment point. The heliostat is then adjusted to align the current alignment point (A0) in the planar view (CP) with the desired alignment point (A00) (see Figure 11d) [68]. Figure 11a,c illustrate 3D and 2D camera views, respectively, of misaligned heliostats, while Figure 11b,d show the same views with aligned heliostats.

This heliostat calibration method eliminates the need for stringent installation requirements and periodic recalibration, exhibiting high accuracy and powerful real-time performance. It automatically calibrates and adjusts errors via a closed-loop control system, utilizes deep learning to extract features for precise identification of the sun and receiver positions, while real-time object detection in the video stream enhances response speeds. The angular accuracy of the camera resolution in the test results ranges from 1.1 mrad per pixel to 0.33 mrad per pixel per radian. Disadvantages include reliance on large amounts of training data, the complexity and time-consuming nature of data processing, the high computational resource demands of deep learning models, which limit their application on resource-constrained devices, and the potential for environmental factors such as dust and clouds to affect precision; moreover, the model requires fine-tuning under different conditions.

Coquand et al. proposed an automatic heliostat orientation calibration system based on artificial vision. The system utilizes a black and white CCD camera to capture images of the reflected solar light from the heliostats and corrects deviations by calculating the difference between the centroid position of the solar beam and the target center using simple image processing algorithms. The correction process minimizes beam offset by adjusting the azimuth and elevation angles of the motors, simultaneously updating the heliostat orientation in the database [69]. This methodology facilitates the detection of a slope error within a margin of ±3 mrad through the reconstruction of the slope of the mirror reflection wavefront, representing a typical range commonly observed across numerous heliostat arrays. This system effectively reduces the time consumption of traditional manual calibration processes, enhances calibration consistency and stability, and has a particularly positive impact on reducing receiver temperature gradient fluctuations, ensuring system safety, and optimizing energy distribution. The system, being based on simple image processing, may require more complex algorithms and hardware support for large-scale power plants and cannot fully guarantee automatic correction of heliostats with specific errors, still requiring manual intervention.

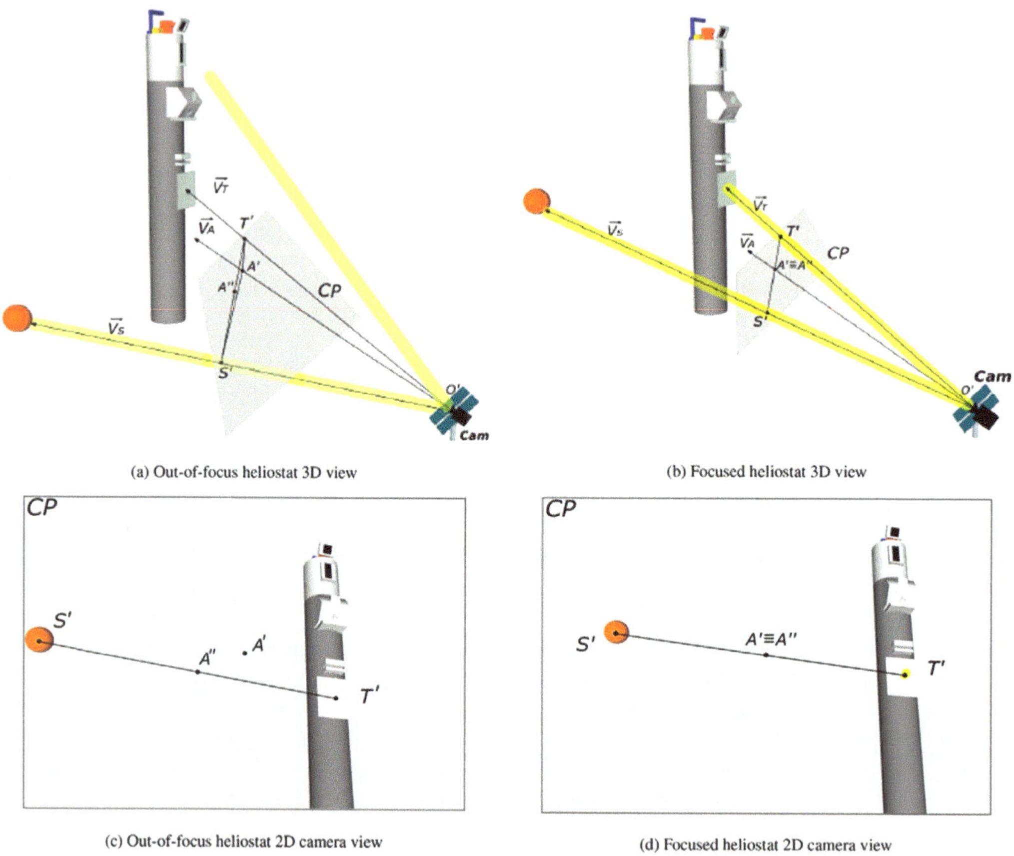

Figure 11. Heliostat tracking principle [68].

3.3.2. Data-Driven Approach

In solar power systems, data-driven calibration methods utilize real-time monitoring data and employ machine learning and statistical analysis techniques to automatically identify and correct tracking errors of the heliostat [70]. These methods rely on sensor networks, data fusion, and algorithm optimization, dynamically adjusting control parameters based on extracted error models from historical data to minimize errors and enhance system performance. This method requires no additional hardware, is capable of adapting to environmental changes, and continuously optimizes system accuracy, thus improving heliostat tracking precision and power generation efficiency. However, data-driven methods exhibit a strong dependence on high-quality data and algorithm optimization and may encounter challenges related to data processing and storage in large-scale applications [71]. In addition, the accuracy of the system is limited by the quality of the input data and the effectiveness of the algorithms. Therefore, although this method provides a highly automated calibration solution, further validation and optimization are required in practical applications.

Hayat et al. proposed a numerical calibration method using mathematical modeling to calculate theoretical values and employing MATLAB software (Version R2008b) to compare the theoretical values with experimental data [72]. The mathematical model incorporates thermal radiation and convection losses, predicting system behavior under various scenarios, enhancing system design precision, validating the reliability of the theoretical model, and is suitable for other similar solar power systems. This method is dependent on the experimental conditions, with experimental accuracy affected by instrumentation and controls. The complexity of the mathematical model increases the computational costs, and climate and geographical factors may influence the experimental results.

García et al. proposed a method to address transient response issues caused by fluctuations in direct normal irradiance (DNI) by calibrating the heat flux and fluid temperature within a solar central receiver; the position of the DNI sensor in the solar field is shown in Figure 12. This strategy incorporates three primary calibration or control methods: internal flow valve control, aiming strategy adjustments, and feedforward control. Firstly, lateral mass flow valves are added at the outlet of each panel to independently regulate the flow through each panel, enabling a rapid response to variations in solar irradiance and avoiding global thermal lag associated with single inlet valve adjustments. Secondly, an aiming strategy similar to valve grouping is employed to distribute heat flux by adjusting the aiming points of the reflectors, ensuring that the heat flux density on the receiver does not exceed design limits. Additionally, feedforward control uses real-time solar field enthalpy data to compensate for DNI fluctuations caused by cloud cover, proactively adjusting the system to minimize its impact on the system [3]. This method can maintain stable output under varying irradiance conditions, adapting to diverse environmental changes including overcast skies. However, the method relies on mathematical models of the solar field and receiver; therefore, the accuracy of the model directly impacts its performance. Furthermore, the implementation of feedforward control and the aiming strategy requires the collection of high-precision, real-time data, increasing system integration and data processing challenges.

Rodríguez proposed a novel, empirical direct method—the superposition method—for estimating solar flux distribution and intensity on the surface of a central receiver, particularly when heliostats are degraded and numerical simulations cannot accurately predict incident solar flux. This calibration method achieves precise estimation of heliostat solar flux distribution through image acquisition and processing, image library creation, and a superposition process [73]. Initially, the reflected beam from each heliostat is characterized using a CCD camera with a resolution of 2.597 mm per pixel and a passive screen, with beam intensity information extracted through digital image analysis to calculate the concentration ratio distribution. A time-independent image library is created for each heliostat, and the solar flux distribution and intensity on the receiver are predicted by superimposing the concentration ratio distributions. This method offers time independence, high spatial resolution, low cost, and strong flexibility, adapting to various aiming strategies and time periods. However, the method also faces challenges including long image library creation times, accuracy impacted by heliostat state, environmental interference, and difficulty in expanding validation.

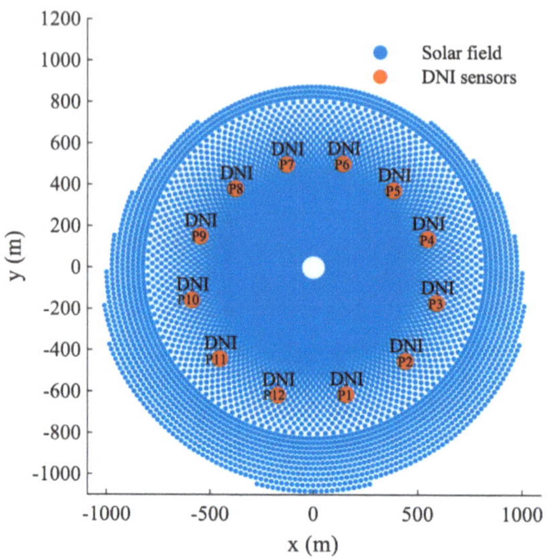

Figure 12. Location of the DNI sensors within the solar field for the feedforward control loop [3].

Berenguel designed a system for the automatic correction of heliostats in solar concentrating power plants using artificial vision techniques. This system employs a common black and white CCD camera to acquire images of the solar light spots reflected by the heliostats, and then corrects deviations based on the offset between the centroid of the solar spot and the center of the target in the image [74]. This calibration method uses the target center and solar spot centroid to achieve automated and precise heliostat calibration. The target area is identified through threshold detection and the center position is calculated. The solar spot centroid is determined by combining grayscale histograms and threshold settings. The azimuth and elevation angles of the heliostat are then adjusted based on the offset between these two, and the calibration results are updated in the control database. This method exhibits a high degree of automation, reducing manual operations; is cost-effective, requiring only common CCD cameras; and is highly flexible, supporting real-time adjustments. However, the method also has limitations, such as encoder and camera resolution limitations impacting accuracy, light variations or mechanical vibrations interfering with reliability, and algorithm performance degrading when the solar spot deviates from the field of view.

Sun proposed an improved method to address heliostat tracking errors in solar power tower systems, using an error correction model in conjunction with the Hartley–Meyer algorithm to calculate six angular parameters: tilt angle, azimuth axis tilt azimuth angle, initial azimuth angle, biaxial non-orthogonality angle, initial elevation angle, and adjustment angle of the mirror surface relative to the elevation axis. The error correction model was validated in tests at the DAHAN solar power plant, demonstrating a significant improvement in heliostat tracking accuracy [75]. The uncorrected root mean square error (RMSE) was 2.97 mrad. Application of a single set of angular correction parameters reduced the RMSE to 1.67 mrad. Further refinement with a second set of parameters yielded a final RMSE of 1.26 mrad. The method further optimizes heliostat tracking performance via a dual correction strategy, combining a tracking angular deviation strategy with the error correction model. Additionally, the model has strong adaptability, considers error distribution across different time periods, and divides the day into two parts based on the time of the solar near-equilibrium point using two sets of regression results to achieve more accurate tracking angles. However, this method has several drawbacks including its dependence on a large amount of test data, potential complexity and tediousness in tracking adjustments for a large number of heliostats during practical operation, the model's complexity involving multiple error parameters and calculation formulas, which requires high technical and computational support, and, although effective in short-term tests, the error correction effect may diminish over time, thus necessitating periodic adjustments and recalibration to maintain accuracy.

Abdallah presented a design and construction solar tracking system controlled by a programmable logic controller (PLC). Comparing the solar energy collected using tracking with a fixed surface tilted at 32 degrees southward, the solar energy collected on the moving surface was significantly greater than that on the fixed surface, with an increase of 41.34%. The PLC control system operates as an open-loop system, adjusting the calculated position of the tracking surface based on pre-calculated solar angles. A photometer was used to measure and record solar irradiance data on various surfaces, and continuous multi-day experiments were conducted to obtain the average daily total solar irradiance values [76]. The study concluded that the two-axis tracking system exhibited a significant increase in efficiency compared to traditional single-axis or fixed systems, with nearly a twofold improvement.

3.3.3. Hybrid Calibration Method

The limitations of individual methods can be mitigated by combining the aforementioned approaches. Data from various sensors and control systems can be integrated to yield more robust and accurate results.

Sánchez-González devised a calibration method using an unmanned aerial system (UAS) equipped with a camera, capturing images of heliostats in operation, detecting images reflected off the back of adjacent heliostats, and comparing these captured images with theoretical images generated via an optical model to detect alignment errors in the heliostat's reflective surface, as shown in Figure 13 [49]. Mounting the camera on a UAV avoids interruptions to power plant operations, allowing flexible detection of heliostat tilt and focusing errors to achieve comprehensive optical calibration. This method can detect tilt errors of 0.25 mrad, making it suitable for fine calibration and adaptable to different solar energy applications, meeting the stringent alignment requirements of high-temperature receivers. However, this method faces challenges including high-precision positioning requirements, substantial equipment needs, high initial investment costs, and the interference of mirror slope errors on the reflected images. To improve sensitivity, multiple shots are typically needed, increasing the complexity of the calibration process.

The utilization of microcontroller technology enables the implementation of robust error correction algorithms on low-cost heliostat controllers, facilitating high tracking accuracy. Malan demonstrated an array of 18 heliostats, proving that this approach effectively reduced tracking errors and that these errors are deterministic, which can be further improved through model optimization. Malan also presented a method for periodically updating the error correction coefficients for each heliostat via camera capture and image processing techniques [30]. Model-based error correction methods reduce the costs and enhance the efficiency of solar concentrating systems.

Leibauer et al. proposed a two-layer hybrid model to improve heliostat calibration accuracy, integrating mechanical rigid-body kinematic models, neural network correction models, and hybrid models to enhance calibration precision. Initially, the rigid-body model is employed for pre-alignment with limited data, as shown in Figure 14; subsequently, a neural network model corrects deviations with minimal data to improve accuracy, as shown in Figure 15; finally, a hybrid model integrates the advantages of both, maintaining high precision even with limited data [77]. Leveraging pretrained neural networks, the model's accuracy was significantly enhanced, achieving robust performance with only 60 data points. This represents a substantial improvement compared to the previous deep learning model, which necessitated 300 data points for comparable results. The hybrid model surpassed the performance of the rigid-body model from the outset of the initial measurements, attaining a peak accuracy of less than 0.7 mrad. This method provides benefits including high precision, low data requirements, robustness, and cost-effectiveness. Data bias may influence the model's performance under specific conditions; inference accuracy may decrease when data diversity is insufficient.

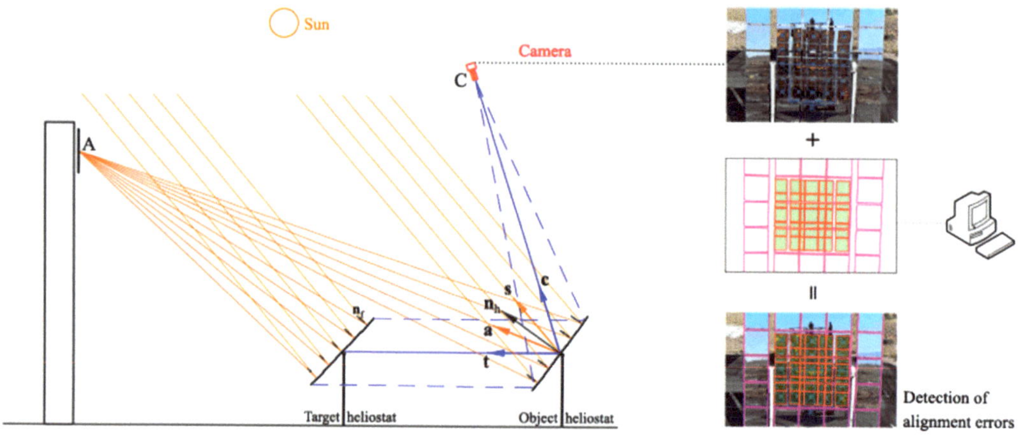

Figure 13. Alignment technique [49].

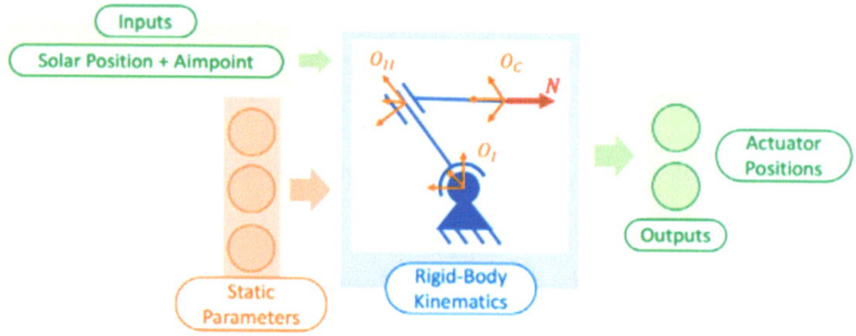

Figure 14. Rigid-body alignment model [77].

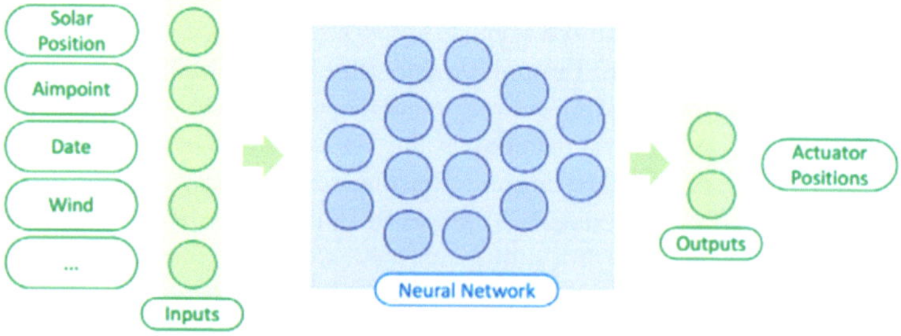

Figure 15. Sketch of a neural network for alignment prediction [77].

Jessen et al. introduced a two-phase methodology for measuring heliostat misalignments in solar power tower systems, leveraging UAVs in conjunction with photogrammetry and deflectometry techniques, thereby enhancing both flexibility and efficiency. The first stage implements photogrammetry, using UAVs to capture multi-angle images of heliostat surface features, facilitating the creation of a preliminary three-dimensional model and ascertaining the initial heliostat orientation. In the first stage of photo measurement, the method was validated using reference data, and the root mean square deviation was 5.9 mrad. The second stage utilizes deflectometry, using a UAV-mounted LED light source for precise heliostat orientation measurements, generating detailed reflected beam data to optimize heliostat alignment, allowing for expeditious and accurate calibration without requiring a completed central receiver tower [78]. This methodology lowers the initial heliostat field calibration time, permitting faster achievement of full operational capacity and reduced power generation costs. Furthermore, this flexible method allows for phased commissioning and calibration in distributed heliostat fields. However, this methodology faces high accuracy requirements for UAV positioning data, complex post-processing algorithms, and potential limitations in measurement efficiency and accuracy for long-distance measurements related to UAV positioning and camera parameters.

4. Future Development and Prospects

The integration of AI and wireless communication technologies provides enhanced possibilities for the intelligent management of solar thermal power systems. Real-time data collected through wireless networks provide substantial inputs for AI models, enabling more precise predictions and optimizations. Concurrently, AI algorithms can intelligently schedule wireless communication networks, ensuring the stability and low latency of data transmission. For example, in large-scale solar thermal power fields, AI can coordinate data transmission across different zones, select optimal transmission paths, avoid network congestion and latency, and ensure the rapid flow of information.

4.1. Technology Integration and Interdisciplinary Research

With the continuous advancement of technologies such as artificial intelligence, wireless communication, and the Internet of Things, future solar thermal power systems will become increasingly intelligent and automated, necessitating greater integration of multidisciplinary and multi-domain technologies. The integration of AI and wireless communication technologies will no longer be limited to single-system optimization, but will progressively enable the intelligentization of the entire system lifecycle, encompassing all stages from design and construction to operation and maintenance.

4.2. Technical Hurdles to Future Development

4.2.1. The Robustness and Generalization Ability of AI Algorithms

Research gap: Current artificial intelligence (AI) algorithms demonstrate proficiency in specific operational environments or power plant conditions. However, their robustness and generalization capabilities—specifically their performance across diverse geographical locations, climatic conditions, and equipment-aging scenarios—require further investigation. For example, an AI model trained for desert environments may exhibit limited applicability in regions characterized by cloud cover or high humidity.

Technical hurdles: The development of adaptive AI algorithms, capable of autonomous adjustment to varying operational conditions and power plant characteristics, necessitates substantial and diverse datasets, coupled with effective transfer learning methodologies. Furthermore, algorithmic interpretability presents a significant challenge, demanding the development of techniques that elucidate AI decision-making processes to facilitate debugging and optimization efforts.

4.2.2. Reliability and Security of Wireless Communication

Research gap: The extensive geographical footprint and complex electromagnetic environments inherent in large-scale CSP power plants present challenges to the reliability of wireless communication systems. Future research should prioritize the investigation of wireless signal attenuation, interference, and multipath effects. Effective channel coding modulation/demodulation techniques, and optimized antenna designs warrant exploration. Furthermore, ensuring the security of wireless communication is paramount, necessitating the development and implementation of robust security protocols and mechanisms to mitigate unauthorized access and malicious cyberattacks.

Technical hurdles: Achieving high-reliability wireless communication necessitates intricate network planning and optimization strategies, alongside efficient management of the wireless spectrum. To fortify the security of these networks, the development of lightweight yet robust encryption algorithms and authentication mechanisms is critical, coupled with the implementation of comprehensive defense strategies to counter a range of network attacks.

4.3. Combination of Edge Computing and Cloud Computing

The integration of edge computing and cloud computing enables a distributed approach to data processing and analysis, leveraging the strengths of both paradigms to enhance the performance and scalability of solar thermal power systems.

4.4. Standardization and Interoperability

Currently, the application of AI and wireless communication in solar thermal power generation still faces challenges related to standardization. The significant variations in equipment, communication protocols, and data formats offered by different manufacturers raise the question of how to ensure interoperability between systems, which will be critical for future development. Promoting a unified standard system and open technology platforms would facilitate seamless integration and collaboration between different technologies.

5. Conclusions

The integration of artificial intelligence (AI) and wireless communication technologies holds significant promise for enhancing the performance of solar power tower systems. AI-driven control strategies, such as optimized heliostat field management, thermal energy storage (TES) management, and predictive maintenance, have the potential to improve overall system efficiency by 2–10% and reduce tracking error by 10–50%. Furthermore, AI can significantly enhance energy dispatchability (5–15%) and equipment maintenance efficiency (reducing maintenance costs and downtime by 10–20%) by compensating for factors like wind and thermal distortion, and by predicting demand-side needs. Concurrently, employing advanced wireless communication technologies can reduce data latency, shortening control response times from hundreds of milliseconds to tens of milliseconds, thereby improving the overall energy capture efficiency. Wireless solutions also lower deployment costs (20–50%), enhance system redundancy, and provide more comprehensive environmental monitoring capabilities. However, the extent of these performance gains is highly dependent on the specific system design, climatic conditions, implementation details, and the maturity of the chosen technologies. Therefore, in practical applications, thorough simulations and field tests are essential, along with careful consideration of system integration complexity, information security, and cost-effectiveness, to fully validate potential performance enhancements. Table 6 summarizes the applicable scenarios of the calibration system mentioned in the article.

Table 6. Applicable scenarios of calibration methods.

Calibration Method Category	Applicable Scenarios
Artificial light source calibration	Night or cloudy environment: in situations where sunlight cannot be utilized, artificial light sources can provide stable light sources for calibrating the heliostat.
Multi-copter calibration	Drones can be equipped with high-precision cameras and visual analysis algorithms to quickly and accurately measure the actual reflection angle of each heliostat and compare it with the target angle. This can help identify and correct angular deviations, ensuring that the beam is accurately focused on the heat absorber.
Traditional camera calibration	After long-term operation, the heliostat may experience positional displacement or deformation due to various factors such as wind force, temperature changes, foundation settlement, etc., which can affect the focusing effect. Regularly using a camera for calibration can detect these changes and provide timely maintenance and adjustments.
Artificial intelligence calibration	Using AI algorithms to accurately calibrate each heliostat, eliminating focusing inaccuracies caused by manufacturing errors, installation deviations, wind disturbances, and other factors.

The reliable operation and efficient power generation of solar power tower systems hinge critically on addressing two key challenges: ensuring the reliability of artificial intelligence (AI) models and mitigating latency in wireless communication networks. Enhancing AI model reliability necessitates a multifaceted approach, encompassing the use of high-quality training data, rigorous pre-deployment testing and validation procedures, and continuous post-deployment monitoring and iterative model updates. Simultaneously, minimizing wireless communication latency, a known performance bottleneck, is essential for maintaining responsive data transmission and remote monitoring capabilities within the system. Resolving these intertwined challenges is paramount for maximizing the overall stability and efficiency of solar power tower operations. The convergence of AI and wireless technologies fundamentally reshapes the design and operation of heliostat systems in concentrating solar power plants. AI enhances calibration accuracy and efficiency, while wireless communication reduces costs and complexity. This integration of AI and wireless technologies holds significant potential for boosting system efficiency, lowering operational costs, and enabling intelligent management in solar thermal power generation. In the future, the profound fusion of AI and wireless communication will elevate the role of solar thermal power in sustainable energy, propelling the technology toward greater efficiency, intelligence, and environmental sustainability, and contributing to the global energy transition.

Author Contributions: Conceptualization, Z.D. and Q.L.; validation, D.W.; writing—original draft preparation, Q.L.; writing—review and editing, Q.L., Y.W., and Y.X.; project administration, Y.W. All authors have read and agreed to the published version of the manuscript.

Funding: This research was funded by the POWERCHINA GROUP (funding number: DJ-HXGG-2022–19).

Acknowledgments: Thanks to the fund for their support, and thanks to my team.

Conflicts of Interest: All authors were employed by the company SEPCO3 Electric Power Construction Co., Ltd.

Abbreviations

The following abbreviations are used in this manuscript:

AI	Artificial intelligence
CSP	Concentrated solar power
FFT	Fast Fourier transform
BCS	Beam Characterization Systems
ALC	Artificial light calibration
MPC	Model predictive control

References

1. Hamanah, W.M.; Salem, A.; Abido, M.A.; Qwbaiban, A.M.; Habetler, T.G. Solar Power Tower Drives: A Comprehensive Survey. *IEEE Access* **2023**, *11*, 83964–83982. [CrossRef]
2. Rizvi, A.A.; Yang, D.; Khan, T.A. Optimization of biomimetic heliostat field using heuristic optimization algorithms. *Knowl.-Based Syst.* **2022**, *258*, 110048. [CrossRef]
3. García, J.; Barraza, R.; Soo Too, Y.C.; Vásquez-Padilla, R.; Acosta, D.; Estay, D.; Valdivia, P. Transient simulation of a control strategy for solar receivers based on mass flow valves adjustments and heliostats aiming. *Renew. Energy* **2022**, *185*, 1221–1244. [CrossRef]
4. Stamatellos, G.; Stamatelos, T. Effect of Actual Recuperators' Effectiveness on the Attainable Efficiency of Supercritical CO_2 Brayton Cycles for Solar Thermal Power Plants. *Energies* **2022**, *15*, 7773. [CrossRef]
5. Mehos, M.; Turchi, C.; Vidal, J.; Wagner, M.; Ma, Z.W.; Ho, C.; Kolb, W.; Andraka, C.; Kruizenga, A. *Concentrating Solar Power Gen3 Demonstration Roadmap*; No. NREL/TP-5500-67464; National Renewable Energy Lab.(NREL): Golden, CO, USA, 2017. [CrossRef]
6. Freeman, J.; Keerthi, K.S.; Chandran, L.R. Closed loop control system for a heliostat field. In Proceedings of the 2015 International Conference on Technological Advancements in Power and Energy (TAP Energy), Kollam, India, 24–26 June 2015; pp. 272–277. [CrossRef]
7. Tsiropoulou, E.E.; Rahman, A.B.; Siraj, M.S. HELIOCOMM: A Wireless Revolution in Concentrated Solar Power Systems. *IT Prof.* **2024**, *26*, 73–79. [CrossRef]
8. Maiga, M.; N'Tsoukpoe, K.E.; Gomna, A.; Fiagbe, Y.K. Sources of solar tracking errors and correction strategies for heliostats. *Renew. Sustain. Energy Rev.* **2024**, *203*, 114770. [CrossRef]
9. Pargmann, M.; Leibauer, M.; Nettelroth, V.; Maldonado Quinto, D.; Pitz-Paal, R. Questioning the reliability of open-loop calibration methods: Introducing a robust data sampling for year-round high accuracy. *Sol. Energy* **2025**, *286*, 113094. [CrossRef]
10. Zhu, G.; Augustine, C.; Mitchell, R.; Muller, M.; Kurup, P.; Zolan, A.; Yellapantula, S.; Brost, R.; Armijo, K.; Smentet, J. HelioCon: A roadmap for advanced heliostat technologies for concentrating solar power *Sol. Energy* **2023**, *264*, 111917. [CrossRef]
11. Ashith Shyam, R.B.; Ghosal, A. Path planning of a 3-UPU wrist manipulator for sun tracking in central receiver tower systems. *Mech. Mach. Theory* **2018**, *119*, 130–141. [CrossRef]
12. Liebenberg, A.D.; Smit, W. Wireless communication for a modular heliostat field. In Proceedings of the SOLARPACES 2020: 26th International Conference on Concentrating Solar Power and Chemical Energy Systems, Online, 28 September–2 October 2020.
13. Kubisch, S.; Randt, M.; Buck, R.; Pfahl, A.; Unterschütz, S. Wireless heliostat and control system for large self-powered heliostat fields. In Proceedings of the SolarPACES 2011, SolarPACES Conference 2011, Granada, Spain, 20–3 September 2011.
14. Lim, B.-H.; Lim, C.-S.; Li, H.; Hu, X.-L.; Chong, K.-K.; Zong, J.-L.; Kang, K.; Tan, W.-C. Industrial design and implementation of a large-scale dual-axis sun tracker with a vertical-axis-rotating-platform and multiple-row-elevation structures. *Sol. Energy* **2020**, *199*, 596–616. [CrossRef]
15. Bern, G.; Schöttl, P.; van Rooyen, D.W.; Heimsath, A.; Nitz, P. Parallel in-situ measurement of heliostat aim points in central receiver systems by image processing methods. *Sol. Energy* **2019**, *180*, 648–663. [CrossRef]
16. Sievers, L.T.E.; Pargmann, M.; Maldonado Quinto, D.; Hoffschmidt, B. End-to-end sensitivity analysis of a hybrid heliostat calibration process involving artificial neural networks. *Sol. Energy* **2025**, *287*, 113219. [CrossRef]
17. Gamra, M. Modelling and an adaptive fuzzy logic controller of solar thermal power plant. *Przegld Elektrotechniczny* **2023**, *1*, 259–265. [CrossRef]
18. Nsengiyumva, W.; Chen, S.G.; Hu, L.; Chen, X. Recent advancements and challenges in Solar Tracking Systems (STS): A review. *Renew. Sustain. Energy Rev.* **2018**, *81*, 250–279. [CrossRef]
19. Lopes, F.M.; Conceição, R.; Silva, H.G.; Fasquelle, T.; Salgado, R.; Canhoto, P.; Collares-Pereira, M. Short-Term Forecasts of DNI from an Integrated Forecasting System (ECMWF) for Optimized Operational Strategies of a Central Receiver System. *Energies* **2019**, *12*, 1368. [CrossRef]
20. Hanrieder, N.; Ghennioui, A.; Wilbert, S.; Sengupta, M.; Zarzalejo, L.F. AATTENUATION—The Atmospheric Attenuation Model for CSP Tower Plants: A Look-Up Table for Operational Implementation. *Energies* **2020**, *13*, 5248. [CrossRef]
21. Abreu, E.F.M.; Canhoto, P.; Costa, M.J. Prediction of Circumsolar Irradiance and Its Impact on CSP Systems under Clear Skies. *Energies* **2023**, *16*, 7950. [CrossRef]

22. Guo, M.; Wang, Z.; Wang, X.; Zhang, X.; Wang, N. Performance study of a general azimuth-elevation heliostat tracking formula. In Proceedings of the SOLARPACES 2018: International Conference on Concentrating Solar Power and Chemical Energy Systems, Casablanca, Morocco, 2–5 October 2018.
23. Soo Too, Y.C.; García, J.; Padilla, R.V.; Kim, J.-S.; Sanjuan, M. A transient optical-thermal model with dynamic matrix controller for solar central receivers. *Appl. Therm. Eng.* **2019**, *154*, 686–698. [CrossRef]
24. García, J.; Soo Too, Y.C.; Padilla, R.V.; Beath, A.; Kim, J.-S.; Sanjuan, M.E. Dynamic performance of an aiming control methodology for solar central receivers due to cloud disturbances. *Renew. Energy* **2018**, *121*, 355–367. [CrossRef]
25. Sarr, M.P.; Thiam, A.; Dieng, B. ANFIS and ANN models to predict heliostat tracking errors. *Heliyon* **2023**, *9*, e12804. [CrossRef] [PubMed]
26. Xie, Q.; Xiao, Y.; Wang, X.; Liu, D.; Shen, Z. Heliostat Cluster Control for the Solar Tower Power Plant Based on Leader-Follower Strategy. *IEEE Access* **2019**, *7*, 135031–135039. [CrossRef]
27. Röger, M.; Herrmann, P.; Ulmer, S.; Ebert, M.; Prahl, C.; Göhring, F. Techniques to measure solar flux density distribution on large-scale receivers *J. Sol. Energy Eng.* **2014**, *136*, 031013. [CrossRef]
28. Ahlbrink, N.; Alexopoulos, S.; Andersson, J.; Belhomme, B.; Boura, C.T.; Gall, J. vICERP—The virtual institute of central receiver power plants: Modeling and simulation of an open volumetric air receiver power plant. In Proceedings of the 6th Vienna Conference on Mathematical Modelling, Vienna, Austria, 11–13 February 2009.
29. Rahman, A.B.; Siraj, M.S.; Tsiropoulou, E.E. Wireless Communications for Concentrated Solar Power Fields. *IEEE Trans. Green Commun. Netw. Early Access.* **2025**. [CrossRef]
30. Benitez, V.H.; Armas-Flores, R.V.; Pacheco-Ramirez, J.H. Experimental Study for the Development of a Wireless Communication System in a Solar Central Tower Facility *Int. J. Inf. Commun. Eng.* **2016**, *10*, 365–371.
31. Pfahl, A.; Randt, M.; Meier, F.; Zaschke, M.; Geurts, C.P.W.; Buselmeier, M. A Holistic Approach for Low Cost Heliostat Fields. *Energy Procedia* **2015**, *69*, 178–187. [CrossRef]
32. Malan, K.; Gauché, P. Model based Open-loop Correction of Heliostat Tracking Errors. *Energy Procedia* **2014**, *49*, 2118–2124. [CrossRef]
33. Shariff, F.; Rahim, N.A.; Hew, W.P. Zigbee-based data acquisition system for online monitoring of grid-connected photovoltaic system. *Expert Syst. Appl.* **2015**, *42*, 1730–1742. [CrossRef]
34. Paredes-Parra, J.M.; García-Sánchez, A.J.; Mateo-Aroca, A.; Molina-García, Á. An Alternative Internet-of-Things Solution Based on LoRa for PV Power Plants: Data Monitoring and Management. *Energies* **2019**, *12*, 881. [CrossRef]
35. Kabalci, E.; Kabalci, Y. A wireless metering and monitoring system for solar string inverters. *Int. J. Electr. Power Energy Syst.* **2018**, *96*, 282–295. [CrossRef]
36. Salgado-Plasencia, E.; Carrillo-Serrano, R.V.; Rivas-Araiza, E.A.; Toledano-Ayala, M. SCADA-Based Heliostat Control System with a Fuzzy Logic Controller for the Heliostat Orientation. *Appl. Sci.* **2019**, *9*, 2966. [CrossRef]
37. Hu, L.; Li, C. Wireless Instrument and Wireless Control Network for Multi-tower CSP Power Plant. In Proceedings of the 2nd International Conference on Civil, Materials and Environmental Sciences, London, UK, 13–15 March 2015; pp. 396–398. [CrossRef]
38. Malan, K.J.; Gauché, P. A locally developed 40 m^2 heliostat array wireless control system. In Proceedings of the Southern African Solar Energy Conference (SASEC'14), Summerset, South Africa, 13–15 November 2024.
39. Sharma, H.; Haque, A.; Jaffery, Z.A. Solar energy harvesting wireless sensor network nodes: A survey. *J. Renew. Sustain. Energy* **2018**, *10*, 023704. [CrossRef]
40. Liebenberg, A.D.; Smit, W.J. Wireless Communication For A Modular Heliostat Field. In Proceedings of the International SAUPEC/RobMech/PRASA Conference, Cape Town, South Africa, 29–31 January 2020; pp. 1–6. [CrossRef]
41. Tsiropoulou, E.E.; Rahman, A.B.; Siraj, M.S. *HELIOCOMM: Wireless Controls State-of-the-Art Report*; National Renewable Energy Lab.(NREL): Golden, CO, USA, 2024.
42. Zhu, R.; Ni, D.; Yang, T.; Yang, J.; Chen, J.; Xiao, G. Heliostat field aiming strategy optimization with post-installation calibration. *Appl. Therm. Eng.* **2022**, *202*, 117720. [CrossRef]
43. Andraka, C.E.; Yellowhair, J.E. AIMFAST for heliostats: Canting tool for long focal lengths. In Proceedings of the SOLARPACES 2018: International Conference on Concentrating Solar Power and Chemical Energy Systems, Casablanca, Morocco, 2–5 October 2018.
44. Elsayed, A.A.; Khalil, E.E.; Kassem, M.A.; Huzzayin, O.A. A novel mechanical solar tracking mechanism with single axis of tracking for developing countries. *Renew. Energy* **2021**, *170*, 1129–1142. [CrossRef]
45. Ren, L.; Wei, X.; Lu, Z.; Yu, W.; Xu, W.; Shen, Z. A review of available methods for the alignment of mirror facets of solar concentrator in solar thermal power system. *Renew. Sustain. Energy Rev.* **2014**, *32*, 76–83. [CrossRef]
46. Iriarte-Cornejo, C.; Arancibia-Bulnes, C.A.; Salgado-Transito, I.; Waissman, J.; Cabanillas, R.E.; Estrada, C.A. Compensation of heliostat drift by seasonal sampling. *Sol. Energy* **2014**, *105*, 330–340. [CrossRef]
47. Koikari, S.; Amano, T.; Onomura, T.; Iemoto, M.; Yoshida, K. Field-aligned heliostats and their application to central receiver system. *Sol. Energy* **2014**, *105*, 575–589. [CrossRef]

48. Sattler, J.C.; Schneider, I.P.; Angele, F.; Atti, V.; Teixeira Boura, C.; Herrmann, U. Development of Heliostat Field Calibration Methods. In Proceedings of the SolarPACES 2022, 28th International Conference on Concentrating Solar Power and Chemical Energy Systems, Albuquerque, NM, USA, 27–30 September 2022. [CrossRef]
49. Sánchez-González, A.; Yellowhair, J. Reflections between heliostats: Model to detect alignment errors. *Sol. Energy* **2020**, *201* 373–386. [CrossRef]
50. Arancibia-Bulnes, C.A.; Peña-Cruz, M.I.; Mutuberría, A.; Díaz-Uribe, R.; Sánchez-González, M. A survey of methods for the evaluation of reflective solar concentrator optics. *Renew. Sustain. Energy Rev.* **2017**, *69*, 673–684. [CrossRef]
51. Mitchell, R.A.; Zhu, G. A non-intrusive optical (NIO) approach to characterize heliostats in utility-scale power tower plants Methodology and in-situ validation. *Sol. Energy* **2020**, *209*, 431–445. [CrossRef]
52. Mitchell, R.A.; Zhu, G. A non-intrusive optical (NIO) approach to characterize heliostats in utility-scale power tower plants Sensitivity study. *Sol. Energy* **2020**, *207*, 450–457. [CrossRef]
53. Les, I.; Peña-Lapuente, A.; Mutuberria, A.; Sanchez, M.; Heras, C.; Salinas, I. Innovative instrumentation and methodology to characterize long distance heliostat beam quality in commercial solar power tower plants. In Proceedings of the SOLARPACES 2018 International Conference on Concentrating Solar Power and Chemical Energy Systems, Casablanca, Morocco, 2–5 October 2018
54. Zavodny, M.; Slack, M.; Huibregtse, R.; Sonn, A. Tower-based CSP Artificial Light Calibration System. *Energy Procedia* **2015**, *69*, 1488–1497. [CrossRef]
55. Burisch, M.; Sanchez, M.; Olarra, A.; Villasante, C. Heliostat calibration using attached cameras and artificial targets. In Proceedings of the SOLARPACES 2015: International Conference on Concentrating Solar Power and Chemical Energy Systems, Cape Town, South Africa, 13–16 October 2015.
56. Burisch, M.; Olano, X.; Sanchez, M.; Olarra, A.; Villasante, C.; Olasolo, D.; Monterreal, R.; Enrique, R.; Fernández, J. *Scalable Heliostat Calibration System (SHORT)—Calibrate a Whole Heliostat Field in a Single Night*; National Renewable Energy Centre: Blyth, UK, 2018.
57. Les, I.; Peña-Lapuente, A.; Sanchez, M.; Olasolo, D.; Villasante, C.; Enrique, R.; Fernandez-Reche, J. Validation of a low-cost camera for Scalable HeliOstat calibRation sysTem (SHORT). In Proceedings of the SOLARPACES 2020: 26th International Conference on Concentrating Solar Power and Chemical Energy Systems, Online, 28 September–2 October 2020.
58. Morales-Sánchez, R.; Lozano-Cancelas, A.; Sánchez-González, A.; Castillo, J.C. Detecting the reflection of heliostat facets through computer vision. In Proceedings of the International Conference on Battery for Renewable Energy and Electric Vehicles (Icb-Rev) 2022, South Tangerang, Indonesia, 21–23 June 2022.
59. Yellowhair, J.; Apostolopoulos, P.A.; Small, D.E.; Novick, D.; Mann, M. Development of an aerial imaging system for heliostat canting assessments. In Proceedings of the SOLARPACES 2020: 26th International Conference on Concentrating Solar Power and Chemical Energy Systems, Online, 28 September–2 October 2020.
60. Lombard, S.J.; Smit, W.J. Practical challenges to calibrate a heliostat with a multi-copter. In Proceedings of the SOLARPACES 2020: 26th International Conference on Concentrating Solar Power and Chemical Energy Systems, Online, 28 September–2 October 2020.
61. Milidonis, K.; Abate, D.; Blanco, M.J. Heliostat geometrical characterization by UAV-assisted, close-range photogrammetry. *Sol. Energy* **2024**, *280*, 112849. [CrossRef]
62. Milidonis, K.; Blanco, M.J.; Grigoriev, V.; Panagiotou, C.F.; Bonanos, A.M.; Constantinou, M.; Pye, J.; Asselineau, C.-A. Review of application of AI techniques to Solar Tower Systems. *Sol. Energy* **2021**, *224*, 500–515. [CrossRef]
63. Cruz, N.C.; Redondo, J.L.; Álvarez, J.D.; Berenguel, M.; Ortigosa, P.M. A parallel Teaching–Learning-Based Optimization procedure for automatic heliostat aiming. *J. Supercomput.* **2016**, *73*, 591–606. [CrossRef]
64. Xu, F.; Li, C.; Sun, F. On-Line Measurement of Tracking Poses of Heliostats in Concentrated Solar Power Plants. *Sensors* **2024**, *24*, 6373. [CrossRef] [PubMed]
65. Li, Q.; Lin, T.; Yu, Q.; Du, H.; Li, J.; Fu, X. Review of Deep Reinforcement Learning and Its Application in Modern Renewable Power System Control. *Energies* **2023**, *16*, 4143. [CrossRef]
66. Pargmann, M.; Maldonado Quinto, D.; Schwarzbözl, P.; Pitz-Paal, R. High accuracy data-driven heliostat calibration and state prediction with pretrained deep neural networks. *Sol. Energy* **2021**, *218*, 48–56. [CrossRef]
67. Pargmann, M.; Quinto, D.M. How can deep learning be used to improve the heliostat field calibration, even with small data sets?—A transfer learning comparison study. In Proceedings of the SOLARPACES 2020: 26th International Conference on Concentrating Solar Power and Chemical Energy Systems, Online, 28 September–2 October 2020.
68. Carballo, J.A.; Bonilla, J.; Berenguel, M.; Fernández, J.; García, G. Solar tower power mockup for the assessment of advanced control techniques. *Renew. Energy* **2020**, *149*, 682–690. [CrossRef]
69. Coquand, M.; Caliot, C.; Hénault, F. Backward-gazing method for heliostats shape errors measurement and calibration. In Proceedings of the SOLARPACES 2016: International Conference on Concentrating Solar Power and Chemical Energy Systems, Abu Dhabi, United Arab Emirates, 11–14 October 2016.

70. Pargmann, M.; Ebert, J.; Götz, M.; Maldonado Quinto, D.; Pitz-Paal, R.; Kesselheim, S. Automatic heliostat learning for in situ concentrating solar power plant metrology with differentiable ray tracing. *Nat. Commun.* **2024**, *15*, 6997. [CrossRef] [PubMed]
71. Bernius, Z.; Danielson, C.; Harper, H.; Armijo, K. Tuning of Real-Time Optimization of Heliostat Concentrated Solar Power. *IEEE Control Syst. Lett.* **2024**, *8*, 2559–2564. [CrossRef]
72. Hayat, H.M.A.; Hussain, S.; Ali, H.M.; Anwar, N.; Iqbal, M.N. Case studies on the effect of two-dimensional heliostat tracking on the performance of domestic scale solar thermal tower. *Case Stud. Therm. Eng.* **2020**, *21*, 100681. [CrossRef]
73. Rodríguez-Sánchez, M.R.; Leray, C.; Toutant, A.; Ferriere, A.; Olalde, G. Development of a new method to estimate the incident solar flux on central receivers from deteriorated heliostats. *Renew. Energy* **2019**, *130*, 182–190. [CrossRef]
74. Berenguel, M.; Rubio, F.R.; Valverde, A.; Lara, P.J.; Arahal, M.R.; Camacho, E.F.; López, M. An artificial vision-based control system for automatic heliostat positioning offset correction in a central receiver solar power plant. *Sol. Energy* **2004**, *76*, 563–575. [CrossRef]
75. Sun, F.; Guo, M.; Wang, Z.; Liang, W.; Xu, Z.; Yang, Y.; Yu, Q. Study on the heliostat tracking correction strategies based on an error-correction model. *Sol. Energy* **2015**, *111*, 252–263. [CrossRef]
76. Abdallah, S.; Nijmeh, S. Two axes sun tracking system with PLC control. *Energy Convers. Manag.* **2004**, *45*, 1931–1939. [CrossRef]
77. Pargmann, M.; Leibauer, M.; Nettelroth, V.; Maldonado Quinto, D.; Pitz-Paal, R. Enhancing heliostat calibration on low data by fusing robotic rigid body kinematics with neural networks. *Sol. Energy* **2023**, *264*, 111962. [CrossRef]
78. Jessen, W.; Röger, M.; Prahl, C.; Pitz-Paal, R. A two-stage method for measuring the heliostat offset. In Proceedings of the SOLARPACES 2020: 26th International Conference on Concentrating Solar Power and Chemical Energy Systems, Online, 28 September–2 October 2020.

energies

MDPI

Article

Advanced Performance Prediction of Triple-Junction Solar Cell Structures Using MATLAB/Simulink Under Variable Conditions

Olfa Bel Hadj Brahim Kechiche * and Habib Sammouda

Laboratory of Energies and Materials (LR11ES34), University of Sousse, ESSTHSousse, Rue Abbassi Lamine, Hammam Sousse 4011, Tunisia
* Correspondence: olfa.belhadjbrahim@essths.u-sousse.tn; Tel.: +216-23882186

Abstract: Raising the efficiency of triple-junction cells such as (GaInP/GaInAs/Ge) is an important goal for designing high-concentration photovoltaic systems. This purpose can be achieved by facing cell obstacles and acting on their configurations to sustain under highly concentrated sunlight and high operating temperatures. In this paper, a prediction performance study of triple-junction solar cells with four types of structures is proposed under variable conditions. The results show that the series structure is well-validated with experimental data under standard test conditions and is presented against those under variable conditions. Then, the triple-junction cells are compared and discussed in terms of photovoltaic cell open circuit voltage, photovoltaic cell electrical efficiency, fill factor, and temperature coefficients. Consequently, the results show that the cells can be separated into two categories that are useful for Low Concentration Systems and High Concentration Systems. The Low Concentration Systems present high efficiency at 20 suns. For the High Concentration Systems, the Hybrid 2 type demonstrates an optimal efficiency of 38.48% at 118 suns with a high *FF* (0.873) and shows a lower temperature coefficient than the series type. So, Hybrid 2 presents a good candidate for high-concentration systems with a performance better than the conventional triple-junction cells.

Keywords: single diode; double diode; triple-junction cell (TJC); structure; variable conditions performance; MATLAB/Simulink

Citation: Kechiche, O.B.H.B.; Sammouda, H. Advanced Performance Prediction of Triple-Junction Solar Cell Structures Using MATLAB/Simulink Under Variable Conditions. *Energies* **2024**, *17*, 5943. https://doi.org/10.3390/en17235943

Academic Editor: Francesco Calise

Received: 25 October 2024
Revised: 15 November 2024
Accepted: 18 November 2024
Published: 26 November 2024

1. Introduction

There has been a significant increase in energy consumption, in particular, in fossil fuels such as natural oil, natural gas, coal, etc. However, our planet has a finite amount of fossil resources based on an initial quantity, and this quantity is limited and does not allow for the durability of our system. So, solar energy presents a renewable, clean, and efficient source that is capable of holding human needs in large quantities in the long term. Sunlight energy can be converted into electricity by several advanced technologies, such as photovoltaic PV, which is primarily constructed of semiconductor materials.

Nowadays, the widely available semiconductor consists of a single PN junction. At this junction, only photons that have an energy equal to or greater than the material band gap energy (denoted Eg in ev) are capable of creating pairs of electrons–holes. The low efficiencies of single-junction solar cells explain the limited exploited solar spectrum. In the aim of exploiting the total incident sunlight, a multi-junction solar cell has been proposed with n number of PN junctions from different materials, stacked on top of each other. Thus, the popular triple-junction solar cell with a combination of (GaInP/GaInAs/Ge) absorbs the quasi-totality of the solar spectrum and has achieved over 40% efficiency since 2006 [1–4]. This technology has been motivated primarily for spatial applications and secondly for terrestrial requirements, which have been used on concentration photovoltaic (CPV) systems [5,6]. At present, the conventional triple-junction cells designed by a lattice matched structure of (GaInP/GaInAs/Ge) sub-cells on germanium substrate are produced by several companies [7–9].

As research progresses, photovoltaic (PV) cells are continually advancing toward higher performance levels by addressing multiple factors. These factors include the selection of materials, which significantly influence the efficiency and longevity of the cells. Additionally, the number of junctions within the cells plays a crucial role in improving energy conversion by enabling more efficient electron transport and reducing energy losses. The arrangement of the devices, such as the stacking of layers or the integration of different semiconductor materials, further enhances the overall performance. Innovations in these areas, as well as other optimization techniques, have led to significant strides in increasing the efficiency of PV cells in recent years [10–13]. Therefore, many scientists focus on advancing layers of architecture and materials. D. C. Law et al. [14] reported the preliminary results of two technical approaches; the first is the multi-junction cells on the new procedure of wafer-bonded and layers transferred epitaxial templates. The second is a metamorphic 1ev GaInAs sub-cells in conjunction with inverted growth. They used these techniques to achieve high efficiency, to reduce the cost of PV system components such as glass, encapsulation materials, and metal support structures, and to open the opening wide market areas for CPV systems.

R. R. King et al. [15] predicted the development of future multi-junction solar cell architectures by increasing the superimposed junctions up to six. They obtained optimal efficiency by using empirical models based on the electrical characteristics of solar cell materials. In the previous papers [14,15], they consider limited conditions, such as the variable solar spectrum, during day illumination, and concentration factors at a fixed operating temperature (25 °C). But none of them consider the temperature variation effect on the performance of multi-junction solar cells.

P. T. Chiu et al. [16] realized and performed an experimental evaluation of five junctions' cells (2.2/1.7/1.4/1.05/0.73 eV) with an efficiency of 35.1% for 1 sun and AM 0. Here, "1 Sun" refers to the standard solar irradiance of 1000 watts per square meter (W/m^2), representing the intensity of sunlight under ideal conditions at noon on a clear day. It is commonly used to test and compare the performance of solar panels. In their work, Chiu and colleagues employed a novel and expensive technology known as Semiconductor Wafer Bonding (SWB). They detailed the bonding process and presented the performance of the SWB cells under concentrated light conditions. However, the cells were found to withstand no more than 10 suns, making them unsuitable for high-concentration systems.

In the case of cell arrangement, it constitutes one of the main underlying factors limiting its efficiencies because the conventional triple-junction solar cell is based on serial sub-cell connections [7,17,18]. This allows many drawbacks, including the dependency of the tunnel junction, the lattice mismatching between materials band gap, and the current mismatching produced by the sub-cells [19,20] To minimize these obstacles, the challenge associated is the optimal arrangement choice under conditions of CPV systems. So, various solar cell structures have been cited in the literature [21,22]. T. W. Hsiech et al. [23] utilized a series-parallel solar cell arrangement with diverse bandgaps under different spectrum and concentration factors to improve the current matching restrictions. The performance results are compared to those of conventional triple cells. Y. Ahn et al. [24] presented a theoretical efficiency and the optimum band gap combination of a new structure called the Hybrid Connected Triple Junction (HCTJ) under high illumination at a fixed operating temperature. Unfortunately, these papers that consider the arrangement of solar cells are only developed by the theoretical method of Shockley.

So, the modeling of triple-junction cells requires an electrical model based on single- or/and double-diode equivalent circuits that describe J-V solar cell characteristics. In fact, many papers developed these electrical models for PN junction solar cells [24], but few of them treated the case of triple-junction solar cells [25].

In this work, two models are developed to predict the electrical performance produced by concentrator solar cells with varying concentration intensities and high temperatures.

The models are applied to a comparative study of the single- and double-diode for four proposed solar cell structures, such as series, Hybrid 1, Hybrid 2, and parallel. The four TJC structures comprise the same materials (GaInP/GaInAs/Ge) and the same thickness. Then, the J-V and P-V characteristics are determined using MATLAB/Simulink software (version9.0.0.341360 (R2016a)) through the cells' input parameters. The series structures are well-validated with experimental data under variable conditions of standard, high concentration, and temperatures. Then, the four TJC are compared and discussed in terms of open circuit voltage V_{oc}, efficiency η, fill factor FF, and temperature coefficient. Based on the current results, a new structure will be adopted and can be indicated as a promising direction for further creation of the cell performance model. This paper is organized as follows: Section 1 provides the introduction, while Section 2 covers the modeling of single- and double-diode sub-cells. Section 3 discusses the development of various structures and electrical models. The implementation of triple-junction cells using MATLAB/Simulink is presented in Section 4. Section 5 presents the results and discussion. Finally, Section 6 concludes this paper with the main findings and observations.

2. Triple-Junction Cell Models

To examine the performance of different triple-junction cell structures, two electrical models are selected, which are based firstly on the single-diode model and secondly on the double-diode model for each sub-cell. The single-diode model, frequently referenced in photovoltaic literature, is chosen for its simplicity and capacity to capture the fundamental electrical characteristics of photovoltaic cells. However, under non-standard conditions such as high concentrations or elevated temperatures, the double-diode model proves superior, offering enhanced sensitivity to resistive and recombination losses that are significant in TJC systems. This model is particularly effective in representing efficiency trends in high-concentration environments, where the single-diode model often falls short due to its limited capacity to capture these losses.

2.1. Single-Diode Model of Triple-Junction Cell

The Ji-Vi and Pi-Vi characteristics of each junction of a triple-junction cell are determined by the single-diode model because it is the most used in the literature [26,27]. Figure 1 presents the equivalent single-diode circuit model of each sub-cell.

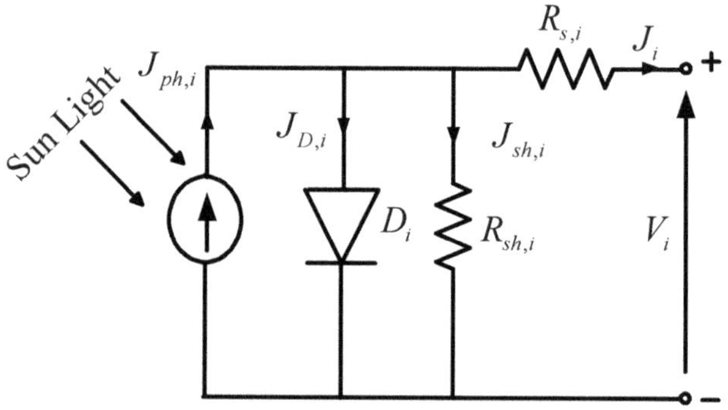

Figure 1. The single-diode equivalent circuit sub-cell model.

Based on Figure 1 and by applying the Kirchhoff law, the sub-cell output current density is described by Equation (1):

$$J_i = J_{ph,i} - J_{D,i} - J_{sh,i} \tag{1}$$

The photocurrent density of the sub-cell $J_{ph,i}$ is proportional to the absorbed concentration factor. It is defined by Equation (2):

$$J_{ph,i} = CJ_{sc,i(1=sun)} \tag{2}$$

where

$$C = \frac{G}{G_{ref}} \tag{3}$$

The sub-cell diode current density $J_{D,i}$ is given by Equation (4) [28]:

$$J_{D,i} = J_{s,i} \exp\left[\left(\frac{(V_i + J_i S R_{s,i})}{n_i V_{th}}\right) - 1\right] \tag{4}$$

The diode reverse current density of each sub-cell $J_{sh,i}$ is highly dependent on the operating cell temperature. It is given by Equation (5) [29]:

$$J_{s,i} = \kappa_i T^{(3+\gamma/2)} \exp\left(\frac{-E_{g,i}}{n_i K_B T}\right) \tag{5}$$

where κ_i and γ are constants, and n_i is the sub-cell diode ideality factor (typically between 1 and 2).

The sub-cell band gap energy $E_{g,i}$ is inversely proportional to the cell operating temperature. It is obtained by Equation (6) using the Varshni relation [30]:

$$E_{g,i}(T) = E_{g,i}(0) + \frac{\alpha_i T^2}{T + \beta_i} \tag{6}$$

where $E_{g,i}(0)$ is the sub-cell band gap energy at 0 K, and α_i, β_i are constants parameters of Varshni.

When the sub-cell in a triple-junction cell is made from an alloy composition selected by the manufacturer (e.g., GaInP, GaInAs). The sub-cell band gap energy is expressed by Vergad's law [31]:

$$E_{g,i}(A_{1-x}B_x) = (1-x)E_{g,i}(A) + xE_{g,i}(B) - x(1-x)P_i \tag{7}$$

where $A_{1-x}B_x$ represents the semiconductors' alloy composition, and P_i is an alloy-dependent parameter that accounts for the deviation from linear approximation.

The thermal voltage of the cell can be defined by Equation (8) [32]:

$$V_{th} = \frac{K_B T}{q} \tag{8}$$

The sub-cell shunt current density $J_{sh,i}$ is given by Equation (9):

$$J_{sh,i} = \frac{V_i + J_i S R_{s,i}}{S R_{sh,i}} \tag{9}$$

Then, the sub-cell output current, resulting from Equations (1), (4) and (9), is given by Equation (10):

$$J_i = J_{ph,i} - J_{s,i} \exp\left[\left(\frac{(V_i + J_i S R_{s,i})}{n_i V_{th}}\right) - 1\right] - \frac{V_i + J_i S R_{s,i}}{S R_{sh,i}} \tag{10}$$

2.2. Double-Diode Model of Triple-Junction Cell

The single-diode model, although widely used in the photovoltaic literature for its simplicity, only captures the basic electrical characteristics of photovoltaic cells, making it suitable for standard test conditions (STCs). However, it is less accurate when modeling advanced structures like triple-junction cells (TJCs) under high concentration or elevated temperature conditions, where more complex phenomena such as recombination losses become more significant. To overcome these limitations, the double-diode model is often employed, as it introduces a second parallel diode that improves the model's sensitivity to resistive and recombination losses, particularly in high-concentration environments.

The double-diode model is particularly advantageous for TJCs because it provides a more detailed representation of the internal phenomena occurring in each sub-cell of the triple junction structure. By adding a second diode, this model accounts for recombination in the space charge region, which is crucial for accurately capturing the behavior of these cells under varying conditions [33,34]. As illustrated in Figure 2 and by applying Kirchhoff's law, the output current density for each sub-cell in the double-diode model is given by Equation (11):

$$J_i = J_{ph,i} - J_{D1,i} - J_{D2,i} - J_{sh,i} \tag{11}$$

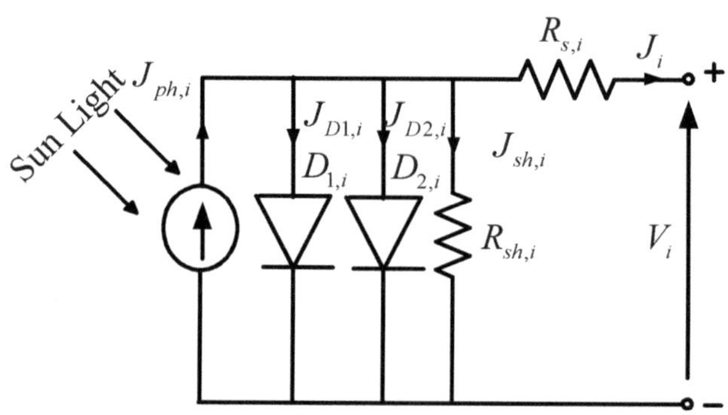

Figure 2. The double-diode equivalent circuit sub-cell model.

The double-diode current density can be written as [35]:

$$J_{D1,i} = J_{s1,i} \exp\left[\left(\frac{(V_i + J_i S R_{s,i})}{n_1 V_{th,i}}\right) - 1\right] \tag{12}$$

$$J_{D2,i} = J_{s2,i} \exp\left[\left(\frac{(V_i + J_i S R_{s,i})}{n_2 V_{th,i}}\right) - 1\right] \tag{13}$$

where the idealities factors n_1 and n_2 are fixed at values of 1 and 2, respectively.

The two-diode saturation current densities as a function of operating triple-junction cell temperature are given as follows [36,37]:

$$J_{s1,i} = \kappa_{1,i} T^3 \exp\left(\frac{-E_{g,i}}{n_1 K_B T}\right) \tag{14}$$

$$J_{s2,i} = \kappa_{2,i} T^{(5/2)} \exp\left(\frac{-E_{g,i}}{n_2 K_B T}\right) \tag{15}$$

where $\kappa_{1,i}$ and $\kappa_{2,i}$ are constants.

Finally, the current density equation of each sub-cell of the triple-junction cell, resulting in Equations (9) and (11)–(13), is given by Equation (16):

$$\begin{aligned} J_i = J_{ph,i} - J_{s1,i} \exp\left[\left(\frac{(V_i + J_i S R_{s,i})}{n_1 V_{th}}\right) - 1\right] \\ - J_{s2,i} \exp\left[\left(\frac{(V_i + J_i S R_{s,i})}{n_2 V_{th}}\right) - 1\right] - \frac{V_i + J_i S R_{s,i}}{S R_{sh,i}} \end{aligned} \tag{16}$$

3. Structures of Triple-Junction Cells

For comparison purposes, we have designed four types of triple-junction cells with different arrangements that are based on series, Hybrid 1, Hybrid 2, and parallel structure, as shown in Figures 3–6, respectively.

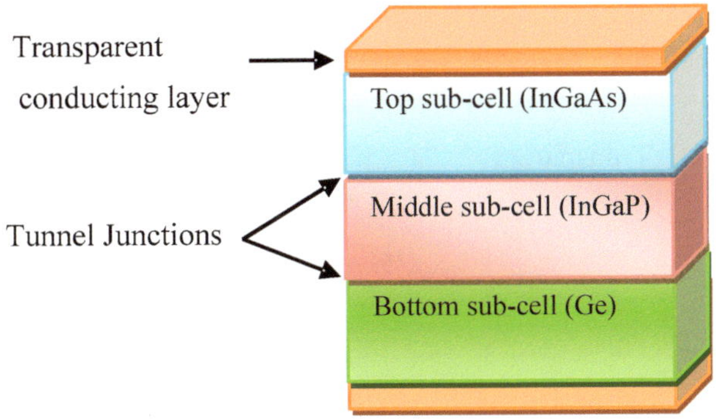

Figure 3. Schematic structure of a series TJC.

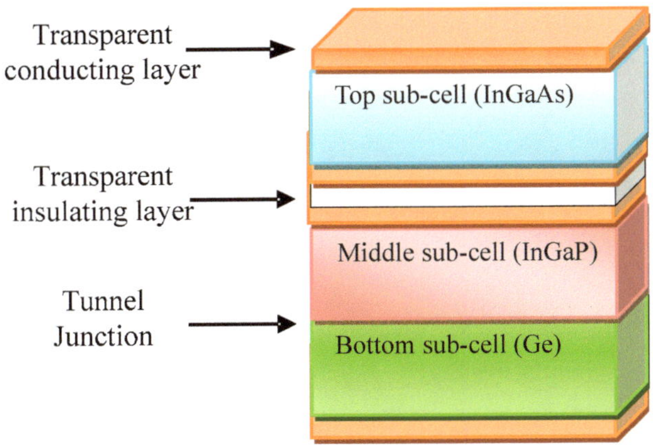

Figure 4. Schematic structure of a Hybrid 1 TJC.

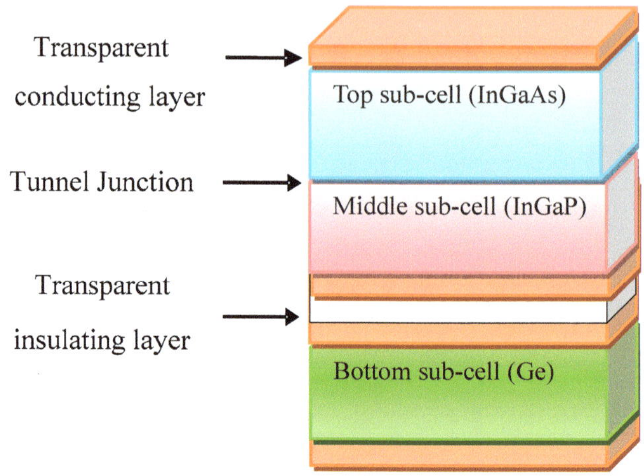

Figure 5. Schematic structure of a Hybrid 2 TJC.

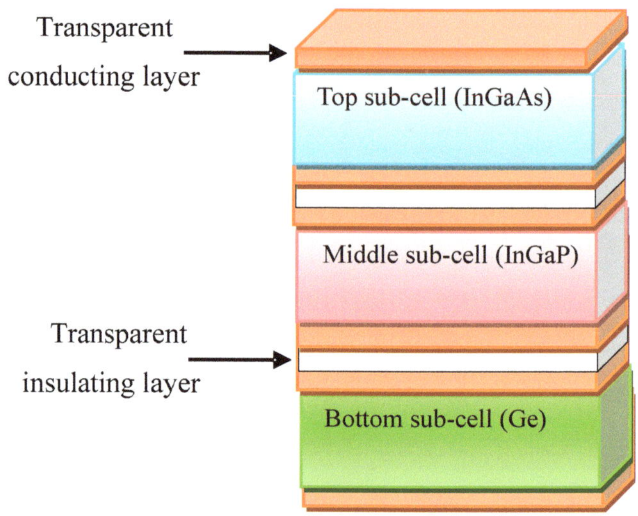

Figure 6. Schematic structure of a parallel TJC.

Therefore, each type is described by electrical models of single and double equivalent circuits in Figures 7–10.

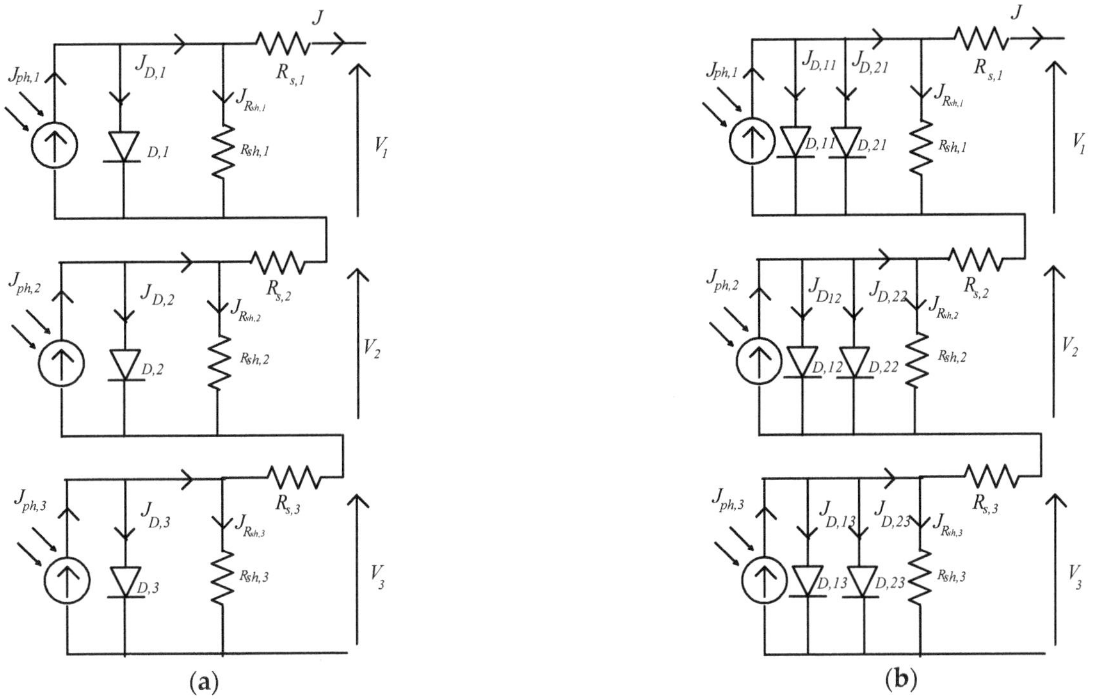

Figure 7. Electrical equivalent circuits of a series TJC. (**a**) Single equivalent circuit of a series TJC; (**b**) double equivalent circuit of a series TJC.

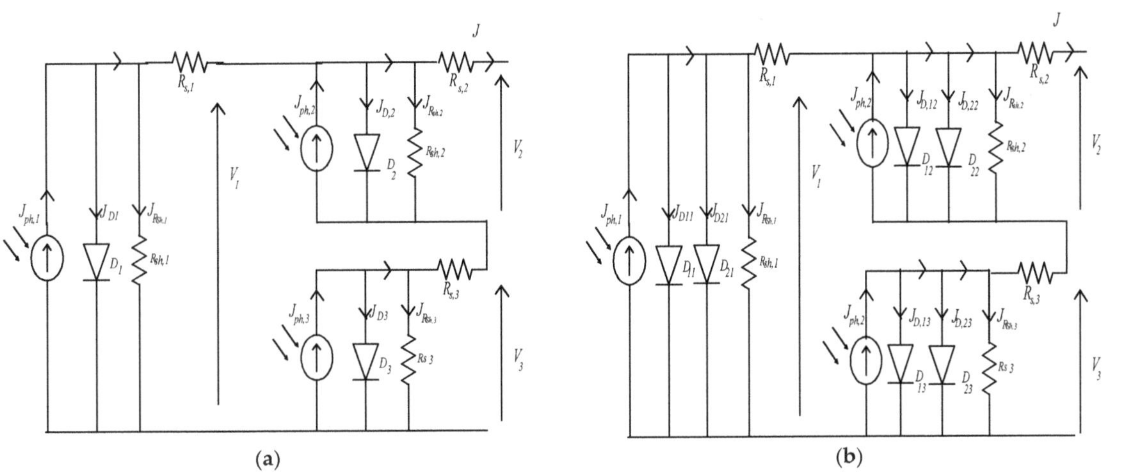

Figure 8. Electrical equivalent circuits of a Hybrid 1 TJC. (**a**) Single equivalent circuit of a Hybrid 1 TJC; (**b**) double equivalent circuit of a Hybrid 1 TJC.

The first type of triple-junction cell denoted "series" is presented in Figure 3 and described by the single and double equivalent circuit, as shown in Figure 7, respectively. The series TJC consists of a series of connected sub-cells in order to obtain high conversion efficiency. Each layer is separated by a special junction called a tunnel junction. From experiment [18], this series structure presents a conventional type of TJC that is composed of three sub-cells (In0.49Ga0.51P/In0.01Ga0.99As/Ge) with a matching lattice. Two tunnel junctions consist of p-AlGaAs/n-InGaP, and p-GaAs/n-GaAs placed between InGaP/InGaAs and InGaAs/Ge, respectively, which are modeled by series resistors as part of $R_{s,i}$.

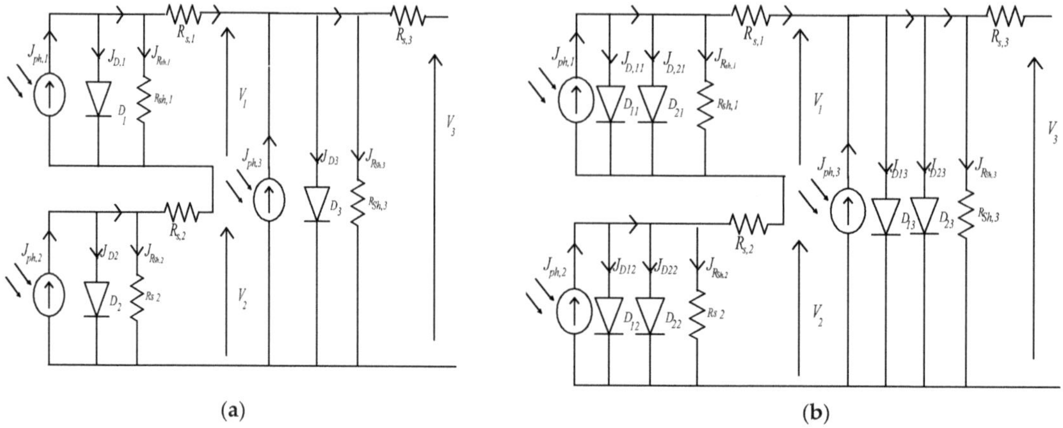

Figure 9. Electrical equivalent circuits of a Hybrid 2 TJC. (**a**) Single equivalent circuit of a Hybrid 2 TJC; (**b**) double equivalent circuit of a Hybrid 2 TJC.

The second type defined as "Hybrid 1" is composed of series and parallel sub-cells which are presented in Figure 4. In this structure, the top sub-cell (InGaAs) is electrically connected in parallel to the middle (InGaP) and the bottom (Ge) sub-cells by changing the current matching to voltage matching and the replacement of the tunnel junction to a simple transparent layer. Thus, only one tunnel junction is presented for the carrier path through the middle and bottom sub-cells. Two Transparent Conducting Oxide (TCO) layers for the electrode between the (InGaAs) and (InGaP) sub-cells are separated by a transparent insulating layer. Figure 8 presents the electrical circuit of this structure.

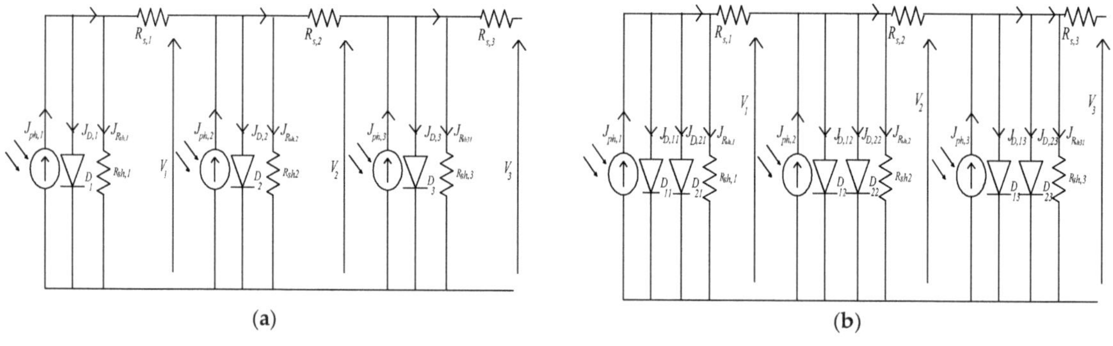

Figure 10. Electrical equivalent circuits of a parallel TJC. (**a**) Single equivalent circuit of a parallel TJC; (**b**) double equivalent circuit of a parallel TJC.

The third type defined as "hybrid 2" is also established by series and parallel sub-cells, which are presented in Figure 5, and electrically modeled by single- and double-diode models, as shown in Figure 9. The arrangement indicates that the bottom sub-cell (Ge) is electrically connected in parallel to the middle (InGaP) and the top (InGaAs) sub-cells. In this case, the junction replacement was inverse to that affected in Hybrid 1.

The fourth type is considered a parallel junction, as presented in Figure 6. The sub-cells are laterally aligned to eliminate the requirement of current matching. In this structure, tunnel junctions are replaced by simple, transparent conducting layers, and the voltage matching rule is followed. In two TCO layers, transparent insulating layers are placed to be electrically isolated from each other, each receiving a certain part of the solar spectrum. Consequently, this structure is electrically modeled by three parallel circuits as illustrated in Figure 10.

4. Implantation of Triple-Junction Cells on MATLAB/ Simulink

The behavior of different structures of triple-junction cells under various conditions, such as concentration factors and a high temperature, were investigated and compared. MATLAB/Simulink are accurate and accessible simulation research tools based on modeling and providing consistent results with a simple structure such as a single-junction solar cell [37,38]. They are suited for investigating triple-junction cell structures. Thus, we created numerical simulation models for the four types of triple-junction cells based on the theories and equations cited in the previous section. The Simulink models considered single- and double-diode equivalent circuits.

To keep consistency with the reference publication, the values of standard test conditions (STCs) of conventional triple-junction cells are considered where 1 sun, temperature, and spectral irradiance are defined, respectively, at 1 KW/m^2, 25 °C, and AM = 1.5. The main parameters of triple-junction cell for single- and double-diode models are indicated in Table 1.

Table 1. Cell input parameters.

	GaInP [18,30]		GaInAs [18,30]		Ge [30]
	GaP	InP	GaAs	InAs	–
E_g (T = 0K)	2.857	1.41	1.519	0.42	0.7347
$\alpha \times 10^{-4}$ (eV/K)	5.771	3.63	5.405	4.19	4.774
β (K)	372	162	204	271	235
Alloy composition	$In_{0.49}Ga_{0.51}P$		$In_{0.01}Ga_{0.99}As$		–
P (eV)	1.018		1.192		–
J_{sc} (mA/cm^2) for C = 1, T = 25 °C	13.78		15.74		20.60
n	1.97		1.75		1.96

The TJC structures with single- and double-diode models have been implanted in Simulink to generate the J-V and P-V electrical characteristics at different values of STCs, concentration factors, and high temperatures. Founded on the simulation results, performance parameters can be determined in terms of open circuits voltages V_{oc}, efficiencies η, fill factors *FF*, and temperature coefficients as a function of the variable conditions treated with the aim of investigating the behavior of TJC structures.

So, we have adopted from the available experimental data [17,18] the range of temperature and concentration factors, which are, respectively, 25 < T < 125 °C and 1 < C < 300.

Figure 11 shows the Simulink model of the top sub-cell (GaInP) using a single diode. In particular, (a) the band gap energy Eg calculation, (b) the reverse saturation current density J1 calculation, and (c) the basic model of a (GaInP) sub-cell. This model presents one sub-system that can join the others to create a series TJC Simulink model.

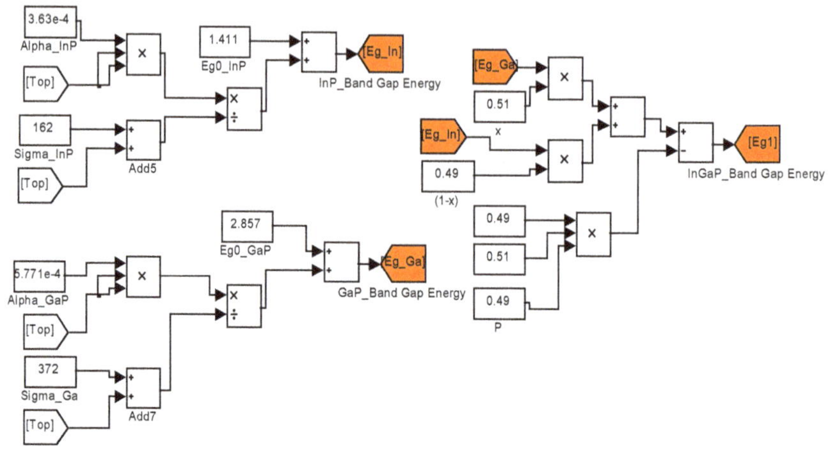

(a)

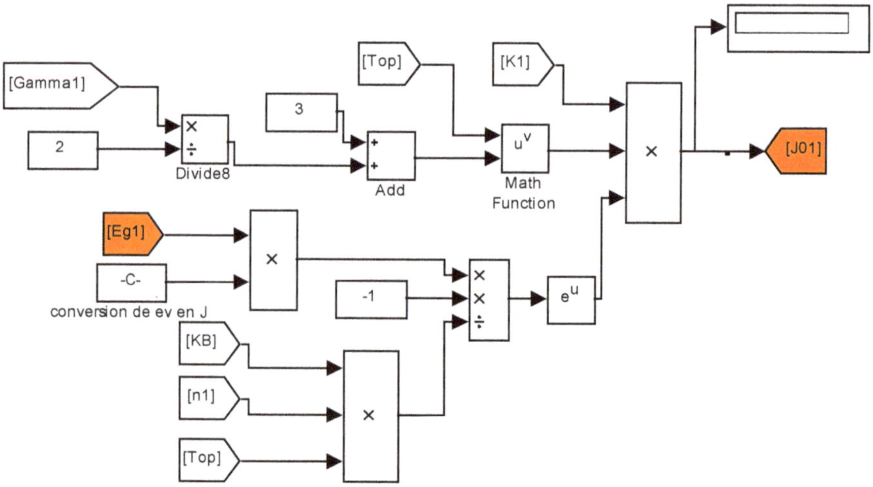

(b)

Figure 11. *Cont.*

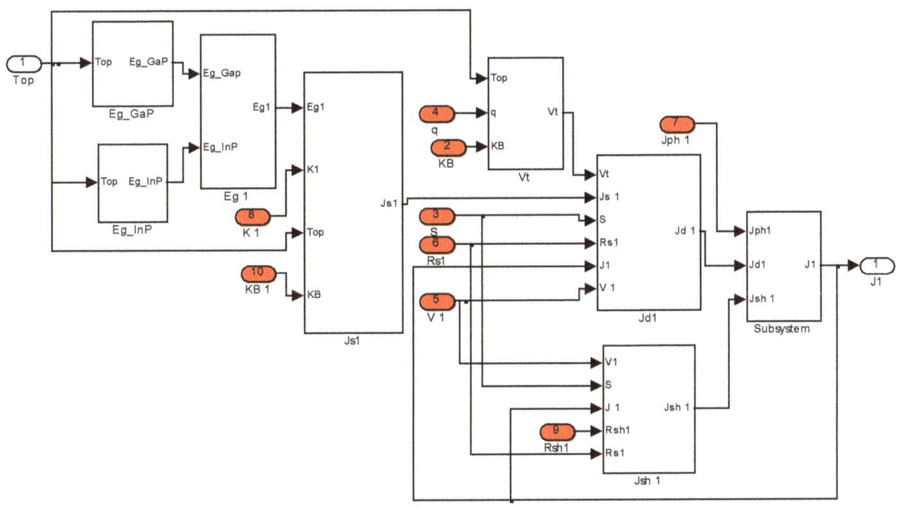

(c)

Figure 11. Simulink models of series TJC (GaInP) sub-cell with (**a**) Eg calculation, (**b**) J1 calculation, (**c**) basic model of (GaInP) sub-cell.

5. Results and Discussion

5.1. Single-Diode Model

Based on Simulink models, we have plotted the current density–voltage (J-V) and the power density–voltage (P-V) characteristics of the series TJC under STCs. The numerical results are indicated below in Figure 12.

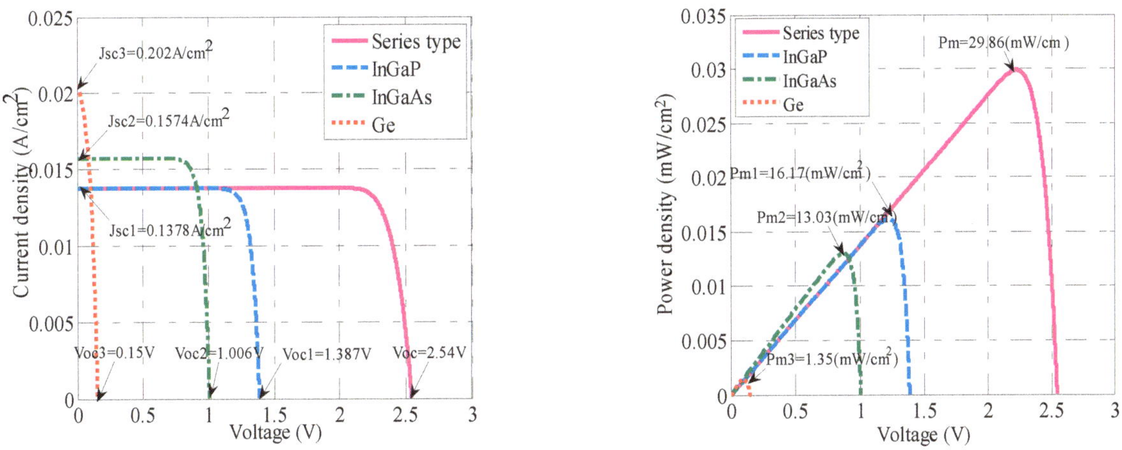

(a) (b)

Figure 12. Electrical characteristics of series TJC at STC (C = 1, T = 25 °C, and AM = 1.5). (**a**) J-V characteristics of series TJC and its sub-cells at STCs; (**b**) P-V characteristics of series TJC and its sub-cells at STCs; by using the single-diode model.

After obtaining the electrical characteristics, it was noted that the current density output due to the series connection is given by the minimum sub-cell current density (J = min (J1, J2, J3)). For the voltage output V, it is the sum of the sub-cell voltage (V ≈ V1 + V2 + V3).

In the aim of extracting and calculating the performance parameters from their J-V and P-V characteristics indicated in Figure 12, we defined these parameters by short circuit density output J_{sc}, open circuit voltage output V_{oc}, fill factor FF, and cell efficiency η.

The fill factor FF is a parameter that indicates the quality of the solar cell. It is defined by the maximum power density P_m divided by the V_{oc} and J_{sc}:

$$FF = \frac{P_m}{J_{sc} \cdot V_{oc}} \tag{17}$$

The efficiency cell η is defined using the following equation:

$$\eta = \frac{J_{sc} \cdot V_{oc} \cdot FF}{C \cdot G_{ref}} \tag{18}$$

Therefore, the model of the series TJC is validated by an experimental study by [18] and cited in Table 2. The comparison study shows good correlations between our results and those found in the literature [18] in terms of I_{sc}, V_{oc}, FF, and η under STCs.

Table 2. Performances parameters of series TJC–single-diode model.

	V_{oc} (V)	I_{sc} (mA)	FF	η (%)
Experimental results [18]	2.53	6.74	0.849	29.5
Numerical results	2.54	6.755	0.852	29.86

5.1.1. Concentration Effect

As presented in Equation (18), the performance parameters are strongly influenced by the concentration factors. Consequently, the J-V and P-V are affected by the illumination level. The effect of concentration factor variation from 1 to 300 suns is shown in Figure 13 for the series TJC and in Figures 14 and 15 for different TJC structures.

Figure 13 presents the variation of predicted electrical efficiency using a single-diode model compared with the experimental data [7,39] under variable concentration factors at a fixed temperature of 25 °C. We remarked that the predicted efficiency is higher and more perfect than the experimental data. The η of the series TJC increased linearly with the concentration factor and reached a maximal value of 40.59% at 300 suns compared to experimental data which achieved 36.44% and 38.78% for 206 and 210 suns, respectively. This can be explained by the logarithmic (weakly) and linear increase of V_{oc} and J_{sc} correspondingly with the increase in concentration factors.

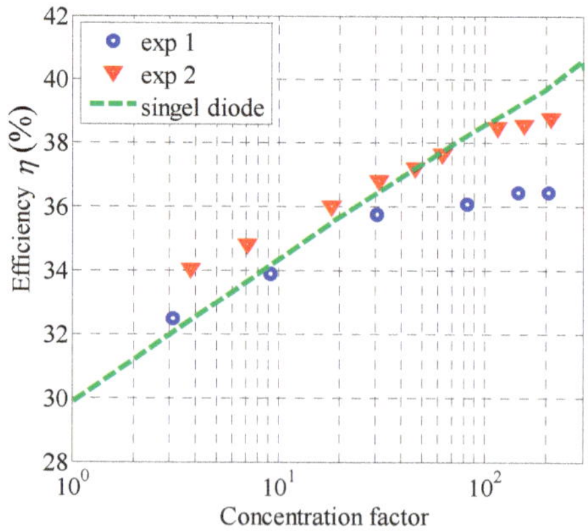

Figure 13. η prediction of series TJC against experimental data under variable concentration factors at T = 25 °C; by using the single-diode model.

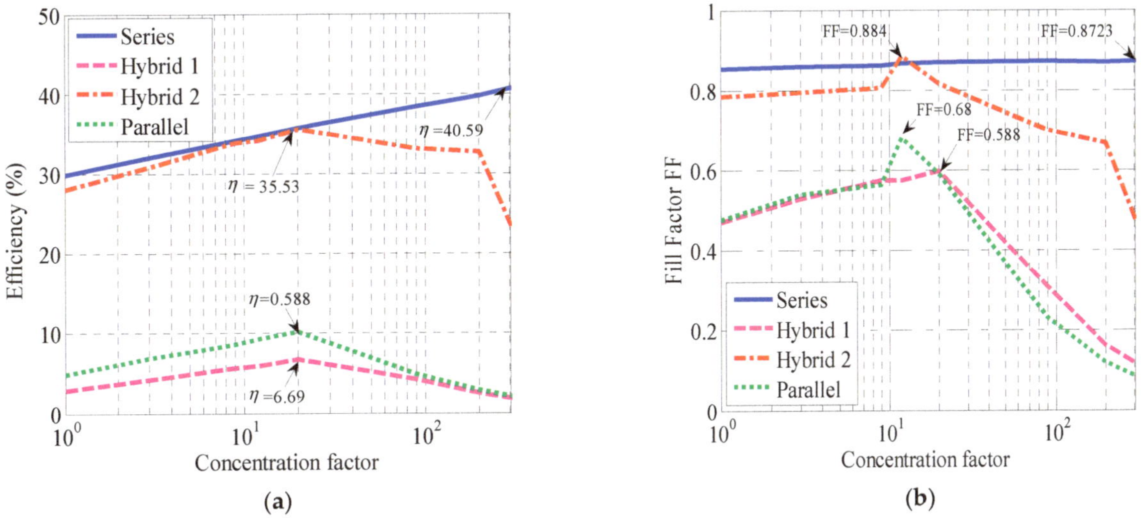

Figure 14. Performance parameters variation of different TJC structures under variable concentration factors at T = 25 °C: (**a**) efficiency η and (**b**) fill factor *FF*; by using the single-diode model.

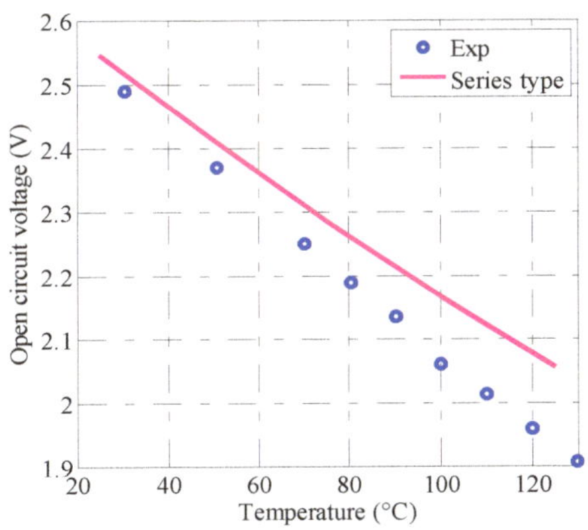

Figure 15. V_{oc} prediction of series TJC against experimental data under variable temperatures at C = 1; by using the single-diode model.

Figure 14a depicts the η of the predicted variations of different TJC structures under variable concentration factors at T = 25 °C using a single-diode model. We noticed that the TJC structures can be divided into two categories; the first is the TJC with high efficiency (series and Hybrid 1), and the second is the TJC with low efficiency (Hybrid 2 and parallel). We observed that the series TJC achieves the highest efficiency with a maximum suns number compared to the other TJC structures. This structure increased to a maximal value of 40.59% at 300 suns compared to Hybrid 1, Hybrid 2, and the parallel type, which achieved 6.69%, 35.53%, and 10.17% at 20 suns, respectively.

Figure 14b shows the *FF*-predicted variations of different TJC structures under variable concentration factors using a single-diode model. The series TJC obtained the higher *FF* with a maximum suns number compared to the other TJC structures. The series TJC increased to a maximal value of 0.87 at 300 suns compared to Hybrid 1, Hybrid 2, and parallel, which achieved 0.59, 0.68, and 0.58 at 20, 12, and 12 suns, respectively. We concluded that Hybrid 2 can hold a good performance under a high concentration factor, and its *FF* values are in the range of [0.6–0.8].

5.1.2. Operating Temperature Effect

As we know, the operating temperature is considered an important factor that affects all kinds of solar cells. In our case, we remarked that the operating temperature of a TJC has a dominant effect on the J-V and P-V characteristics as well as their performances. From these results, firstly, we have plotted and compared the V_{oc} predicted values of the series type against those of the experimental conditions [18], as presented in Figure 15. Secondly, we have determined the various parameters of the four types in terms of their open-circuit voltages V_{oc}, cell efficiency η, and fill factors FF, as shown in Figures 16 and 17.

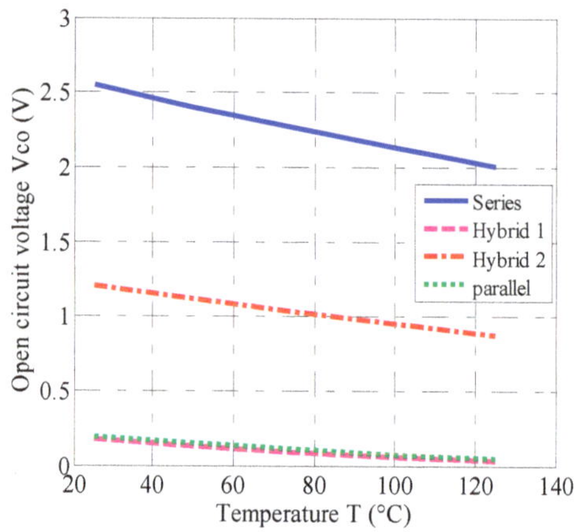

Figure 16. V_{oc} variation of different TJC under variable temperatures at C = 1; by using the single-diode model.

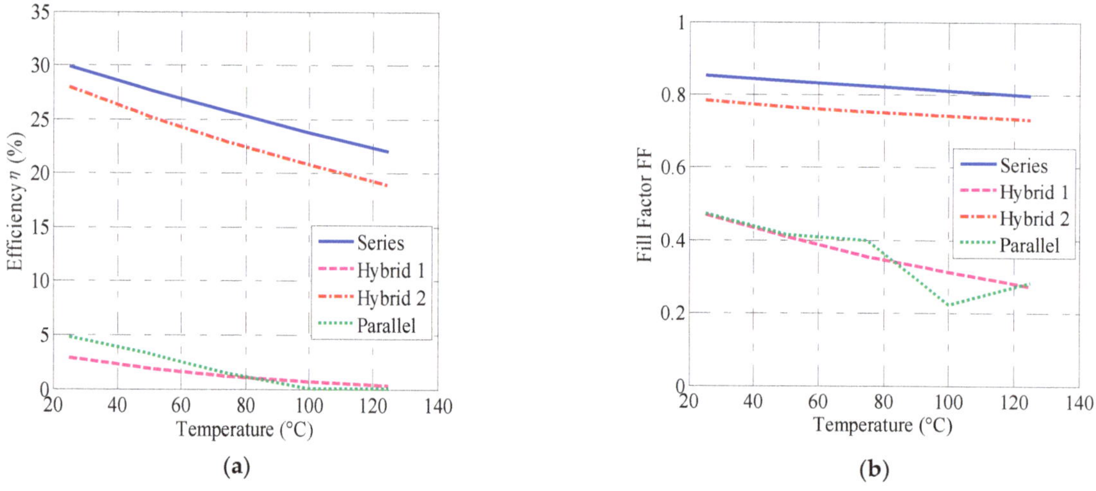

Figure 17. Performance parameters variation of different TJC structures under variable temperature at C = 1: (**a**) efficiency η and (**b**) fill factor FF; by using the single-diode model.

Figure 15 presents the variation of V_{oc} predicted electrical open circuit voltage of conventional TJC (series type) using a single-diode model compared with the experimental data [18] under variable temperature at a concentration factor of 1 sun. We noted that the experimental and the predicted values of V_{oc} have a linear decrease with an increase in cell operating temperature. Also, we considered that the V_{oc} predicted values are nearly estimated for low operating temperatures (lower than 80 °C) and overestimated for high temperatures. We observed that the predicted V_{oc} temperature coefficient dV_{oc}/dt is -0.049 V/°C compared to the measured data 0.06 V/°C [18].

Figure 16 illustrates the V_{oc} variation of different TJC structures under variable temperatures and at C = 1 using a single-diode model. We have observed that the V_{oc} temperature coefficient of series TJC with operating cells temperature was higher than the other types. The voltage decreases with -5.4×10^{-3} V/°C compared to -1.4×10^{-3}, -3.3×10^{-3}, and -1.4×10^{-3} V/°C in Hybrid 1, Hybrid 2, and the parallel type, respectively.

Figure 17 shows the variation in the η and fill factors *FF* of different TJC structures as a function of their operating temperatures and at C = 1 using a single-diode model.

More explication is in the following:

- From Figure 17a, the increase in temperature has separated the structures into two categories as the effect of concentration factors. Therefore, the TJC with high performance decreased linearly in the range of [30–20%].
- From Figure 17b, the variation of *FF* decreased but it does not maintain 0.7. On the other hand, the TJCs with low performance presents poor efficiencies, which decreased less than 5%, and their *FF* shows a linear decrease in the range of [0.2, 0.5].

5.2. Double-Diode Model

After evaluating the single-diode model results of series TJCs, we have also investigated the variation of their electrical characteristics under STCs by using the double-diode model. The simulation results are presented in Figure 18.

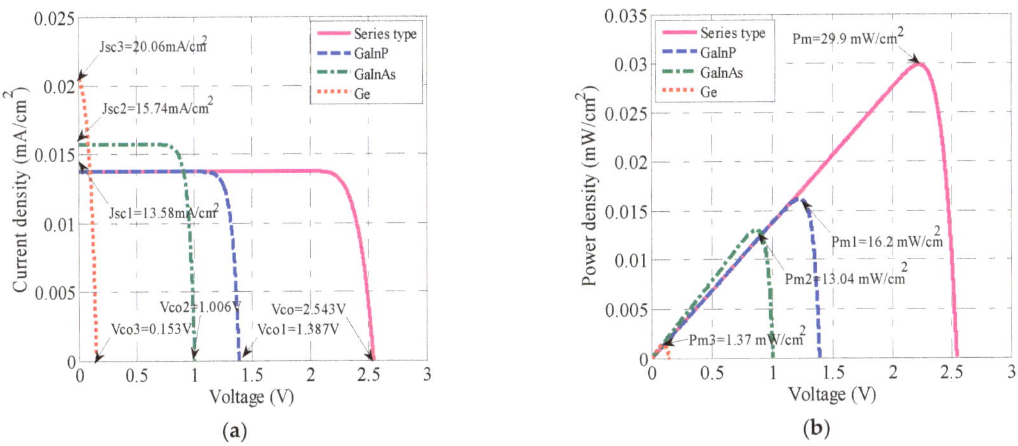

Figure 18. Electrical characteristics of series TJC at STCs (C = 1, T = 25 °C and AM = 1.5): (**a**) J–V characteristics of series TJC and its sub-cells at STCs; (**b**) P–V characteristics of series TJC and its sub-cells at STCs; by using the double-diode model.

In the aim of determining the performance parameters from their J-V and P-V characteristics indicated in Figure 18, the model of the series TJC was validated with the experimental study by [18] and cited in Table 3. The comparison study shows a good correlation between our results and those found in the literature [18] in terms of I_{sc}, V_{oc}, *FF*, and η under STCs.

Table 3. Performances parameters of series TJC–double-diode model.

	V_{oc} (V)	I_{sc} (mA)	*FF*	η (%)
Experimental results [18]	2.53	6.74	0.849	29.5
Numerical results	2.543	6.79	0.848	29.9

5.2.1. Concentration Effect

Figure 19 presents the predicted η variation of series TJC using a double-diode model compared with the experimental data [7] under variable concentration factors at a fixed temperature of 25 °C. We remarked that the predicted efficiency is under the limits of experimental variation [7]. Hence, the model overestimated the efficiency for low concentration (lower than 20). Then, it followed the experimental trend and reached a maximal value of 36.46% at 90 suns compared to experimental data which achieved 36.44% and 38.78% for 206 and 210 suns respectively, after that it decreased. This decrease is explained by power density losses (J2 × Rs) which increase rapidly with concentration intensities.

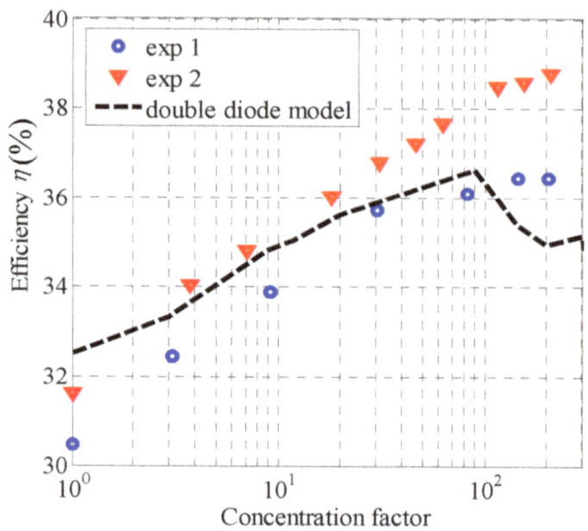

Figure 19. η prediction of series TJC against experimental data under variable concentration factors at T = 25 °C; by using the double-diode model.

Figure 20a depicts the η predicted variations of different TJC structures under variable concentration factors using a double-diode model. As in the previous case of the single diode model, we remarked that the TJC structures can be divided into two categories; the first is the TJC with high efficiency (series and Hybrid 1), and the second is the TJC with low efficiency (Hybrid 2 and parallel). In this model, we observed that the Hybrid 2 of the TJC obtained higher efficiency with a maximum suns number compared to the other TJC structures. This structure increased to a maximal value of 38.48% at 118 suns compared to the series, Hybrid 1, and parallel types, which achieved 36.64%, 13.2%, and 8.58% at 90, 20 and 20 suns, respectively.

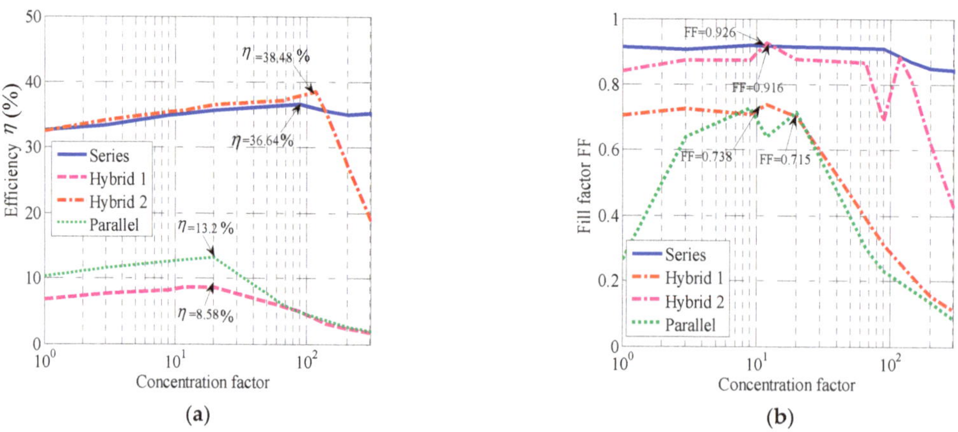

Figure 20. Performance parameters variation of different TJC structures under variable concentration factors at T = 25 °C: (**a**) efficiency η and (**b**) fill factor *FF*; by using the double-diode model.

Figure 20b shows the *FF*-predicted variations of different TJC structures under variable concentration factors using a double-diode model. The Hybrid 2 of the TJC obtains a higher *FF* with a maximum number of suns compared to the other TJC structures. The Hybrid 2 of the TJC increased to a maximal value of 0.926 at 12 suns compared to the series, Hybrid 1, and parallel types, which achieved 0.916, 0.738, and 0.715 at 12, 12, and 20 suns, respectively. As a result, we concluded from the second model that the Hybrid 2 can offer a high η under a high concentration factor, and its *FF* values are in the range of [0.6–0.8], which is an optimal performance quality.

5.2.2. Operating Temperature Effect

From the simulation results, we have first plotted and compared the V_{oc} predicted values of the series type against those of the experimental conditions, as presented in Figure 21. Secondly, we have determined the various parameters of the four types in terms of their open-circuit voltage V_{oc}, cell efficiency η, and fill factor *FF*, as shown in Figures 22 and 23.

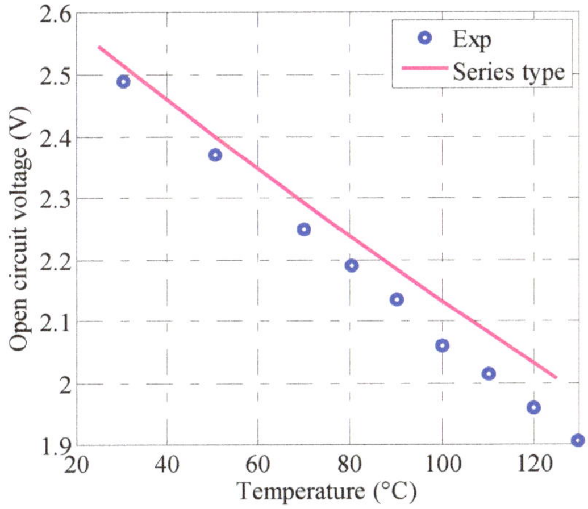

Figure 21. V_{oc} prediction of series TJC against experimental data under variable temperatures at C = 1; by using the double-diode model.

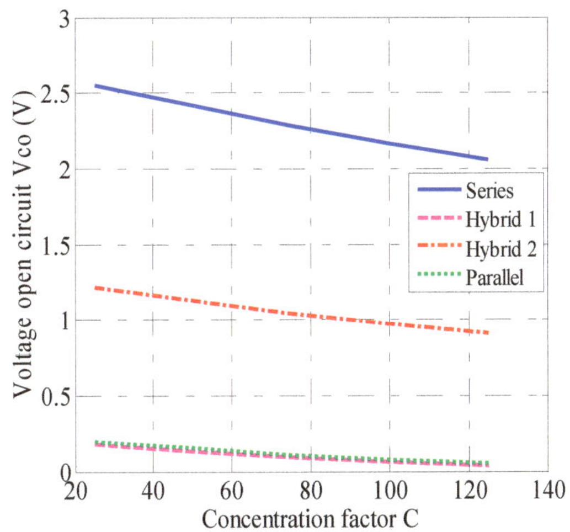

Figure 22. V_{oc} variation of different TJC under variable temperatures at C = 1; by using the double-diode model.

Figure 21 presents the variation in the predicted electrical open-circuit voltage V_{oc} of the conventional TJC (series type) using a double-diode model compared with the experimental data [18] under variable temperature at a concentration factor of 1 sun. We noted that the experimental and the predicted values of V_{oc} have a linear decrease with an increase in cell operating temperature. Also, we considered that the V_{oc} predicted values are nearly estimated for low operating temperatures (lower than 80 °C) and overestimated for high temperatures. We observed that the predicted V_{oc} temperature coefficient dV_{oc}/dt is -0.055 V/°C compared to the -0.0049 V/°C calculated with the single-diode model (the measured datum is -0.06 V/°C [18]).

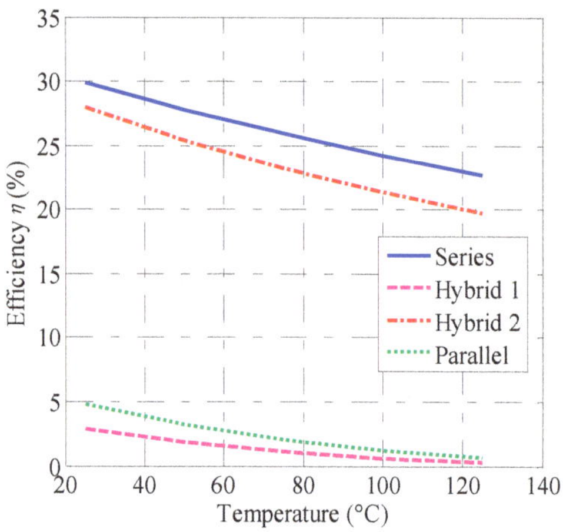

(a)

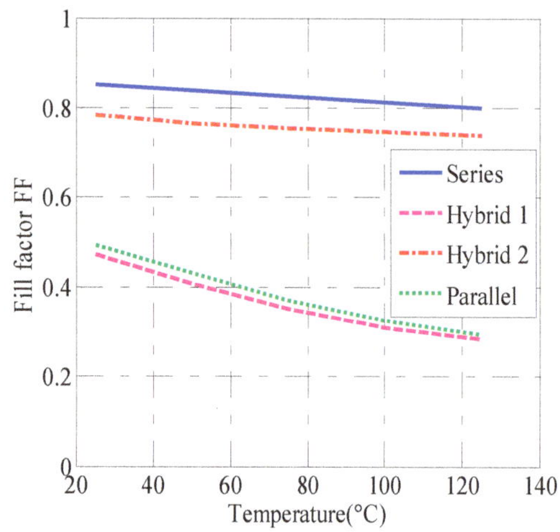

(b)

Figure 23. Performance parameters variation of different TJC structures under variable temperature at C = 1: (**a**) efficiency η and (**b**) fill factor *FF*; by using the double-diode model.

Figure 22 illustrates the V_{oc} variation of different TJC structures under variable temperatures and at C = 1 using a double-diode model. We have observed that the V_{oc} temperature coefficient of series TJC with operating cells temperature was higher than the other types. The voltage decreases with -5.4×10^{-3} V/°C compared to -1.4×10^{-3}, -3.1×10^{-3}, and -1.5×10^{-3} V/°C in Hybrid 1, Hybrid 2, and the parallel type, respectively.

Figure 23 shows the variation of the η and fill factors *FF*s of different TJC structures as a function of their operating temperatures and at C = 1 using a double-diode model.

For Figure 23a, the predicted efficiency temperature coefficient of the series TJC is -0.072%/°C compared to -0.077%/°C by the single-diode model (the measured is 0.073%/°C [18]). Again, the trend in the double-diode model follows the trend in the single-diode model. The increase in temperature has separated the structures into two categories as the effect of concentration factors. Therefore, the TJC with high performance decreased linearly in the range of [30–20%].

For Figure 23b, the predicted *FF* temperature coefficient of type 1 is -0.00055/°C compared to -0.005/°C by the single-diode model (the measured is 0.0006/°C [19]). In the case of the high-performance TJC, the *FF* decreased, but it does not maintain 0.7. On the other hand, the TJCs with the low-performance TJC present poor efficiencies, which decreased less than 5%, and their *FF* shows a linear decrease in the range of [0.2, 0.5].

5.3. Comparison of Single- and Double-Diode Model Results

This study utilizes both single- and double-diode models to simulate the behavior of triple-junction cells (TJCs) in MATLAB Simulink, offering valuable insights into TJC performance under varying concentration factors and temperature conditions. Selecting the appropriate diode model is essential, as each model provides distinct advantages in terms of simplicity, accuracy, and computational efficiency, which facilitate effective simulations within the MATLAB/Simulink environment.

We developed equivalent circuit models for four TJC configurations using both single- and double-diode models, analyzing their electrical performance across different temperature and concentration intensities. To compare the impact of each model on TJC performance, Figure 24 illustrates the predicted efficiency of the series TJC under various concentration factors at a fixed temperature. This comparison highlights the double-diode model's improved alignment with observed efficiency trends, especially at higher concentration levels where resistive losses are more pronounced.

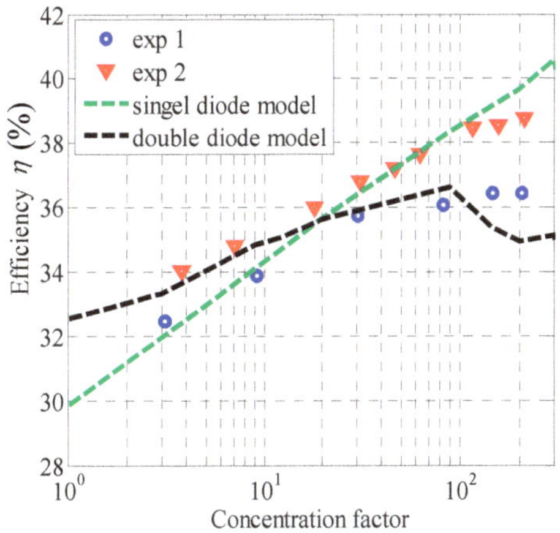

Figure 24. Efficiency prediction of series TJC under variable concentration factors.

As shown in Figure 24, the results from both models align well with experimental data trends [7]. However, the double-diode model demonstrates greater sensitivity, especially near the optimal concentration efficiency, and accurately reflects a decline in efficiency beyond 200 suns—a behavior not captured by the single-diode model. This decline is primarily due to a series of resistive losses, which represent cumulative power losses in the contact and neutral regions of the semiconductor.

In photovoltaic cell modeling, particularly for triple-junction cells (TJCs), the Shockley–Queisser limit defines a theoretical efficiency cap based on detailed balance principles for single-junction cells. Initially set at 33.7% under one sun illumination, this limit can be exceeded in multi-junction architectures like TJCs, where layered structures capture a broader range of the solar spectrum. However, surpassing this limit in practical applications relies on advanced material properties and quantum effects—such as photon absorption in photoactive layers—especially in designs incorporating quantum dots or plasmonic enhancements.

This study employs a simplified MATLAB Simulink model to simulate TJCs under varying concentration and temperature conditions. While effective for modeling general performance trends, this model does not incorporate quantum mechanical effects, limiting its ability to capture certain quantum efficiency characteristics seen in more complex designs. Consequently, while our model accurately reflects general trends, it does not account for phenomena such as photon recycling, multi-exciton generation, or photon management via plasmonic structures. These aspects represent potential areas for enhancement in future modeling efforts.

This context defines the current model's limitations and places it within the Shockley–Queisser framework, highlighting its computational simplicity. The model is well-suited for studies under standard conditions, but it has limitations when applied to advanced photovoltaic architectures.

The observed lower efficiencies of the Hybrid 1 and parallel configurations, compared to the series and Hybrid 2 configurations in both the single- and double-diode models, can be attributed to several factors:

1. Electrical Configuration Differences: The series and Hybrid 2 configurations allow for efficient current matching across the sub-cells, optimizing energy conversion. However, the Hybrid 1 and parallel configurations encounter current mismatches, especially in high-concentration scenarios, which leads to greater resistive losses and subsequently lower efficiencies.
2. Temperature Coefficients: The series and Hybrid 2 configurations exhibit lower temperature coefficients for V_{oc}, FF, and efficiency, which enables them to sustain higher performance under varying concentration and temperature conditions. By contrast, the Hybrid 1 and parallel configurations are more sensitive to temperature changes, resulting in performance degradation, particularly under high-concentration conditions.
3. Limitations in Current Matching: Effective current balancing across sub-cells is crucial for maintaining high efficiency across different diode models. The series and Hybrid 2 configurations achieve this balance more effectively, while the Hybrid 1 and parallel configurations face inherent limitations in current matching, resulting in efficiency losses, as observed in Figures 14a and 20a.

Moreover, the efficiency of conventional TJCs (series type) is highly dependent on the series resistance (R_s) of the solar cell, particularly under high-concentration intensities. Thus, the efficiency parameter can be expressed in relation to R_s, as shown in the following equation:

$$\eta(x)_{Rs} = \eta(1)\left[1 + \frac{nkT}{qV_{oc}(1)}Ln(x) - \frac{R_S x S J_{sc}^2}{G}\right] \quad (19)$$

The calculated temperature coefficients of the series type are compared to the experimental parameters in Table 4. In both diode models, we remarked a good agreement between the predicted values and the experimental data [18] in terms of V_{oc}, FF, and η, whereas the double-diode model results are close to the experimental values.

Table 4. Series temperature coefficient: experiment and models.

	dV_{oc}/dT	dFF/dT	$d\eta/dT$
Exp [6]	−0.0062	−0.0073	−0.006
Single diode	−0.0049	−0.0077	−0.0051
Double diode	−0.0054	−0.0072	−0.0055

Also, the temperature coefficients of the TJCs, such as V_{oc}, FF, and η, are calculated using single- and double-diode models and compared in Table 4. In both diode models, we observed that all the temperature coefficients are converged. Moreover, we have remarked that dV_{oc}/dT and $d\eta/dT$ have a similar dependency. This can be explained by the approximation shown in Equation (20):

$$\frac{1}{\eta}\frac{d\eta}{dT} \approx \frac{1}{V_{oc}}\frac{dV_{oc}}{dT} \quad (20)$$

However, it should be noted that this approximation may not be valid with an increase in concentration factors due to the strong decrease in dV_{oc}/dT, as discussed in the literature.

As can be seen from Table 5, the Hybrid 2 type has lower temperature coefficients in terms of V_{oc}, FF, and η than the series type. Therefore, we concluded that this type can be a good candidate for triple-junction cells under a high variation in operating temperature. We also encourage elaborating on this type because the Hybrid 2 type demonstrates optimal efficiency under a high concentration factor with a high quality of performance (fill factor).

Table 5. TJC temperature coefficients: single- and double-diode models.

		Single Diode	Double Diode
dV_{oc}/dT	Series	−0.0049	−0.0054
	Hybrid 1	−0.0014	−0.0014
	Hybrid 2	−0.0031	−0.0033
	parallel	−0.0015	−0.0014
dFF/dT	Series	−0.072	−0.0775
	Hybrid 1	−0.025	−0.025
	Hybrid 2	−0.082	−0.09
	parallel	−0.041	−0.051
$d\eta/dT$	Series	−0.0051	−0.0055
	Hybrid 1	−0.0019	−0.0019
	Hybrid 2	−0.0042	−0.0051
	parallel	−0.002	−0.0014

In this study, relative errors for performance parameters such as open-circuit voltage V_{oc}, fill factor FF, and efficiency η were computed and compared to experimental results to validate the single- and double-diode models, as seen in Equation (21).

$$\text{Relative Error}(\%) = \frac{\text{Experimental Value-Simulated Value}}{\text{Experimental Value}} \times 100 \quad (21)$$

The low relative errors obtained (mostly below 5%), seen in Table 6, confirm the validity of the models, although larger discrepancies were observed under certain conditions, such as at high temperatures or extreme concentration factors. This analysis not only quantifies the robustness of the models but also highlights areas for improvement to better reflect cell performance under varying conditions.

Table 6. Comparison of experimental and simulated values with relative errors.

Parameter	Experimental Value	Single-Diode Model	Relative Error Single (%)	Double-Diode Model	Relative Error Double (%)
V_{oc} (V)	2.53	2.54	0.39	2.543	0.51
I_{sc} (mA)	6.74	6.755	0.22	6.79	0.74
FF	0.849	0.852	0.35	0.848	0.12
η (%)	29.5	29.86	1.22	29.9	1.35

However, while the relative errors provide a strong indication of model accuracy under standard and moderate conditions, they also point to certain limitations of the MATLAB/Simulink simulation method when applied to high-concentration environments. In these scenarios, resistive and recombination losses become significant, with the single-diode model underestimating power losses and the double-diode model offering limited accuracy under extreme thermal variations. Furthermore, the models do not fully account for complexities such as non-uniform current effects or rapid junction temperature increases, which restrict their ability to faithfully reflect TJC behavior in extreme conditions. Thus, while robust for low to moderate concentrations, these models require further refinement to enhance their applicability under high-concentration scenarios.

6. Conclusions

In this work, we presented single- and double-diode equivalent circuit models for four triple-junction cell structures—series, Hybrid 1, Hybrid 2, and parallel—under varying conditions of high concentration and high temperature. These structures, composed of identical materials with realistic parameters, were evaluated using both models. Validation of the series structure's performance against experimental data [7,19,39] confirmed that both models yield consistent results, with the double-diode model offering slightly higher accuracy. Importantly, the results revealed that these configurations can be categorized into two performance groups: low efficiency (Hybrid 1 and parallel) and high efficiency (series and Hybrid 2). The low-efficiency group demonstrates relatively high efficiency at 20 suns making these configurations suitable for low-concentration systems. In contrast, Hybrid 2 shows optimal performance, achieving an efficiency of 38.48% at 118 suns with a high fill factor (FF = 0.873) and lower temperature coefficients for V_{oc}, FF, and efficiency η compared to the series type. This makes Hybrid 2 a strong candidate for high-concentration systems with a performance surpassing that of the conventional series configuration.

Furthermore, our study observed that both Hybrid 2 and series configurations demonstrate superior efficiency, especially in high-concentration conditions, as illustrated in Figures 14 and 20. While Hybrid 1 and parallel configurations may be effective in low concentration applications, series and Hybrid 2 consistently deliver enhanced efficiency across both low- and high-concentration levels. This distinction provides valuable guidance for selecting configurations based on specific concentration requirements, with Hybrid 1 and parallel suited for low-concentration applications and series and Hybrid 2 performing well at higher concentrations.

Author Contributions: O.B.H.B.K.: methodology, original draft, conceptualization, editing—final draft, methodology, investigation, formal analysis; H.S.: project administration, visualization writing—supervision, and editing. All authors have read and agreed to the published version of the manuscript.

Funding: This research received no external funding.

Data Availability Statement: The original contributions presented in the study are included in the article, further inquiries can be directed to the corresponding author.

Acknowledgments: This research project was supported by the Tunisian Ministry of Higher Education and Scientific Research under Grant LabEM—ESSTHSousse—LR11ES34—University of Sousse.

Conflicts of Interest: The authors declare no conflict of interest.

Nomenclature

C	Concentration factor
TJC	Triple-junction cell
i	Sub-cell number
E_g	Band gap energy (ev)
FF	Fill factor
G	Incident flux irradiance on the cell area (W/cm^2)
G_{ref}	Reference flux irradiance on the cell area (=10 W/cm^2)
J	PV cell ouput current density (mA/cm^2)
J_{ph}	Photocurrent density (mA/cm^2)
J_D	Diode current density (mA/cm^2)
J_s	Diode reverse saturation current density (mA/cm^2)
$J_{sc,ref}$	Short circuit current density of PV cell (mA/cm^2)
J_i	Sub-cell current density (mA/cm^2)
K_B	Boltzmann constant (1.38 $\times$ 10^{-23} J/K)
S	PV cell area (cm^2)
n	Diode ideality factor
P_m	Maximum power density (W/cm^2)
P_i	Sub-cell power density (W/cm^2)
R_s	Cell series resistor (Ω)
R_{sh}	Cell shunt resistor (Ω)
$STCs$	Standard test conditions
T	Cell operating temperature (K)

T_{ref}	Cell temperature at $STCs$ (=298.18 K)
V	PV cell output voltage (V)
V_{oc}	PV cell open circuit voltage (V)
V_{th}	Thermal voltage of the cell (V)
V_i	Sub-cell voltage (V)
x	Suns number
q	Electronic charge
η	PV cell electrical efficiency (%)

References

1. Schygulla, P.; Müller, R.; Lackner, D.; Höhn, O.; Hauser, H.; Bläsi, B.; Predan, F.; Benick, J.; Hermle, M.; Glunz, S.W.; et al. Two-terminal III–V//Si triple-junction solar cell with power conversion efficiency of 35.9% at AM1.5g. *Prog. Photovolt. Res. Appl.* **2022**, *30*, 869–879. [CrossRef]
2. Barrutia, L.; Garcia, I.; Barrigón, E.; Ochoa, M.; Lombardero, I.; Hinojosa, M.; Cano, P.; Bautista, J.; Cifuentes, L.; Rey-Stolle, I.; et al. Development of the Lattice Matched GaInP/GaInAs/Ge Triple Junction Solar Cell with an Efficiency Over 40%. In Proceedings of the 2018 12th Spanish Conference on Electron Devices (CDE), Salamanca, Spain, 14–16 November 2018; pp. 1–4. [CrossRef]
3. King, R.R.; Law, D.C.; Edmondson, K.M.; Fetzer, C.M.; Kinsey, G.; Yoon, H.; Sherif, R.A.; Karam, N.H. 40% efficient metamorphic GaInP/GaInAs/Ge multijunction solar cells. *Appl. Phys. Lett.* **2007**, *90*, 183516. [CrossRef]
4. Mousa, M.; Amer, F.Z.; Mubarak, R.I.; Saeed, A. Simulation of Optimized High-Current Tandem Solar-Cells With Efficiency Beyond 41%. *IEEE Access* **2021**, *9*, 49724–49737. [CrossRef]
5. Gomez-San-Juan, A.M.; Cubas, J.; Pindado, S. On the Thermo-Electrical Modeling of Small Satellite's Solar Panels. *IEEE Trans. Aerosp. Electron. Syst.* **2021**, *57*, 1672–1684. [CrossRef]
6. Kuo, M.-T.; Lo, W.-Y. A Combination of concentrator photovoltaics and water cooling system to improve solar energy utilization. *IEEE Trans. Ind. Appl.* **2014**, *50*, 2818–2827. [CrossRef]
7. Kinsey, G.S.; Edmondson, K.M. Spectral response and energy output of concentrator multijunction solar cells. *Prog. Photovolt. Res. Appl.* **2009**, *17*, 279–288. [CrossRef]
8. Hassan, S. *Four-Terminal Mechanically Stacked GaAs/Si Tandem Solar Cells*; Univesity of Michigan: Ann Arbor, MI, USA, 2015; pp. 2–5.
9. Essig, S.; Allebé, C.; Geisz, J.F.; Steiner, M.A.; Barraud, L.; Descoeudres, A.; Ward, J.S.; Schnabel, M.; Young, D.L.; Despeisse, M.; et al. Mechanically stacked 4-terminal III-V/Si tandem solar cells. In Proceedings of the 2017 IEEE 44th Photovoltaic Specialist Conference (PVSC), Washington, DC, USA, 25–30 June 2017; pp. 25–30.
10. Wang, Y.; Wang, Y.; Gao, F.; Yang, D. Efficient Monolithic Perovskite/Silicon Tandem Photovoltaics. *Energy Environ. Mater.* **2023**, *7*, e12639. [CrossRef]
11. France, R.M.; Geisz, J.F.; Garcia, I.; Steiner, M.A.; McMahon, W.E.; Friedman, D.J.; Moriarty, T.E.; Osterwald, C.; Ward, J.S.; Duda, A.; et al. Quadruple-Junction Inverted Metamorphic Concentrator Devices. *IEEE J. Photovolt.* **2015**, *5*, 432–437. [CrossRef]
12. Dimroth, F.; Grave, M.; Beutel, P.; Fiedeler, U.; Karcher, C.; Tibbits, T.N.; Oliva, E.; Siefer, G.; Schachtner, M.; Wekkeli, A.; et al. Wafer bonded four-junction GaInP/GaAs//GaInAsP/GaInAs concentrator solar cells with 44.7% efficiency. *Prog. Photovolt. Res. Appl.* **2014**, *22*, 277–282. [CrossRef]
13. Trupke, T.; Würfel, P. Improved spectral robustness of triple tandem solar cells by combined series/parallel interconnection. *J. Appl. Phys.* **2004**, *96*, 2347–2351. [CrossRef]
14. Law, D.C.; King, R.; Yoon, H.; Archer, M.; Boca, A.; Fetzer, C.; Mesropian, S.; Isshiki, T.; Haddad, M.; Edmondson, K.; et al. Future technology pathways of terrestrial III–V multijunction solar cells for concentrator photovoltaic systems. *Sol. Energy Mater. Sol. Cells* **2010**, *94*, 1314–1318. [CrossRef]
15. King, R.R.; Bhusari, D.; Larrabee, D.; Liu, X.; Rehder, E.; Edmondson, K.; Cotal, H.; Jones, R.K.; Ermer, J.H.; Fetzer, C.M.; et al. Solar cell generations over 40% efficiency. *Prog. Photovolt. Res. Appl.* **2012**, *20*, 801–815. [CrossRef]
16. Chiu, P.T.; Law, D.C.; Woo, R.L.; Singer, S.B.; Bhusari, D.; Hong, W.D.; Zakaria, A.; Boisvert, J.; Mesropian, S.; King, R.R.; et al. Direct semiconductor bonded 5J cell for space and terrestrial applications. *IEEE J. Photovolt.* **2014**, *4*, 493–497. [CrossRef]
17. Nishioka, K.; Takamoto, T.; Agui, T.; Kaneiwa, M.; Uraoka, Y.; Fuyuki, T. Evaluation of temperature characteristics of high-efficiency InGaP/InGaAs/Ge triple-junction solar cells under concentration. *Sol. Energy Mater. Sol. Cells* **2005**, *85*, 429–436. [CrossRef]
18. Nishioka, K.; Takamoto, T.; Agui, T.; Kaneiwa, M.; Uraoka, Y.; Fuyuki, T. Annual output estimation of concentrator photovoltaic systems using high-efficiency InGaP/InGaAs/Ge triple-junction solar cells based on experimental solar cell's characteristics and field-test meteorological data. *Sol. Energy Mater. Sol. Cells* **2006**, *90*, 57–67. [CrossRef]
19. Paulauskas, T.; Pačebutas, V.; Strazdienė, V.; Geižutis, A.; Devenson, J.; Kamarauskas, M.; Skapas, M.; Kondrotas, R.; Drazdys, M.; Rudzikas, M.; et al. Performance assessment of a triple-junction solar cell with 1.0 eV GaAsBi absorber. *Discov. Nano* **2023**, *18*, 86. [CrossRef]
20. Das, N.; Wongsodihardjo, H.; Islam, S. Modeling of multi-junction photovoltaic cell using MATLAB/Simulink to improve the conversion efficiency. *Renew. Energy* **2015**, *74*, 917–924. [CrossRef]

21. Xing, Y.; Han, P.; Wang, S.; Fan, Y.; Liang, P.; Ye, Z.; Li, X.; Hu, S.; Lou, S.; Zhao, C.; et al. Performance analysis of vertical multi-junction solar cell with front surface diffusion for high concentration. *Sol. Energy* **2013**, *94*, 8–18. [CrossRef]
22. Ahn, Y.; Kim, Y.-H.; Kim, S.-I. Detailed balance calculation of a novel triple-junction solar cell structure. *IEEE J. Photovolt.* **2013**, *3*, 1403–1408. [CrossRef]
23. Hsieh, T.W.; Yu, P. Design optimization of series-parallel triple-junction solar cells. In Proceedings of the 2010 35th IEEE Photovoltaic Specialists Conference (PVSC), Honolulu, HI, USA, 20–25 June 2010; pp. 002924–002927.
24. Hamza, M.; Kechiche, O.B.H.B.; Sammouda, H. Dependence of Novel Triple Junction Solar Cell Parameters on Cell's Temperature. In Proceedings of the 2019 10th International Renewable Energy Congress (IREC), Sousse, Tunisia, 26–28 March 2019; pp. 1–6.
25. Houssein, E.H.; Nassef, A.M.; Fathy, A.; Mahdy, M.A.; Rezk, H. Modified search and rescue optimization algorithm for identifying the optimal parameters of high efficiency triple-junction solar cell/module. *Int. J. Energy Res.* **2022**, *46*, 13961–13985. [CrossRef]
26. Hamza, M.; Sammouda, H. Modeling and optimizing of non-imaging disc concentrator (NIDC) photovoltaic system performance under non-uniform illumination. *Optik Int. J. Light Electron Opt.* **2020**, *203*, 163906. [CrossRef]
27. Vinod; Kumar, R.; Singh, S. Solar photovoltaic modeling and simulation: As a renewable energy solution. *Energy Rep.* **2018**, *4*, 701–712. [CrossRef]
28. Bellia, H.; Youcef, R.; Fatima, M. A detailed modeling of photovoltaic module using MATLAB. *NRIAG J. Astron. Geophys.* **2014**, *3*, 53–61. [CrossRef]
29. Richard, O.; Jaouad, A.; Bouzazi, B.; Arès, R.; Fafard, S.; Aimez, V. Simulation of a through cell via contacts architecture for HCPV multi-junction solar cells. *Sol. Energy Mater. Sol. Cells* **2016**, *144*, 173–180. [CrossRef]
30. O'donnell, K.P.; Chen, X. Temperature dependence of semiconductor band gaps. *Appl. Phys. Lett.* **1991**, *58*, 2924–2926. [CrossRef]
31. Nahory, R.E.; Pollack, M.A.; Johnston, W.D.; Barns, R.L. Band gap versus composition and demonstration of Vegard's law for In 1−xGaxAsyP1−y lattice matched to InP. *Appl. Phys. Lett.* **1978**, *33*, 659–661. [CrossRef]
32. Kechiche, O.B.H.B.; Hamza, M. Enhancement of a commercial PV module performance under Low Concentrated Photovoltaic (LCPV) conditions: A numerical study. *Renew. Energy Focus* **2022**, *41*, 258–267. [CrossRef]
33. Yadav, D.; Singh, N.; Bhadoria, V.S.; Vita, V.; Fotis, G.; Tsampasis, E.G.; Maris, T.I. Analysis of the Factors Influencing the Performance of Single- and Multi-Diode PV Solar Modules. *IEEE Access* **2023**, *11*, 95507–95525. [CrossRef]
34. Aidoud, M.; Feraga, C.-E.; Bechouat, M.; Sedraoui, M.; Kahla, S. Development of photovoltaic cell models using fundamental modeling approaches. *Energy Procedia* **2019**, *162*, 263–274. [CrossRef]
35. Elbaset, A.A.; Ali, H.; Sattar, M.A.-E. Novel seven-parameter model for photovoltaic modules. *Sol. Energy Mater. Sol. Cells* **2014**, *130*, 442–455. [CrossRef]
36. Lv, H.; Sheng, F.; Dai, J.; Liu, W.; Cheng, C.; Zhang, J. Temperature-dependent model of concentrator photovoltaic modules combining optical elements and III–V multi-junction solar cells. *Sol. Energy* **2015**, *112*, 351–360. [CrossRef]
37. Badi, N.; Khasim, S.; Al-Ghamdi, S.A.; Alatawi, A.S.; Ignatiev, A. Accurate modeling and simulation of solar photovoltaic panels with simulink-MATLAB. *J. Comput. Electron.* **2021**, *20*, 974–983. [CrossRef]
38. Premkumar, M.; Kumar, C.; Sowmya, R. Mathematical modelling of solar photovoltaic cell/panel/array based on the physical parameters from the manufacturer's datasheet. *Int. J. Renew. Energy Dev.* **2020**, *9*, 7–22. [CrossRef]
39. Kinsey, G.S.; Hebert, P.; Barbour, K.E.; Krut, D.D.; Cotal, H.L.; Sherif, R.A. Concentrator multijunction solar cell characteristics under variable intensity and temperature. *Prog. Photovolt. Res. Appl.* **2008**, *16*, 503–508. [CrossRef]

technologies

MDPI

Review

Overview of Recent Solar Photovoltaic Cooling System Approach

Yaareb Elias Ahmed [1,2,*], Mohammad Reza Maghami [1], Jagadeesh Pasupuleti [3,*], Suad Hassan Danook [4,*] and Firas Basim Ismail [1]

1 Department of Electrical and Electronics Engineering, College of Engineering, Universiti Tenaga Nasional, Kajang 43000, Selangor, Malaysia; reza.maghami@uniten.edu.my (M.R.M.); firas@uniten.edu.my (F.B.I.)
2 Technical Institute of Hawija, Northern Technical University, Kirkuk 36001, Iraq
3 Institute of Sustainable Energy, Universiti Tenaga Nasional, Kajang 43000, Selangor, Malaysia
4 Department of Mechanical Engineering, IJSU Imam Jaafer AL-Sadiq University, Kirkuk 36001, Iraq
* Correspondence: yaarub_hwj@ntu.edu.iq (Y.E.A.); jagadeesh@uniten.edu.my (J.P.); suad@sadiq.edu.iq (S.H.D.)

Abstract: In recent years, research communities have shown significant interest in solar energy systems and their cooling. While using cells to generate power, cooling systems are often used for solar cells (SCs) to enhance their efficiency and lifespan. However, during this conversion process, they can generate heat. This heat can affect the performance of solar cells in both advantageous and detrimental ways. Cooling cells and coordinating their use are vital to energy efficiency and longevity, which can help save energy, reduce energy costs, and achieve global emission targets. The primary objective of this review is to provide a thorough and comparative analysis of recent developments in solar cell cooling. In addition, the research discussed here reviews and compares various cooling systems that can be used to improve cell performance, including active cooling and passive cooling. The outcomes reveal that phase-change materials (PCMs) help address critical economic goals, such as reducing the cost of PV degradation, while enhancing the lifespan of solar cells and improving their efficiency, reliability, and quality. Active PCMs offer precise control, while passive PCMs are simpler and more efficient in terms of energy use, but they offer less control over temperature. Moreover, an innovative review of advanced cooling methods is presented, highlighting their potential to improve the efficiency of solar cells.

Keywords: solar energy; water colling; solar photovoltaic; active cooling; passive cooling

Citation: Ahmed, Y.E.; Maghami, M.R.; Pasupuleti, J.; Danook, S.H.; Basim Ismail, F. Overview of Recent Solar Photovoltaic Cooling System Approach. *Technologies* **2024**, *12*, 171. https://doi.org/10.3390/technologies12090171

Academic Editor: Dongran Song

Received: 20 July 2024
Revised: 10 September 2024
Accepted: 12 September 2024
Published: 19 September 2024

1. Introduction

Today, one of the primary challenges for photovoltaic (PV) systems is overheating caused by intense solar radiation and elevated ambient temperatures [1–4]. To prevent immediate declines in efficiency and long-term harm, it is essential to utilize efficient cooling techniques [5]. Each degree of cooling of a silicon solar cell can increase its power production by 0.4–0.5%. Therefore, achieving additional cooling of a cell by more than 1.5 °C beyond the existing standard module practices in any location could be beneficial. The primary goal of lowering the temperature of PV modules is to increase the energy yield of solar panel systems. Both air- and water-based cooling methods are employed to reduce the operational temperatures of PV modules. Solar cell cooling plays a crucial role in optimizing the performance, reliability, and longevity of solar panel systems. Effective strategies maximize energy production and reduce temperature stress, making solar energy systems more reliable and cost-effective. Researchers have evaluated cooling system techniques and intelligent control systems, focusing on solar cell cooling systems and phase-change materials (PCMs) [6]. Cooling systems are essential for regulating the temperature of PV modules in large installations, and it is crucial that these methods are cost-effective The following paragraph provides some reasons as to why cooling solar cells is necessary.

Solar cells are temperature-sensitive, and their efficiency decreases as the temperature rises. Most solar cell technologies experience a diminution in performance of roughly 0.5% to 0.8% for every 1-degree Celsius rise in temperature. By cooling solar cells, their operating temperature can be lowered, allowing them to maintain higher efficiency. The operational temperature of a PV module affects its electrical effectiveness and power generation, demonstrating a strong correlation between temperature and the power conversion technique. According to the authors of [7], solar cells capture sunlight and transform it into electrical energy. In this conversion process, some of the absorbed energy is altered into heat. In the case that this heat is not dissipated effectively, it can accumulate within the solar cells, leading to increased operating temperatures and reduced efficiency. Cooling solar cells helps dissipate excess heat, preventing performance degradation. In [8], the solar-based refrigeration system was shown to effectively dissipate heat, reducing resistance and enhancing power output by keeping solar cells cooler. Cooler temperatures help reduce resistive losses and allow the solar cells to operate closer to their optimal voltage and current levels, maximizing their electrical generation capacity and the dissipation of energy [9] as heat during peak sunlight, which diminishes the power output and effectiveness of a PV module. High temperatures can accelerate the degradation of solar cell materials, reducing their lifespan. By cooling the solar cells, the overall operating temperature is lowered, reducing the stress on the materials and prolonging their lifespan. Lower operational temperatures additionally augment the long-standing reliability of the solar cells [10] The photovoltaic industry enhances the efficiency and durability of polymer-based modules through high-speed and high-resolution surface inspection for extended longevity, superior quality, and increased product yield. Concentrated solar power (CSP) systems distillate sunlight by utilizing lenses or mirrors to generate high temperatures. Cooling mechanisms are crucial in these systems to prevent overheating, maintain cell efficiency, and protect the system components [11]. In [12], researchers extensively studied thermal regulation techniques for PV modules, with a particular focus on PCMs for regulating PV system temperature.

The Web of Science portal has published an updated review on PCM technologies, highlighting the increased amount of research in the domain of cooling solar PV systems over the past five years [13]. Figure 1 illustrates the annual number of relevant publications and citations, showcasing the growing importance of this topic among researchers. Last year, the topic peaked, with over 600 articles published. However, in the 2020–2021 period, the publication growth rate decreased, likely due to the COVID-19 pandemic, while citations increased dramatically. Therefore, it is estimated that, by the end of 2024, the number of research publications, as well as citations, will reflect the trends in and importance of this area, indicating future attention from researchers. This trend underscores the resilience and continued relevance of the topic within the scholarly community.

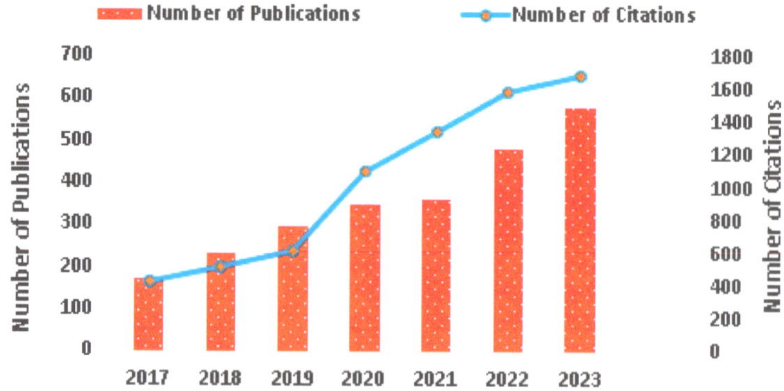

Figure 1. Number of articles in the area of solar cooling published in WOS.

This review is organized into nine sections. Section 1 provides an overview of both solar energy and the cooling systems used to increase efficiency. Section 2 describes the factors that influence the efficiency of solar cells. Section 3 describes an overview of cooling technologies. Section 4 provides an overview of active cooling and a table of the most important studies in the literature that used this system. Section 5 explains passive cooling and provides a table of the most important studies in the literature that used this system. Section 6 provides an overview of the cooling of phase-changing materials and includes a table of the most important studies in the literature that used this system. Section 7 provides an overview of cooling by active phase-changing materials and includes a table of the most important studies in the literature that has used this type. Section 8 provides a comparison of the cooling systems mentioned in this research paper in terms of usefulness, limitations, and impact. Section 9 provides the conclusion and suggestions for future studies. Figure 2 shows the structure of this review paper.

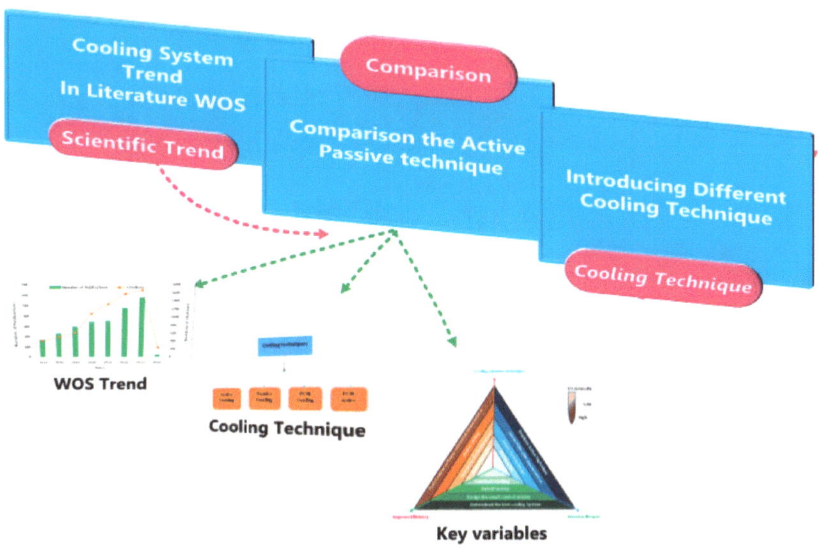

Figure 2. The review paper structure and steps.

2. The Factors Affecting Cooling Performance

Various factors affect the cooling system that enhances the productivity of PV panels, which is essential for maximizing their efficiency and productivity, as presented in Figure 3 [14,15]. Key factors include temperature management, dust, materials, design, environmental challenges, and long-term performance [16,17]. Solar cells are sensitive to temperature changes; higher temperatures can decrease their efficiency, leading to reduced energy generation. The cooling system helps maintain optimal temperatures, thereby enhancing the efficiency and lifespan of the PV panels [18]. Additionally, another important factor affecting the productivity of solar panels is dust accumulation on their surfaces, which can significantly reduce light transmission. The cooling system also aids in the regular cleaning of panels to prevent dust buildup and maintain optimal performance. One of the primary factors influencing cooling system performance is the materials used in its construction. The choice of materials directly impacts the heat transfer rate, durability, and overall effectiveness of the cooling system.

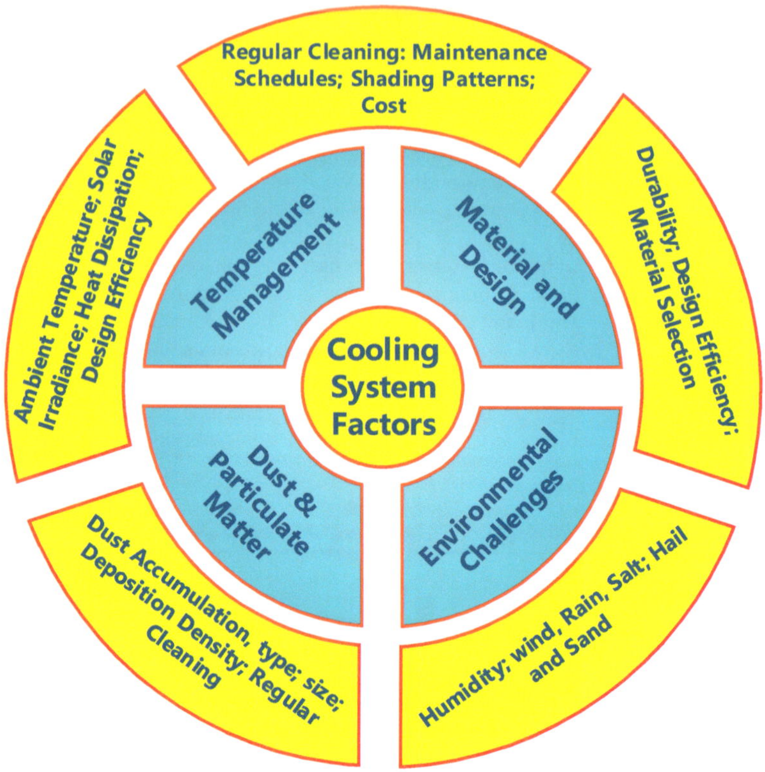

Figure 3. Factors affecting cooling systems.

Long-term performance is another critical factor that must be considered when evaluating the effectiveness of a cooling system. Factors such as material degradation, fouling, and wear can impact the system's efficiency over time. Implementing regular maintenance and monitoring procedures can help mitigate these issues and ensure the long-term reliability of the cooling system. Environmental challenges such as fluctuating temperatures, humidity, wind, rain, hail, sand, and salt can reduce the efficiency and lifespan of PV modules. The cooling system helps in mitigating these effects by maintaining the cleanliness and integrity of the panels. For instance, high humidity can lead to condensation and corrosion, while low humidity can cause overheating. Wind can aid in heat dissipation but also poses challenges in dust accumulation. Rain can clean panels but also lead to temporary drops in performance. Hail and sand can cause physical damage, and salt accumulation in coastal areas can corrode materials [19]. The cooling system must be designed to withstand and adapt to these varied conditions. Figure 3 shows the most common factors that affect cooling performance.

3. Overview of Cooling System Technique

Various cooling techniques can be employed to cool solar cells, including passive cooling methods, such as natural convection and radiation, and active cooling methods involving the use of a water-spray cooling technique (Figure 4) [20]. Figure 5 shows the immersion of polycrystalline solar cells in water [21]. Figure 6 shows the process of active air cooling [22]. Figure 7 shows the cooling process with PCM [23]. The choice of cooling method is contingent on elements such as the specific solar cell technology, system design, and environmental conditions. In summary, the cooling of solar cells is essential in maintaining their efficiency, preventing performance degradation, increasing power output, extending their lifespan, and ensuring the reliable operation of solar energy systems. Overall, and by reviewing the literature, we can see that water-active cooling systems offer more precise control and higher cooling capacities, making them suitable for applications used in solar cell cooling. Figure 8 shows the most common cooling techniques.

Figure 4. Water-spray cooling technique.

Figure 5. Panel immersed in water.

Figure 6. Air-based cooling technique.

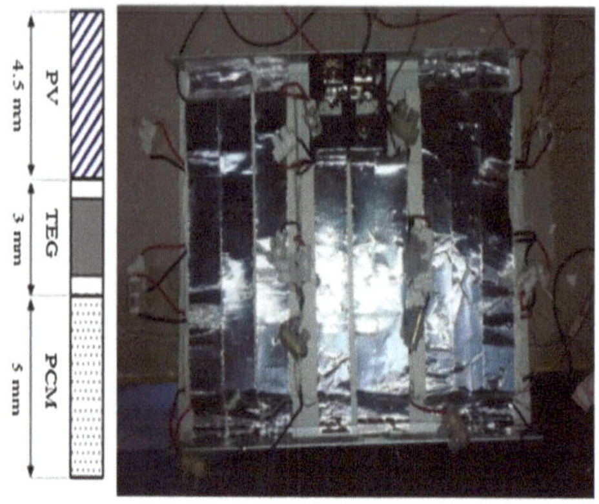

Figure 7. PV/TEG/PCM layout.

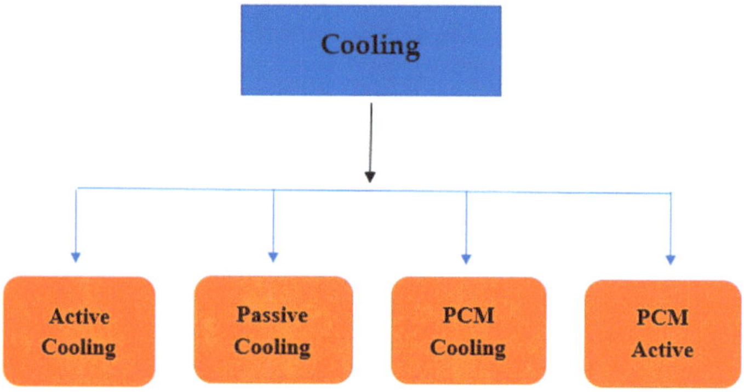

Figure 8. Most common cooling techniques.

The detrimental effect of increasing the surface temperature of PV solar systems, particularly in terms of cooling, is a significant concern for researchers [24]. Passive cooling systems lessen the temperature of PV modules by 6–20 °C, leading to a maximum boost in electrical efficiency of up to 15.5%. Active cooling solutions enhance performance by lowering the temperature of PV modules by up to 30 °C. In [24], the researchers suggested various cooling techniques for photovoltaic panels. The aluminum fins and PCM thermoelectric (TE) were selected for cooling. In [25], the specialists devised a pulsed-spray water cooling system for PV panels that aimed to enhance the efficiency of solar systems while conserving water usage for cooling purposes. The water-spraying approach involves applying a spray of water over the surfaces of PV panels as an alternative method. Another cooling technique involves simultaneously cooling both sides of the PV panel. In [26], the primary performance metrics were detailed for each specific coolant type analyzed, including air, water, and nanofluids. Less-explored cooling methods, namely those associated with concentrated photovoltaic (CPV) systems, have received limited attention. A small number of studies have explored the use of nanofluids in a cooling method for PV systems, highlighting their potential in improving efficiency and longevity.

4. Active Cooling

Active cooling refers to a cooling mechanism that actively removes heat from a system or device. A notable rise in the operating temperature of a cell during the absorption of solar radiation adversely affects its electrical efficiency [27]. Active cooling systems aid in preventing solar cells from reaching elevated operating temperatures, which may poorly affect their performance and efficiency. Active cooling is especially beneficial in regions with high ambient temperatures or in situations where solar panels experience higher heat loads due to factors like concentrated sunlight or limited airflow. It allows for better control and management of the solar cell temperature, ensuring optimal performance and maximizing energy generation. The specific implementation of active cooling methods can vary based on system design, available resources, and cooling requirements. Direct water cooling is a method in which water flows directly over the solar cells' surface, either in contact with the cells or through a separate heat sink. The water absorbs heat from the cells and carries it away, dissipating it through a heat exchanger or a cooling tower.

Another way to active cooling is spray cooling; spray cooling involves the use of nozzles or atomizers which can spray a fine mist or droplets of water onto the solar cell surface. The water droplets evaporate, absorbing heat from the cells and cooling them down. This method provides effective cooling while minimizing water usage. Some researchers have used microchannel cooling; microchannel cooling systems consist of small channels embedded within the solar panel structure. Water flows through these channels, absorbing heat from the solar cells and transferring it away. Microchannel cooling offers high heat transfer efficiency and effective temperature control Others have used heat pipe cooling; here, heat pipes are sealed, closed-loop systems containing a working fluid that transfers heat through evaporation and condensation. Heat pipes are crucial for temperature regulation in solar panels, ensuring efficient heat transfer and the dissipation of heat from cells to the panel structure. To sum up, active cooling is vital for averting overheating and sustaining ideal operational states across various applications. The selection of cooling techniques relies on factors such as device characteristics, efficiency demands, space availability, and cost factors. Table 1 summarizes the findings and details of recent studies in the area of active cooling techniques.

Table 1. Details of several studies on the cooling of PV cells through active cooling techniques.

Year	References	Techniques	Type	Result	Objective
(2019)	[28]	Active cooling	Forced-water cooling; forced-water cooling with buried water	The study demonstrated a 12.20% enhancement in the relative levelized cost of energy as a result of the suggested cooling system. Additionally, this system contributed to a reduction of around 49,209 g CO_2/summer sea in global average CO_2 emissions.	To evaluate the efficiency of a photovoltaic cooling mechanism that combines a V-trough configuration with an underground water heat exchanger.
(2020)	[29]	Active cooling	Forced-water cooling	The results indicated that augmenting the mass flow rate of cooling water has negligible effects. However, under ideal conditions, with a solar irradiance of 1000 W m^{-2}, an ambient temperature of 45 °C, and a water velocity of 0.9 m s^{-1}, the cell's efficiency improved by 17.12%.	To explore how changes in ambient temperature, the flow rate of cooling water, and solar irradiance affect cell efficiency and temperature.
(2017)	[30]	Active cooling	Forced-water cooling	Improves electrical efficiency; yet, with a steady flow rate, it cannot achieve maximal efficiency. If you want maximum efficiency, it is best to modify the flow rate when the temperature changes.	To reduce the negative effects of increased temperature using various techniques.
(2018)	[31]	Active cooling	Forced-water cooling	Experiments were conducted to determine the ideal cooling cycle, which was 20%/80%. The cooling system had an initial startup temperature of 30 °C. We anticipate a two-year payback period for the system.	To cool many photovoltaic chains simultaneously, both on and off, and to consider different facets of its potential as a product for commercial use.
(2020)	[32]	Active cooling	Forced-water cooling	The research demonstrated that the peak exergy of the hybrid renewable system amounts to 872.06 kWh, representing a 2.6% increase compared to the maximum exergy achieved using the Taguchi standard orthogonal array (849.9 kWh).	To maximize the overall exergy, with the integration of an advanced optimization algorithm.
(2012)	[27]	Active cooling	Hybrid photo-voltaic/thermal (PV/T)	Solar cell efficiency went up from 12% to 14% as a result of the precipitous drop in temperature.	To connect a series of ducts to the back of the panel, each with its inlet and output manifold to distribute airflow evenly to cool the PV cells efficiently.
(2017)	[33]	Active cooling	Thermoelectric module (TEM)	Research indicated that the temperature of photovoltaic panels decreased by 6–26% when a temperature-based "Maximum Power Point Tracking (MPPT)" controller was employed under solar insolation levels ranging from 0.8 to 1 kW/m^2 and temperatures between 25 and 45 °C.	To improve the effectiveness and durability of photovoltaic systems. By utilizing thermoelectric technologies to construct and model the PV module
(2020)	[29]	Active cooling	Water flow (forced)	The study revealed that this cooling technique is most efficient in conditions characterized by elevated ambient temperatures and intense sun radiation.	To scrutinize how the mass flow rate of cooling water, ambient room temperature, and variations in solar irradiance affect both the temperature and efficiency of the cell.

Table 1. *Cont.*

Year	References	Techniques	Type	Result	Objective
(2016)	[34]	Active cooling	Forced-water heat exchanger	The effectiveness of the PV panel augmented by 57%, going from 7 W to 11 W, and the module temperature decreased by 32%, from 50 °C.	To study the cooling impact of the PV panel, a forced water heat exchanger will be incorporated through numerical simulation and experimental investigation.
(2016)	[35]	Active cooling	Water flow (forced), reflectors	The cooling system reduces the working temperature of the PV module to 30–35 °C, resulting in an 18.5% increase in power output for water-cooled CPV and an 8% increase for CPV.	To utilize a technique that focuses on and lowers the temperature of sunlight to enhance the electrical performance of the photovoltaic (PV) module.

5. Passive Cooling

Our thorough examination of the literature showed that most investigated passive cooling solutions incorporate PCM, with air-based and liquid-based methods (such as water, nanofluids, etc.) following closely behind [26]. In [36], passive cooling using an aluminum heat sink was studied to evaluate its effect on silicon solar cell performance. The outcomes exhibited a notable improvement in the efficiency of power conversion, exergy, and energy of the solar cell with this cooling method. In [37], researchers used the design of experiment (DOE) methodology to find the best design parameters for fins, such as height, pitch, thickness, number, and tilt angle. Passive fin heat sinks were evaluated in real-world conditions using their optimal design parameters. In [38], the mechanism of passive cooling was devised to tackle the overheating issue of photovoltaic modules. This system involves the utilization of the capillary action of hessian fabric attached to the rear surface of the module and water evaporation to enhance its performance. Air that is static and air that is ventilated are used to cool the modules that have fins. The authors of [39] introduce an innovative passive cooling method for PV modules harnessing the natural flow of cooling water. The system includes a segmented fin heatsink designed to lower the operational temperature of solar modules while maintaining their efficiency intact. The performance of this heatsink under different wind-attack situations is superior to that of the typical continuous fin profile heatsink design [40]. A proposal was made to enhance passive cooling for a solar module by placing it in a heat sink designed as a finned container. This research used palm wax as a PCM, based on the findings of [41], as it costs significantly less than rival commercial PCMs, it exhibits selective spectral cooling, and because passive radiative cooling relies on the PV module's natural ability to reduce heat [42]. In [43], a novel PV panel passive cooling solution is introduced. A segmented aluminum sheet was suggested as a way of enhancing cooling in high-irradiation environments through enhanced airflow. In [44], passive cooling was implemented by adding perforated aluminum fins to the back of the PV panel, resulting in a synergistic design approach when combined with PV systems. The integration of these technologies not only improves energy efficiency and performance but also contributes to a greener and more resilient energy future. Table 2 summarizes the findings and details of recent studies in the area of passive cooling techniques.

Table 2. Details of several studies on the cooling of PV cells through passive cooling techniques..

Year	References	Techniques	Type	Result	Objective
(2018)	[45]	Passive cooling	Convection or conduction	Passive technology is more advanced than other technologies but requires additional maintenance. Passive technology is more economically efficient when considering the factors that limit it.	To enhance the efficiency of a photovoltaic panel by employing various methods to lower its temperature.
(2021)	[38]	Passive cooling	Cooling by natural water evaporation	An evaporative cooling system helped achieve a substantial 26% drop in temperature. The operational temperature of the reference module fluctuated between 54.2 °C and 76.4 °C, averaging at approximately 66.4 °C.	To scrutinize the efficacy of water-based passive cooling mechanisms in enhancing the efficiency of photovoltaic (PV) modules in hot, arid environments.
(2020)	[39]	Passive cooling	Natural water circulation	The introduction of nano-composed oil resulted in the greatest augmentation of the maximum generated power compared to the reference condition. The percentage enhancements were 44.74%, 46.63%, and 48.23% at radiation intensities of 410, 530, and 690 W/m^2, respectively.	To insert nano-composed oil to increase the maximum generated power compared to the baseline condition. The percentage improvements were 44.74%, 46.63%, and 48.23% at radiation intensities of 410, 530, and 690 W/m^2, respectively.
(2022)	[46]	Passive cooling	Floating photovoltaic system	The research demonstrated that the improved floating PV system featuring a finned heatsink outperforms the conventional floating photovoltaic system, decreasing the operating temperature by roughly 19.07%.	To improve the efficacy of a floating PV system, a novel partially floating system is combined with a passive arrangement of finned heatsinks to lower the operating temperature and sustain the module's productivity.
(2019)	[47]	Passive cooling	Evaporative cooling and natural water mass	Raising the water mass from 0 kg to 600 kg (heat capacity) leads to a 4.67% increase in electrical efficiency, attributed to a 4.79 °C decrease in solar cell temperature at midday.	To comprehend the impacts of different passive cooling methods on integrated semitransparent photovoltaic thermal systems.
(2014)	[48]	Passive cooling	Rainwater, gas expansion device	On a design day, the research indicates a reduction in cell temperature along with an 8.3% upsurge in the electrical efficiency of the PV panel attributed to the passive cooling system.	To employ rainwater for passive cooling in a solar system to enhance its performance by distributing it through a gas expansion mechanism.
(2021)	[36]	Passive cooling	Plant cooling, greenhouse cooling, coir pith	Examples of net cooling within greenhouses and cooling plants using greenhouse structures demonstrate that temperature reduction does not consistently lead to increased power. Nations in tropical regions with agriculture-based economies stand to gain the most from employing this cooling technique.	To measure a 50 W polycrystalline photovoltaic module's power generation and temperature drop.
(2011)	[49]	Passive cooling	Aluminum heat sink	Under conditions of 800 W/m^2 radiation, the PV cell's power production increases by around 20%. The cooling effect is most pronounced at an intensity level of 600 W/m^2. Photovoltaic cell efficiency increases when temperature decreases, regardless of the presence of fins.	To understand how passive cooling with an aluminum heat sink affects the performance of silicon photovoltaic systems under various radiation settings.

Table 2. *Cont.*

Year	References	Techniques	Type	Result	Objective
(2013)	[37]	Passive cooling	Cotton wick	A 30% enhanced cooling system has a 1.4% increase in module efficiency, resulting in a 15.61% increase in PV module output power and a module temperature.	To propose a passive cooling system for flat PV modules using cotton wick structures.
(2021)	[50]	Passive cooling	Fin heat sinks namely	The payback period for different types of PV modules was determined to be 4.2 years for longitudinal fins, 5 years for lapping fins, and 8.4 years for exposed PV modules. The electrical efficiency and power output achieved were 10.	To evaluate the efficiency of passive cooling in a concentrated solar module experimentally by using two different types of passive fin heat sinks: pounding and long-term.
(2020)	[51]	Passive cooling	Lapping fin, wind speed	Increasing the fin pitch from 20 to 60 mm reduced the number of fins from 20 to 10 and raised the PV module temperature from 44.13 to 54.01 °C.	To improve the efficiency of the PV module, they incorporated a planar reflector and expand the surface of the back plate.
(2021)	[38]	Passive cooling	Heat sink, aluminum fins, ultrasonic humidifier	The study employed a cooling technique that reduced the temperature of the panel by an average of 14.61 °C. This lessening resulted in a 6.8% upgrade in the electrical efficiency of the module.	To improve the electrical output of the photovoltaic module by employing an aluminum fin heat sink and an ultrasonic humidifier.
(2021)	[52]	Passive cooling	Water evaporation, capillary action, and burlap fabric	The research demonstrated that the proposed evaporative cooling system efficiently lessened the temperature of the PV module by 20 degrees Celsius, marking a 26% reduction. Consequently, there was a significant 14.7% increase in electricity efficiency.	To create a passive cooling system aimed at averting overheating of solar modules while enhancing their efficiency.

6. Passive PCM Cooling System

Recently, there has been a significant increase in the installed capacity of solar photovoltaic cells, particularly crystal silicon cells. Research has focused on enhancing the photovoltaic (PV) conversion efficiency of the cells by exploring methods to cool PV systems, as elevated PV temperatures can reduce conversion efficiency. The efficiency of cooling photovoltaic cells relies on phase-change materials (PCMs) with high latent heat capacities [23]. In fact, PCMs are being studied as a solution for reducing the surface temperature of PV cells during sunlight exposure, with a goal of improving the electrical efficiency of the cells. PCMs can control temperatures by absorbing and releasing thermal energy when they change from one phase to another. This allows them to act as a thermal buffer, maintaining a stable temperature within a desired range. PCM cooling effectively manages heat by absorbing and dissipating excess thermal energy. It acts as a heat sink, preventing overheating and protecting sensitive components or equipment [53]. PCM cooling improves energy efficiency by stabilizing temperatures and reducing reliance on energy-intensive cooling systems. It leads to lower electricity consumption and operating costs. PCMs are effective in storing and releasing large quantities of latent heat during phase transitions, making them valuable for thermal energy storage. This stored energy can be utilized for either cooling or heating purposes [54]. PCM cooling is environmentally friendly as it reduces reliance on energy-intensive cooling methods, leading to lower greenhouse gas emissions. Additionally, PCMs can be derived from renewable or bio-based sources, making them a sustainable cooling option. Figure 9a [55] and Figure 9b [56] show some uses of PCMs. Table 3 summarizes the findings and details of recent studies in the area of PCM cooling techniques.

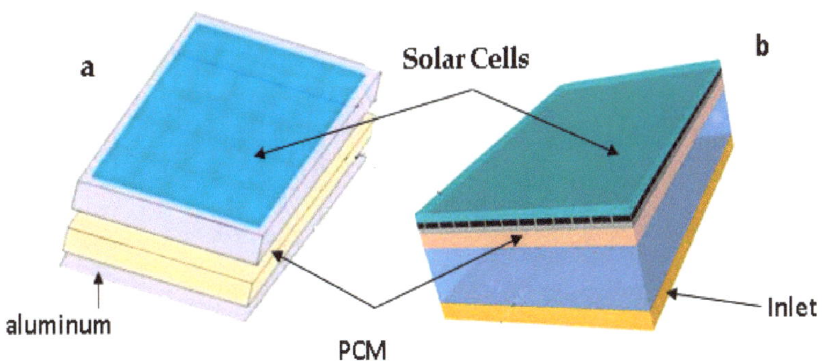

Figure 9. Three-dimensional illustration of PV/PCM configurations featuring aluminum (**a**); schematic representation of an air-based PV/T collector incorporating PCM (**b**).

Table 3. Details of several studies on the cooling of PV cells through composite PCM cooling.

Year	References	Techniques	Type	Result	Objective
(2016)	103	PCM cooling	PV/CS6P-M, PCM	The simulation's final results indicate that the power generation of the PV–PCM panel in Ljubljana exceeded the previous year's output by 7.3%.	To boost the electrical efficiency and power generation of a photovoltaic panel by integrating a phase-change material (PCM).
(2021)	[57]	PCM cooling	PV–PCM cooling systems	The power generated by the PV system rose by 2.5% when utilizing a full PCM container in contrast to a typical PV panel. The innovative PV–PCM passive cooling technology, featuring numerous dissimilar PCM containers, elevated performance by 10.7% compared to the use of a solitary PCM container.	To review the traditional passive cooling method frequently utilized in photovoltaic systems by integrating phase-change material for cooling (PV–PCM cooling systems).
(2021)	[58]	PCM cooling	Water (natural), nano-PCM, PV	The results indicate that utilizing nano-composed PCMs derives superior results compared with using traditional PCMs. The most notable improvement in maximum power output was observed with nano-composed oil, reaching 44.74%, 46.63%, and 48.23% at radiation intensities of 410, 530, and 690 W/m^2, respectively.	To propose a novel passive cooling arrangement for photovoltaic modules that utilizes natural water flow for cooling, supplemented with a nano-enhanced cooling system.
(2019)	[59]	PCM cooling	PV/T-PCM, nanofluid	Increasing coolant concentration boosted electricity and power generation, while higher nanofluid concentration increased pumping power but decreased thermal–electrical equivalent power.	To measure the thermal and electrical efficiency of a photovoltaic solar panel utilizing a nano-suspension containing multi-walled carbon nanotubes in a water/ethylene glycol (50:50) solution.
(2019)	[60]	PCM cooling	PV, natural water, Al_2O_3/PCM mixture	The results show that including Al_2O_3 nanoparticles at a concentration of $\gamma = 1\%$ enhances the effectiveness of the compound approach (Al_2O_3/PCM combination + water) compared to using 100% water for cooling. The compound strategy using Al_2O_3 (=1%)/PCM mixture (thermal conductivity of PCM = 25%) with 75% water yields the highest photovoltaic performance among all cooling techniques examined.	To implement a compound improvement approach to achieve a cooling effect on PV modules.
(2021)	[61]	PCM cooling	Water/ethylene glycol with PCM	By adjusting the coolant concentration, there was an improvement in electricity and power generation. Similarly, increasing the nanofluid concentration led to higher pumping power but a decrease in thermal–electrical equivalent power.	To assess the thermal of cooling photovoltaic solar panels and electrical efficiency using a combination of multi-walled carbon nanotubes, water/ethylene glycol, and phase-change material.
(2019)	[62]	PCM cooling	PCM, natural water circulation	Incorporating phase-change material (PCM) and using natural water circulation improved the performance of a PV panel. A top-to-bottom continuous water supply cooling method was found to be more effective than previous methods, leading to increased electricity generation.	To evaluate the effectiveness of solar panels in cold climates by employing phase-change material (PCM) with natural water circulation.

Table 3. *Cont.*

Year	References	Techniques	Type	Result	Objective
(2017)	[63]	PCM cooling	Paraffin, PCM, melting range	PCM cooling efficiency decreases in extreme temperatures due to incomplete melting and solidification processes, but in hot climates, the PV–PCM system boosts the annual electrical energy output of the PV system by 5.9%.	To measure the sustained energy-saving efficiency through incorporating paraffin-based PCM with a melting range of 38–43 °C behind the PV plate, while monitoring the cooling effects.
(2023)	[45]	PCM cooling	Nano-emulsions PCM	The mean overall thermal-equivalent energy efficiency reached 84.41%, with a peak of 89.23% when employing nano-emulsion within the module, in contrast to 79.95% and 83.23% observed in the water-cooled system. The collective exergy efficiency stood at 10.69% with nano-emulsion, slightly lower than the 11.66% attained with water.	To evaluate the application of nano-emulsion phase-change materials as advanced coolants to improve the overall efficiency of liquid-cooled PV/thermal systems.
(2022)	[64]	PCM cooling	Foam (AMF), PCM	The PV–PCM/AFM system exhibited lower PV surface temperatures by 4%, 7.4%, and 13.12% compared to standard PV, while achieving higher power outputs by 1.85%, 3.38%, and 4.14%.	To develop a method for passively regulating the temperature of a photovoltaic system using aluminum metal foam and PCM
(2022)	[65]	PCM cooling	PV–PCM cooling, (ECM)	Findings show that PCM can reduce PV temperature by up to 17.5 °C when used as a cooling system, leading to less efficiency losses and more power output. The usage of a 40 mm thick coating of $CaCl_2$-$6H_2O$ boosts the electrical generation for Vicuña by 5.8% and for Calame by 4.5% in a 1-year period.	To upsurge efficiency in silicon photovoltaic (PV) systems, this study compares several PV–PCM cooling system configurations according to PCM material

7. Active PCM Cooling System

The integration of PCMs into photovoltaic (PV) cooling systems has emerged as a promising approach for enhancing the performance and longevity of PV modules. PCMs are substances that absorb and release thermal energy during their phase transition, typically between solid and liquid states, at a specific temperature range. This property makes them ideal for stabilizing the temperature of PV cells by absorbing excess heat during peak sunlight hours and releasing it when the ambient temperature drops [66].

Active PCM cooling systems involve the circulation of a heat transfer fluid (HTF) through a network of channels or pipes that are in thermal contact with the PCM. This active approach ensures that the heat absorbed by the PCM is effectively removed and dissipated, preventing the PV cells from overheating. The result is a significant reduction in the PV module temperature (TPV), which has a direct positive impact on the electrical efficiency of the PV cells [67]. The benefits of active PCM cooling are multifaceted. Firstly, by keeping the PV cells at a lower temperature, the systems can mitigate the inherent decrease in conversion efficiency that occurs as PV cells heat up. This is particularly beneficial in regions with high solar irradiance and ambient temperatures, where the efficiency of PV modules can be significantly compromised.

Moreover, active PCM cooling can extend the operational life of PV modules. High temperatures can lead to accelerated aging and degradation of the module components including the encapsulant, back sheet, and even the solar cells themselves. By reducing the thermal stress on these components, PCM cooling can delay the onset of degradation mechanisms such as potential-induced degradation (PID), hot spot formation, and corrosion, thereby enhancing the reliability and durability of the PV system [68]. The economic implications of PCM cooling are also noteworthy. By improving the efficiency and longevity of PV modules, active PCM cooling systems can contribute to a higher return on investment for PV system owners. The initial cost of installing such cooling systems can be offset by the increased power output over the lifetime of the PV system, making it an attractive option for both residential and commercial applications. In addition to these advantages, active PCM cooling systems can contribute to the overall sustainability of PV technology. By reducing the energy losses associated with overheating, these systems can help to maximize the energy yield from renewable sources, aligning with global efforts to reduce carbon emissions and combat climate change [68]. As the PV industry continues to evolve, the development of advanced PCM cooling solutions is likely to play a crucial role in the optimization of PV system performance. Ongoing research is focused on improving the thermal conductivity of PCMs, enhancing their heat exchange properties, and integrating them more seamlessly into PV module designs. The ultimate goal is to create a synergistic cooling system that not only protects the PV cells from thermal stress but also contributes to the aesthetic and structural integrity of the PV installation.

In conclusion, active PCM cooling systems represent a significant advancement in the thermal management of PV modules. By effectively reducing TPV and enhancing performance, these systems are poised to become an integral part of the next generation of PV technologies, driving the industry towards higher efficiency, greater reliability, and increased sustainability. Figure 10 represents one of the types of active PCM cooling [69] Table 4 shows the most common studies in this area.

Table 4. Details of several studies on the cooling of PV cells through composite phase-change material (PCM) systems.

Year	References	Techniques	Type	Result	Objective
(2022)	[70]	PCM active	PV–PCM-TE system	Research findings show that in a PV–PCM-TE system, solar cells cooled from 79.72 °C to 57.39 °C, while in a PV-TE system, the temperature remained at 73.62 °C. The yearly average efficiencies were 17.57% for PVs.	To boost power output through the optimization of thermoelectric (TE) module attachment and phase-change material (PCM) to photovoltaic (PV) cells.
(2022)	[71]	PCM active	Nano-PCM, PV	Using nano-PCM resulted in augmented electrical efficiency and lower panel surface temperature	To expand the productivity of photovoltaic panels by utilizing nano-phase-change material (nano-PCM) to enhance heat radiation absorption.
(2021)	[39]	PCM active	PCM, nanoparticle	The results indicate that incorporating twisted bundles of multi-walled carbon nanotubes (MWCNTs) into the PCM/CFM at a concentration ratio of 0.2% significantly improved the material's heat absorption and rejection capabilities. Throughout the test, the solar panel's average electrical efficiency surged from 4% to 21% owing to the enhanced electrical performance of the cells.	To examine how empirically dispersing nanoparticles within porous materials and phase-change materials (PCMs) affects the electrical efficiency of photovoltaic (PV) panels in hot climate conditions.
(2022)	[53]	PCM active	PV–PCM/AMF system	Experimental results show significant temperature decreases in July for PV/PCM-AMF and PV/PCM solar cells, with reductions of 8.1% and 13.4% respectively. In contrast, November had the smallest temperature decreases at 3.8% and 5% for the same cells. The annual exergy and energy productivities of the PV–PCM.	To compare the energy and exergy efficiency of a proposed PV–PCM/AMF system with conventional PV systems and PV–PCM systems without metal foam.
(2018)	[72]	PCM active	CPV/PCM, fins	Studies show that both vertical and horizontal fins improve the thermal efficiency of PCM systems. Vertical fins are especially effective in controlling the temperature of PV cells and exhibit better performance than horizontal fins, particularly under a solar irradiation level of 670 W/m^2.	To implement aluminum fins of varying thicknesses in both horizontal and vertical orientations to enhance heat transfer within the PCM.
(2021)	[73]	PCM active	Convection (PVT/PCM-III)	Experimental findings from existing literature were used to validate the outcomes of a numerical model. After 120 min, the conversion efficiency of PV cells in different setups was recorded, ranging from 16.84% to 18.98%, with an inflow velocity of 3 m/s.	To securitize four solar module configurations to ensure a consistent cell temperature for optimal efficiency: a standard PV module (PVT module) and a traditional module featuring a PCM layer beneath (PVT/PCM-I).
(2021)	[74]	PCM active	Solar simulator, PCM	Studies have shown that PCM composites can reduce PV cell temperature by 6.8% and recuperate electrical efficiency by 14%. Implementing active cooling techniques involved flowing water through a cooling block underneath the PVT system at flow rates varying from 0.3 to 1 liter per minute.	To find out the effects of the effectiveness of active PCM on solar panels/experimental study.

Table 4. *Cont.*

Year	References	Techniques	Type	Result	Objective
(2019)	[75]	PCM active	Nanofluid, PCM	A new technology reduces the average temperature of concentrated photovoltaic (CPV) systems by 60% compared to traditional cooling methods. With specific settings, the cell temperature remains below 78 °C.	To evaluate and improve system efficiency by employing nanofluid as the heat transfer fluid (HTF) in a PCM active cooling system.
(2023)	[76]	PCM active	Coconut husk and paraffin wax (PCM)	The outcomes display that the rear surface temperature of a PV panel can reach 69.02 °C under an irradiance of 752 W/m^2. The cooling effect of PCM reduces this temperature by 12.83% compared to a standard PV panel.	To recover photovoltaic output performance and mitigate thermal stress in PV panels, the implementation of active fog phase-change material (PCM) will be conducted. A comprehensive applied study will assess its effectiveness.

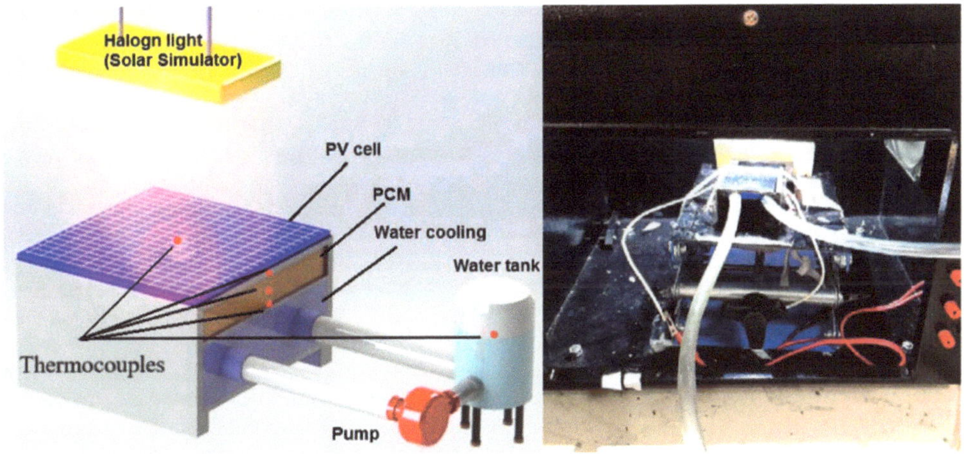

Figure 10. Experimental test facility in PCM active cooling PVT system.

8. Comparison of the Cooling Systems

The comparison of cooling systems in photovoltaic (PV) systems is a critical aspect in undertaking research to enhance the overall efficiency and performance of solar energy conversion. The literature review presented here revealed that cooling methods can significantly affect the temperature regulation of PV modules, which in turn influences their electrical output and longevity. Active cooling techniques, such as those involving water or air circulation, can effectively remove heat from the PV cells, but they often require energy input from pumps or fans, which can offset some of the energy gains. Several cooling techniques are employed for solar PV, and how these technologies impact solar PV is discussed in [61]. In [77], active and passive cooling techniques for a CPV system. In the functioning process, a wide microchannel heat sink (WMCHS) and manifold microchannel heat sink (MMCHS) are used to achieve better thermal management of the CPV system. Passive cooling methods, on the other hand, rely on natural heat transfer processes and do not require external energy input. PCMs are a popular choice for passive cooling as they absorb and release heat during their phase transition, helping to stabilize the temperature of the PV modules. The integration of PCMs with other passive techniques, such as the use of metal fins for heat dissipation, can further enhance the cooling effect. The study of microchannel heat sinks in [77] for concentrated photovoltaic (CPV) systems shows promise due to their ability to efficiently remove heat over a large surface area. The comparison between wide microchannel heat sinks (WMCHSs) and manifold microchannel heat sinks (MMCHS) provides valuable insights into the design considerations for effective thermal management in CPV systems. Moreover, the development of numerical models for proton exchange membrane fuel cells in [78] demonstrates the complexity of thermal management in energy conversion systems. These models can be adapted and applied to PV systems to predict and optimize cooling performance. The experimental investigation in [79] into the cooling of electronic chipsets using both active and passive methods is particularly relevant to PV systems. The findings that the combination of active and passive cooling can significantly improve thermal management are applicable to the design of cooling systems for PV modules.

The impact of operating temperature on the electrical and thermal efficiency of PV panels cannot be overstated. High temperatures can lead to a decrease in power output and accelerated degradation of the PV material. Therefore, research into various cooling methods is essential for the advancement of PV technology. The use of PCMs, as highlighted in [80], is a promising area of research. PCMs can provide a stable temperature environment for PV cells, which is crucial for maintaining high performance. The comparison between systems with and without active PCM cooling in [81] clearly shows the benefits of using active PCMs, such as improved temperature regulation and, consequently, higher energy conversion efficiency. In conclusion, the comparison of cooling systems in photovoltaic systems reveals that there is no one-size-fits-all solution. The choice of cooling method depends on various factors, including the specific type of PV system, the climate in which it is installed, and the balance between cooling efficiency and energy consumption for the cooling process. Ongoing research and development in this area are crucial for the continued improvement of solar PV technologies and their widespread adoption as a sustainable energy source. In summary, the choice between active and passive cooling and active and passive PCMs depends on the specific requirements of a given application. Dynamic systems are suitable for precise control and high heat loads, but they are energy intensive. Passive systems are more energy-efficient but may have limited control or may not handle extreme conditions well. Active PCMs offer precise control, while passive PCMs are simpler and more efficient in terms of energy use but offer less control over temperature.

Figure 11 shows the variables which are important in designing a cooling system. The figure presents a comprehensive overview of cooling system techniques, structured as a pyramid to illustrate the hierarchy and interrelation of different strategies. At the base of the pyramid, the focus is on improving efficiency through foundational methods such as determining the best cooling system and designing a smart control system. These elements are crucial as they provide the groundwork for more advanced cooling strategies. The integration of hybrid systems and techniques for overheat tracking ensures a balance between passive and active cooling methods, aiming to control temperatures effectively and prevent hotspots.

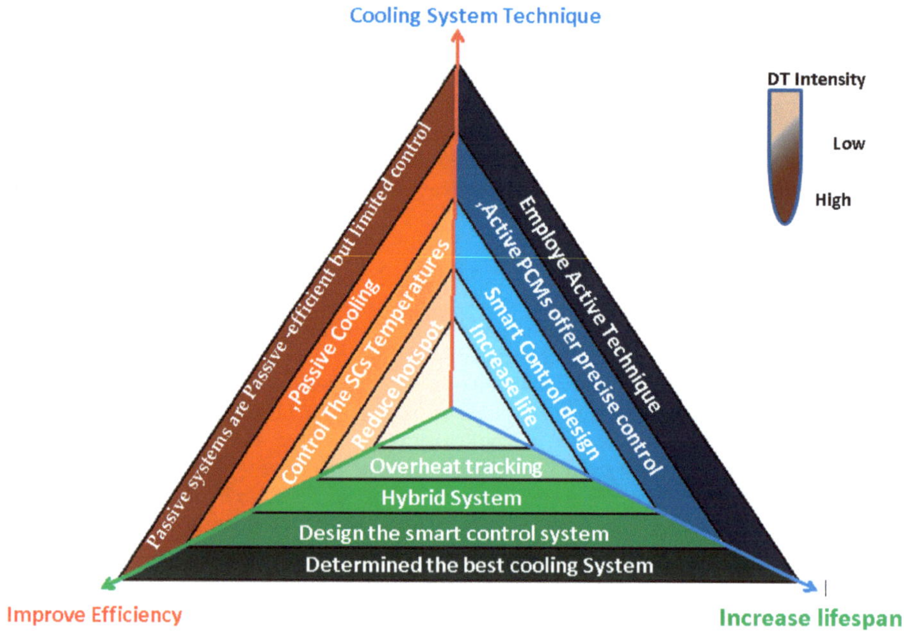

Figure 11. Cooling system strategies.

As we move up the pyramid, the emphasis shifts towards increasing the lifespan of the cooling systems. This involves employing advanced active techniques and smart control designs, which offer precise control over temperature management. The use of PCMs is highlighted for their ability to provide precise thermal regulation. The color gradient from low to high DT intensity underscores the increasing complexity and effectiveness of these methods. Overall, the pyramid visually encapsulates the progression from basic to advanced cooling strategies, emphasizing the importance of both efficiency and longevity in system design.

Table 5 provides a comparative analysis of different cooling techniques for solar PV systems, including both passive and active methods. It delves into the advantages and disadvantages of each technique, shedding light on how these cooling mechanisms impact the overall performance of solar energy systems. Passive cooling, relying on natural heat dissipation, offers simplicity and low cost but may be insufficient in high-temperature environments. Active cooling, on the other hand, utilizes energy to remove heat more effectively, potentially increasing system efficiency, but this comes at the expense of higher operational costs and greater complexity. The table also explores the impact of these techniques on the longevity of PV modules and their environmental footprints, helping to inform decisions on the most suitable cooling approach for various solar energy applications.

Table 5. Comparison of the four cooling systems mentioned in this research paper.

Type	References/ Year	Advantage	Disadvantage	Impact
Active Cooling	[26,82] (2020,2018)	Efficient heat transfer, precise temperature control, suitable for high heat loads, flexibility, higher cooling capacity, adaptability, highly effective in maintaining a specific temperature, suitable for applications where precise temperature control is critical, can handle large heat loads.	High energy consumption, complexity, environmental impact, noise generation, size and space requirements, consumes electricity and can be costly to operate, requires maintenance and can be noisy, may not be ideal for portable or off-grid applications.	Enhanced performance and efficiency heat, dissipation in high-power applications, thermal management in vehicles medical, applications aerospace and space exploration, environmental control in buildings
Passive Cooling	[38,83] (2021,2021)	High effectiveness, energy efficiency, simplicity, reliability, cost-effective, quiet operation, natural ventilation, longevity, suitable for remote or off-grid locations, environmentally friendly.	Slower heat dissipation, limited cooling capacity, lack of precise control, adaptation challenges, insufficient heat dissipation, design complexity, less effective than active cooling for precise temperature control, may not be appropriate for high heat loads or harsh circumstances.	Energy efficiency, cost savings, environmental benefits, low maintenance, resilience, architectural integration, natural ventilation, sustainable design.
PCM Cooling	[84,85] (2023,2023)	Energy-efficient and environmentally friendly, lower operating costs, reduce the average temperature, reduced peak load, thermal inertia, space-efficient design, maintenance reduction, remote and off-grid applications, compatibility with renewable energy, thermal energy storage.	Limited cooling capacity, limited temperature range, thermal cycling fatigue, thermal management, long charging times, volume and weight constraints, thermal conductivity, material compatibility, limited material selection, maintenance complexity, initial cost, limited control over the exact temperature (it stabilizes around the melting/freezing point of the PCM), slower response to temperature changes compared to active systems.	Quiet operation, renewable energy integration, improved thermal comfort, energy efficiency, peak load reduction, temperature regulation, environmental benefits, space efficiency.
PCM Active	[69,86] (2019,2016)	Extract thermal energy stored, thermal energy storage, energy efficiency, temperature regulation, reduced peak loads, thermal stability, space-efficient design, environmental benefits, precise temperature control. Suitable for applications with varying heat loads, can store and release thermal energy efficiently.	Requires energy input for phase change control, complex systems may require maintenance, limited by the specific properties of the pCm used, limited temperature range, thermal conductivity, volume expansion, limited energy density, durability, initial cost, complexity of system, integration, regulatory and safety concerns.	Integration with renewable energy, improved thermal, comfort efficient, thermal storage, energy savings, reduced peak loads, temperature regulation, environmental benefits, space-efficient design.

9. Conclusions and Future Study

In conclusion, this examination of cooling systems in photovoltaic (PV) systems has underscored the importance of effective thermal management in enhancing the efficiency and longevity of solar energy conversion. The literature review has shown that both active and passive cooling methods have their merits and drawbacks. Active systems provide more immediate and controllable cooling at the expense of energy consumption, while passive systems offer a more sustainable and energy-efficient approach, albeit with potentially less cooling capacity. The exploration of microchannel heat sinks and the integration of phase-change materials (PCMs) with other passive techniques have demonstrated innovative strategies for managing heat in PV modules. The development of numerical models for predicting thermal behavior in PV systems has also provided valuable tools for optimizing cooling designs.

As the demand for renewable energy sources continues to grow, further research into PV cooling systems is imperative. Future studies could focus on the following areas: hybridization of cooling methods, combining the strengths of both active and passive systems to achieve optimal thermal management with minimal energy consumption; exploration of new materials for PCMs with improved thermal properties and phase change temperatures more suited to PV applications; environmental impact—conducting a life-cycle analysis of different cooling technologies and assessing their overall sustainability; economic viability, performing cost–benefit analyses, and examining return-on-investment assessments for PV system operators; scalability—addressing the scalability of cooling solutions from small-scale laboratory tests to large-scale commercial PV installations; field studies and long-term performance evaluations—conducting field studies and long-term performance evaluations in various climatic conditions to validate the effectiveness of these cooling technologies in real-world scenarios.

It is important to note that, while currently available research has laid a solid foundation for understanding and improving PV cooling systems, there is still much to be explored. The pursuit of more efficient, sustainable, and cost-effective cooling solutions will be a critical component in the continued evolution of photovoltaic technologies, ensuring their competitiveness in the global energy market. We are committed to continuing this line of research and will be undertaking future studies to address the topics outlined above. Our goal is to contribute to the development of advanced cooling technologies that can significantly enhance the performance and sustainability of photovoltaic systems.

Author Contributions: Investigation, Y.E.A. and J.P.; Methodology, M.R.M.; Project administration, Y.E.A. and F.B.I.; supervision, J.P. and S.H.D.; Writing—original draft, Y.E.A. and S.H.D.; Writing—review and editing, S.H.D. and M.R.M. All authors have read and agreed to the published version of the manuscript.

Funding: This research was funded by Universiti Tenaga Nasional, Malaysia, through UNITEN R&D Sdn Bhd. TNB R&D Seeding Fund U-TD-RD-21-14 and Northern Technical University, Iraq.

Institutional Review Board Statement: No applicable.

Informed Consent Statement: Not applicable.

Data Availability Statement: The data provided in this study can be obtained upon request from the corresponding authors.

Acknowledgments: The authors express their gratitude for the support provided by Universiti Tenaga Nasional, Malaysia, and Northern Technical University, Iraq, in conducting this research.

Conflicts of Interest: The authors declare no conflicts of interest.

References

1. Moharram, K.A.; Abd-Elhady, M.; Kandil, H.; El-Sherif, H. Enhancing the performance of photovoltaic panels by water cooling *Ain Shams Eng. J.* **2013**, *4*, 869–877. [CrossRef]
2. Maghami, M.R.; Hizam, H.; Gomes, C.; Radzi, M.A.; Rezadad, M.I.; Hajighorbani, S. Power loss due to soiling on solar panel: A review. *Renew. Sustain. Energy Rev.* **2016**, *59*, 1307–1316. [CrossRef]
3. Zareian-Jahromi, M.; Fadaeinedjad, R.; Hosseini-Biyouki, M.M.; Askarian-Abyaneh, H. Investigation of Solar Irradiance Impact on Electro-Thermo-Mechanical Characteristics of a Dish-Stirling Engine Power Generation System. In Proceedings of the 2014 IEEE Electrical Power and Energy Conference, Washington, DC, USA, 12–14 November 2014; pp. 196–201.
4. Libra, M.; Petrík, T.; Poulek, V.; Tyukhov, I.I.; Kouřím, P. Changes in the efficiency of photovoltaic energy conversion in temperature range with extreme limits. *IEEE J. Photovolt.* **2021**, *11*, 1479–1484. [CrossRef]
5. Zhu, L.; Boehm, R.F.; Wang, Y.; Halford, C.; Sun, Y. Water immersion cooling of PV cells in a high concentration system. *Sol. Energy Mater. Sol. Cells* **2011**, *95*, 538–545. [CrossRef]
6. An, Q.; Bagheritabar, M.; Basem, A.; Ghabra, A.A.; Li, Y.; Tang, M.; Sabri, L.S.; Sabetvand, R. The effect of size of copper oxide nanoparticles on the thermal behavior of silica aerogel/paraffin nanostructure in a duct using molecular dynamics simulation. *Case Stud. Therm. Eng.* **2024**, *60*, 104666. [CrossRef]
7. Dubey, S.; Sarvaiya, J.N.; Seshadri, B. Temperature dependent photovoltaic (PV) efficiency and its effect on PV production in the world–a review. *Energy Procedia* **2013**, *33*, 311–321. [CrossRef]
8. Lu, S.; Zhang, J.; Liang, R.; Zhou, C. Refrigeration characteristics of a hybrid heat dissipation photovoltaic-thermal heat pump under various ambient conditions on summer night. *Renew. Energy* **2020**, *146*, 2524–2534. [CrossRef]
9. Larciprete, M.C.; Passeri, D.; Michelotti, F.; Paoloni, S.; Sibilia, C.; Bertolotti, M.; Belardini, A.; Sarto, F.; Somma, F.; Lo Mastro, S. Second order nonlinear optical properties of zinc oxide films deposited by low temperature dual ion beam sputtering. *J. Appl. Phys.* **2005**, *97*, 023501. [CrossRef]
10. Horschig, T.; Adams, P.W.; Röder, M.; Thornley, P.; Thrän, D. Reasonable potential for GHG savings by anaerobic biomethane in Germany and UK derived from economic and ecological analyses. *Appl. Energy* **2016**, *184*, 840–852. [CrossRef]
11. Benkahoul, M.; Chaker, M.; Margot, J.; Haddad, E.; Kruzelecky, R.; Wong, B.; Jamroz, W.; Poinas, P. Thermochromic VO2 film deposited on Al with tunable thermal emissivity for space applications. *Sol. Energy Mater. Sol. Cells* **2011**, *95*, 3504–3508. [CrossRef]
12. Browne, M.; Norton, B.; McCormack, S. Phase change materials for photovoltaic thermal management. *Renew. Sustain. Energy Rev.* **2015**, *47*, 762–782. [CrossRef]
13. Maghami, M.R.; Asl, S.N.; Rezadad, M.E.; Ale Ebrahim, N.; Gomes, C. Qualitative and quantitative analysis of solar hydrogen generation literature from 2001 to 2014. *Scientometrics* **2015**, *105*, 759–771. [CrossRef] [PubMed]
14. Maghami, M.R.; Hizam, H.; Gomes, C. Mathematical Relationship Identification for Photovoltaic Systems under Dusty Condition. In Proceedings of the 2015 IEEE European Modelling Symposium (EMS), Madrid, Spain, 6–8 October 2015; pp. 288–292.
15. Rusănescu, C.O.; Rusănescu, M.; Istrate, I.A.; Constantin, G.A.; Begea, M. The effect of dust deposition on the performance of photovoltaic panels. *Energies* **2023**, *16*, 6794. [CrossRef]
16. Maghami, M.; Hizam, H.; Gomes, C.; Hajighorbani, S.; Rezaei, N. Evaluation of the 2013 southeast asian haze on solar generation performance. *PLoS ONE* **2015**, *10*, e0135118. [CrossRef] [PubMed]
17. Xiao, M.; Tang, L.; Zhang, X.; Lun, I.Y.F.; Yuan, Y. A review on recent development of cooling technologies for concentrated photovoltaics (CPV) systems. *Energies* **2018**, *11*, 3416. [CrossRef]
18. Kumari, S.; Pandit, A.; Bhende, A.; Rayalu, S. Thermal management of solar panels for overall efficiency enhancement using different cooling techniques. *Int. J. Environ. Res.* **2022**, *16*, 53. [CrossRef]
19. Maghami, M.; Hizam, H.; Gomes, C.; AG, I. Characterization of dust materials on the surface of solar panel. *Life Sci. J.* **2014**, *11*, 387–390.
20. Nižetić, S.; Čoko, D.; Yadav, A.; Grubišić-Čabo, F. Water spray cooling technique applied on a photovoltaic panel: The performance response. *Energy Convers. Manag.* **2016**, *108*, 287–296. [CrossRef]
21. Mehrotra, S.; Rawat, P.; Debbarma, M.; Sudhakar, K. Performance of a solar panel with water immersion cooling technique. *Int. J. Sci. Environ. Technol.* **2014**, *3*, 1161–1172.
22. Nižetić, S.; Papadopoulos, A.; Giama, E. Comprehensive analysis and general economic-environmental evaluation of cooling techniques for photovoltaic panels, Part I: Passive cooling techniques. *Energy Convers. Manag.* **2017**, *149*, 334–354. [CrossRef]
23. Darkwa, J.; Calautit, J.; Du, D.; Kokogianakis, G. A numerical and experimental analysis of an integrated TEG-PCM power enhancement system for photovoltaic cells. *Appl. Energy* **2019**, *248*, 688–701. [CrossRef]
24. Dwivedi, P.; Sudhakar, K.; Soni, A.; Solomin, E.; Kirpichnikova, I. Advanced cooling techniques of PV modules: A state of art. *Case Stud. Therm. Eng.* **2020**, *21*, 100674. [CrossRef]
25. Hadipour, A.; Zargarabadi, M.R.; Rashidi, S. An efficient pulsed-spray water cooling system for photovoltaic panels: Experimental study and cost analysis. *Renew. Energy* **2021**, *164*, 867–875. [CrossRef]
26. Nižetić, S.; Giama, E.; Papadopoulos, A. Comprehensive analysis and general economic-environmental evaluation of cooling techniques for photovoltaic panels, Part II: Active cooling techniques. *Energy Convers. Manag.* **2018**, *155*, 301–323. [CrossRef]
27. Teo, H.; Lee, P.; Hawlader, M. An active cooling system for photovoltaic modules. *Appl. Energy* **2012**, *90*, 309–315. [CrossRef]

28. Elminshawy, N.A.; El-Ghandour, M.; Elhenawy, Y.; Bassyouni, M.; El-Damhogi, D.; Addas, M.F. Experimental investigation of a V-trough PV concentrator integrated with a buried water heat exchanger cooling system. *Sol. Energy* **2019**, *193*, 706–714. [CrossRef]
29. Maleki, A.; Ngo, P.T.T.; Shahrestani, M.I. Energy and exergy analysis of a PV module cooled by an active cooling approach. *J. Therm. Anal. Calorim.* **2020**, *141*, 2475–2485. [CrossRef]
30. Siecker, J.; Kusakana, K.; Numbi, E.B. A review of solar photovoltaic systems cooling technologies. *Renew. Sustain. Energy Rev.* **2017**, *79*, 192–203. [CrossRef]
31. Castanheira, A.F.; Fernandes, J.F.; Branco, P.C. Demonstration project of a cooling system for existing PV power plants in Portugal. *Appl. Energy* **2018**, *211*, 1297–1307. [CrossRef]
32. Tang, L.; Zhou, Y.; Zheng, S.; Zhang, G. Exergy-based optimisation of a phase change materials integrated hybrid renewable system for active cooling applications using supervised machine learning method. *Sol. Energy* **2020**, *195*, 514–526. [CrossRef]
33. Kane, A.; Verma, V.; Singh, B. Optimization of thermoelectric cooling technology for an active cooling of photovoltaic panel. *Renew. Sustain. Energy Rev.* **2017**, *75*, 1295–1305. [CrossRef]
34. Colţ, G. Performance evaluation of a PV panel by rear surface water active cooling. In Proceedings of the 2016 International Conference on Applied and Theoretical Electricity (ICATE), Craiova, Romania, 6–8 October 2016; pp. 1–5.
35. Zubeer, S.A.; Ali, O.M. Performance analysis and electrical production of photovoltaic modules using active cooling system and reflectors. *Ain Shams Eng. J.* **2021**, *12*, 2009–2016. [CrossRef]
36. Cuce, E.; Bali, T.; Sekucoglu, S.A. Effects of passive cooling on performance of silicon photovoltaic cells. *Int. J. Low-Carbon Technol.* **2011**, *6*, 299–308. [CrossRef]
37. Elbreki, A.; Muftah, A.; Sopian, K.; Jarimi, H.; Fazlizan, A.; Ibrahim, A. Experimental and economic analysis of passive cooling PV module using fins and planar reflector. *Case Stud. Therm. Eng.* **2021**, *23*, 100801. [CrossRef]
38. Dida, M.; Boughali, S.; Bechki, D.; Bouguettaia, H. Experimental investigation of a passive cooling system for photovoltaic modules efficiency improvement in hot and arid regions. *Energy Convers. Manag.* **2021**, *243*, 114328. [CrossRef]
39. Abdollahi, N.; Rahimi, M. Potential of water natural circulation coupled with nano-enhanced PCM for PV module cooling. *Renew. Energy* **2020**, *147*, 302–309. [CrossRef]
40. Hernandez-Perez, J.; Carrillo, J.; Bassam, A.; Flota-Banuelos, M.; Patino-Lopez, L. Thermal performance of a discontinuous finned heatsink profile for PV passive cooling. *Appl. Therm. Eng.* **2021**, *184*, 116238. [CrossRef]
41. Wongwuttanasatian, T.; Sarikarin, T.; Suksri, A. Performance enhancement of a photovoltaic module by passive cooling using phase change material in a finned container heat sink. *Sol. Energy* **2020**, *195*, 47–53. [CrossRef]
42. Li, H.; Zhao, J.; Li, M.; Deng, S.; An, Q.; Wang, F. Performance analysis of passive cooling for photovoltaic modules and estimation of energy-saving potential. *Sol. Energy* **2019**, *181*, 70–82. [CrossRef]
43. Hernandez-Perez, J.; Carrillo, J.; Bassam, A.; Flota-Banuelos, M.; Patino-Lopez, L. A new passive PV heatsink design to reduce efficiency losses: A computational and experimental evaluation. *Renew. Energy* **2020**, *147*, 1209–1220. [CrossRef]
44. Čabo, F.G.; Nižetić, S.; Giama, E.; Papadopoulos, A. Techno-economic and environmental evaluation of passive cooled photovoltaic systems in Mediterranean climate conditions. *Appl. Therm. Eng.* **2020**, *169*, 114947. [CrossRef]
45. Kalaiselvan, S.; Karthikeyan, V.; Rajesh, G.; Kumaran, A.S.; Ramkiran, B.; Neelamegam, P. Solar PV active and passive cooling technologies-a review. In Proceedings of the 2018 International Conference on Computation of Power, Energy, Information and Communication (ICCPEIC), Chennai, India, 28–29 March 2018; pp. 166–169.
46. Elminshawy, N.A.; El-Damhogi, D.; Ibrahim, I.; Elminshawy, A.; Osama, A. Assessment of floating photovoltaic productivity with fins-assisted passive cooling. *Appl. Energy* **2022**, *325*, 119810. [CrossRef]
47. Wu, S.; Xiong, C. Passive cooling technology for photovoltaic panels for domestic houses. *Int. J. Low-Carbon Technol.* **2014**, *9*, 118–126. [CrossRef]
48. Ramkiran, B.; Sundarabalan, C.; Sudhakar, K. Sustainable passive cooling strategy for PV module: A comparative analysis. *Case Stud. Therm. Eng.* **2021**, *27*, 101317.
49. Chandrasekar, M.; Suresh, S.; Senthilkumar, T. Passive cooling of standalone flat PV module with cotton wick structures. *Energy Convers. Manag.* **2013**, *71*, 43–50. [CrossRef]
50. Elbreki, A.; Sopian, K.; Fazlizan, A.; Ibrahim, A. An innovative technique of passive cooling PV module using lapping fins and planner reflector. *Case Stud. Therm. Eng.* **2020**, *19*, 100607. [CrossRef]
51. Agyekum, E.B.; PraveenKumar, S.; Alwan, N.T.; Velkin, V.I.; Shcheklein, S.E.; Yaqoob, S.J. Experimental investigation of the effect of a combination of active and passive cooling mechanism on the thermal characteristics and efficiency of solar PV module. *Inventions* **2021**, *6*, 63. [CrossRef]
52. Gupta, N.; Tiwari, G. Parametric study to understand the effect of various passive cooling concepts on building integrated semitransparent photovoltaic thermal system. *Sol. Energy* **2019**, *180*, 391–400. [CrossRef]
53. Sarafraz, M.; Safaei, M.R.; Leon, A.S.; Tlili, I.; Alkanhal, T.A.; Tian, Z.; Goodarzi, M.; Arjomandi, M. Experimental investigation on thermal performance of a PV/T-PCM (photovoltaic/thermal) system cooling with a PCM and nanofluid. *Energies* **2019**, *12*, 2572. [CrossRef]
54. Sudhakar, P.; Santosh, R.; Asthalakshmi, B.; Kumaresan, G.; Velraj, R. Performance augmentation of solar photovoltaic panel through PCM integrated natural water circulation cooling technique. *Renew. Energy* **2021**, *172*, 1433–1448. [CrossRef]

55. Sharaf, M.; Huzayyin, A.; Yousef, M.S. Performance enhancement of photovoltaic cells using phase change material (PCM) in winter. *Alex. Eng. J.* **2022**, *61*, 4229–4239. [CrossRef]
56. Díaz, F.A.; Moraga, N.O.; Cabrales, R.C. Computational modeling of a PV-PCM passive cooling system during a day–night cycle at arid and semi-arid climate zones. *Energy Convers. Manag.* **2022**, *270*, 116202. [CrossRef]
57. Sharma, A.; Tyagi, V.V.; Chen, C.R.; Buddhi, D. Review on thermal energy storage with phase change materials and applications. *Renew. Sustain. Energy Rev.* **2009**, *13*, 318–345. [CrossRef]
58. Liu, M.; Saman, W.; Bruno, F. Review on storage materials and thermal performance enhancement techniques for high temperature phase change thermal storage systems. *Renew. Sustain. Energy Rev.* **2012**, *16*, 2118–2132. [CrossRef]
59. Sarı, A.; Karaipekli, A. Thermal conductivity and latent heat thermal energy storage characteristics of paraffin/expanded graphite composite as phase change material. *Appl. Therm. Eng.* **2007**, *27*, 1271–1277. [CrossRef]
60. Kyaligonza, S.; Cetkin, E. Photovoltaic System Efficiency Enhancement with Thermal Management: Phase Changing Materials (PCM) with High Conductivity Inserts. *Int. J. Smart Grid* **2021**, *5*, 138–148.
61. Ahmadi, R.; Monadinia, F.; Maleki, M. Passive/active photovoltaic-thermal (PVT) system implementing infiltrated phase change material (PCM) in PS-CNT foam. *Sol. Energy Mater. Sol. Cells* **2021**, *222*, 110942. [CrossRef]
62. Nasef, H.; Nada, S.; Hassan, H. Integrative passive and active cooling system using PCM and nanofluid for thermal regulation of concentrated photovoltaic solar cells. *Energy Convers. Manag.* **2019**, *199*, 112065. [CrossRef]
63. Said, Z.; Ahmad, F.F.; Radwan, A.M.; Hachicha, A.A. New thermal management technique for PV module using Mist/PCM/Husk An experimental study. *J. Clean. Prod.* **2023**, *401*, 136798. [CrossRef]
64. Radwan, A.; Emam, M.; Ahmed, M. Comparative study of active and passive cooling techniques for concentrated photovoltaic systems. In *Exergetic, Energetic and Environmental Dimensions*; Elsevier: Amsterdam, The Netherlands, 2018; pp. 475–505.
65. Atyabi, S.A.; Afshari, E.; Udemu, C. Comparison of active and passive cooling of proton exchange membrane fuel cell using a multiphase model. *Energy Convers. Manag.* **2022**, *268*, 115970. [CrossRef]
66. Aldaghi, A.; Banejad, A.; Kalani, H.; Sardarabadi, M.; Passandideh-Fard, M. An experimental study integrated with prediction using deep learning method for active/passive cooling of a modified heat sink. *Appl. Therm. Eng.* **2023**, *221*, 119522. [CrossRef]
67. Bria, A.; Raillani, B.; Chaatouf, D.; Salhi, M.; Amraqui, S.; Mezrhab, A. Effect of PCM thickness on the performance of the finned PV/PCM system. *Mater. Today Proc.* **2023**, *72*, 3617–3625. [CrossRef]
68. Prakash, K.B.; Amarkarthik, A. Energy analysis of a novel butterfly serpentine flow-based PV/T and PV/T heat pump system with phase change material–an experimental comparative study. *Energy Sources Part A Recovery Util. Environ. Eff.* **2023**, *45*, 5494–5507.
69. Ma, T.; Li, Z.; Zhao, J. Photovoltaic panel integrated with phase change materials (PV-PCM): Technology overview and materials selection. *Renew. Sustain. Energy Rev.* **2019**, *116*, 109406. [CrossRef]
70. Stritih, U. Increasing the efficiency of PV panel with the use of PCM. *Renew. Energy* **2016**, *97*, 671–679.
71. Nižetić, S.; Jurčević, M.; Čoko, D.; Arıcı, M. A novel and effective passive cooling strategy for photovoltaic panel. *Renew. Sustain. Energy Rev.* **2021**, *145*, 111164. [CrossRef]
72. Salem, M.; Elsayed, M.; Abd-Elaziz, A.; Elshazly, K. Performance enhancement of the photovoltaic cells using Al2O3/PCM mixture and/or water cooling-techniques. *Renew. Energy* **2019**, *138*, 876–890. [CrossRef]
73. Velmurugan, K.; Kumarasamy, S.; Wongwuttanasatian, T.; Seithtanabutara, V. Review of PCM types and suggestions for an applicable cascaded PCM for passive PV module cooling under tropical climate conditions. *J. Clean. Prod.* **2021**, *293*, 126065. [CrossRef]
74. Akhtar, M.; Arendt, C.; Das, U. A review on active cooling techniques of photovoltaic modules. *Renew. Sustain. Energy Rev.* **2015**, *50*, 724–742.
75. Kargarian, A.; Bahaidarah, H.M.; Gandhidasan, P. Cooling techniques and design considerations for photovoltaic modules: A review. *Sol. Energy* **2019**, *183*, 278–305.
76. Padullés, J.; Ramírez, L.; Escobar, R.; Roca, J. Analysis of the impact of active cooling strategies on the electrical performance of a photovoltaic module under real working conditions. *Energy* **2018**, *152*, 206–216.
77. Luo, Z.; Zhu, N.; Hu, P.; Lei, F.; Zhang, Y. Simulation study on performance of PV-PCM-TE system for year-round analysis. *Renew. Energy* **2022**, *195*, 263–273. [CrossRef]
78. Stalin, P.M.J.; Prasad, K.S.; Kumar, K.P.; Hemadri, G.; Rajesh, M.; Kumar, K.P. Performance improvement of solar PV through the thermal management using a nano-PCM. *Mater. Today Proc.* **2022**, *50*, 1553–1558.
79. Abdulmunem, A.R.; Samin, P.M.; Rahman, H.A.; Hussien, H.A.; Ghazali, H. A novel thermal regulation method for photovoltaic panels using porous metals filled with phase change material and nanoparticle additives. *J. Energy Storage* **2021**, *39*, 102621. [CrossRef]
80. Sharaf, M.; Yousef, M.S.; Huzayyin, A. Year-round energy and exergy performance investigation of a photovoltaic panel coupled with metal foam/phase change material composite. *Renew. Energy* **2022**, *189*, 777–789. [CrossRef]
81. Lu, W.; Liu, Z.; Flor, J.-F.; Wu, Y.; Yang, M. Investigation on designed fins-enhanced phase change materials system for thermal management of a novel building integrated concentrating PV. *Appl. Energy* **2018**, *225*, 696–709. [CrossRef]
82. Ahmed, A.; Shanks, K.; Sundaram, S.; Mallick, T.K. Theoretical investigation of the temperature limits of an actively cooled high concentration photovoltaic system. *Energies* **2020**, *13*, 1902. [CrossRef]

83. Chandavar, A.U. Quantifying the performance advantage of using passive solar air heater with chimney for photovoltaic module cooling. *Int. J. Energy Res.* **2021**, *45*, 1576–1586. [CrossRef]
84. Gad, R.; Mahmoud, H.; Ookawara, S.; Hassan, H. Impact of PCM type on photocell performance using heat pipe-PCM cooling system: A numerical study. *J. Energy Syst.* **2023**, *7*, 67–88. [CrossRef]
85. Al Miaari, A.; Ali, H.M. Technical method in passive cooling for photovoltaic panels using phase change material. *Case Stud. Therm. Eng.* **2023**, *49*, 103283. [CrossRef]
86. Elarga, H.; Goia, F.; Zarrella, A.; Dal Monte, A.; Benini, E. Thermal and electrical performance of an integrated PV-PCM system in double skin façades: A numerical study. *Sol. Energy* **2016**, *136*, 112–124. [CrossRef]

energies

Article

Prediction of Global Solar Irradiance on Parallel Rows of Tilted Surfaces Including the Effect of Direct and Anisotropic Diffuse Shading

Sara Pereira [1,*], Paulo Canhoto [1,2] and Rui Salgado [1,3]

1 Institute of Earth Sciences, University of Évora, Rua Romão Ramalho 59, 7000-671 Évora, Portugal
2 Department of Mechatronics Engineering, University of Évora, Rua Romão Ramalho 59, 7000-671 Évora, Portugal
3 Physics Department, University of Évora, Rua Romão Ramalho 59, 7000-671 Évora, Portugal
* Correspondence: spereira@uevora.pt

Abstract: Solar photovoltaic power plants typically consist of rows of solar panels, where the accurate estimation of solar irradiance on inclined surfaces significantly impacts energy generation. Existing practices often only account for the first row, neglecting shading from subsequent rows. In this work, ten transposition models were assessed against experimental data and a transposition model for inner rows was developed and validated. The developed model incorporates view factors and direct and circumsolar irradiances shading from adjacent rows, significantly improving global tilted irradiance (GTI) estimates. This model was validated against one-minute observations recorded between 14 April and 1 June 2022, at Évora, Portugal (38.5306, −8.0112) resulting in values of mean bias error (MBE) and root-mean-squared error (RMSE) of -12.9 W/m^2 and 76.8 W/m^2, respectively which represent an improvement of 368.3 W/m^2 in the MBE of GTI estimations compared to the best-performing transposition model for the first row. The proposed model was also evaluated in an operational forecast setting where corrected forecasts of direct and diffuse irradiance (0 to 72 h ahead) were used as inputs, resulting in an MBE and RMSE of -33.6 W/m^2 and 169.7 W/m^2, respectively These findings underscore the potential of the developed model to enhance solar energy forecasting accuracy and operational algorithms' efficiency and robustness.

Keywords: solar radiation; solar energy; transposition model; solar power plant; forecast

Citation: Pereira, S.; Canhoto, P.; Salgado, R. Prediction of Global Solar Irradiance on Parallel Rows of Tilted Surfaces Including the Effect of Direct and Anisotropic Diffuse Shading. *Energies* **2024**, *17*, 3444. https://doi.org/10.3390/en17143444

Academic Editors: Annamaria Buonomano, Jayanta Deb Mondol and Biplab Das

Received: 6 June 2024
Revised: 5 July 2024
Accepted: 10 July 2024
Published: 12 July 2024

1. Introduction

In recent decades, there has been a significant surge in the installed capacity of PV systems. In 2022 alone, solar photovoltaics comprised two-thirds of the new renewable energy capacity added to the grid, totaling 239 GW [1]. This growth is driven by global efforts to decarbonize the economy and achieve net-zero greenhouse gas emissions by 2050 The International Energy Agency (IEA) estimates that PV systems will continue to expand with new capacity projected to exceed 700 GW by 2028 [2]. With the rapid increase in the installed capacity of solar energy systems worldwide and the characteristic variability of solar radiation, accurate estimation and forecasting of power output is becoming increasingly important. The factor that most affects this output is the irradiance incident on the solar energy collectors. However, weather stations, satellites, and numerical weather prediction (NWP) models usually provide global (GHI) and diffuse (DIF) irradiance data on the horizontal plane and direct irradiance on a plane normal to the sun's rays (DNI) Therefore, transposition models that can compute the irradiance incident on a tilted surface from the available variables are essential.

Various transposition models have been developed and studied over the years. They can be divided into analytical, semi-empirical, and empirical models [3]. Analytical models are based on the laws of physics and only require the geometric characteristics and position data of the surface, while semi-empirical and empirical models also rely on observation data.

Physical models are usually based on the sum of the direct, diffuse, and reflected irradiances on the tilted surface. The computation of the direct and reflected components is often the same for all models, with the computation of the diffuse irradiance being the distinct factor [4]. Among these models, some assume a uniform sky dome radiance with the same intensity in all directions, termed isotropic. Anisotropic models, on the other hand, define indices representing the irradiance intensity from different regions of the sky dome, such as the circumsolar region (an annular region surrounding the sun disk), the horizon-brightening region (a band along the horizon), and the background or remaining region often called the isotropic region.

Transposition models also tend to show varying degrees of performance depending on sky conditions, whether clear, cloudy, or overcast [5]. The complexity and variability of the shape and position of clouds make the accurate instantaneous evaluation of the diffuse component extremely difficult without knowing the distribution of sky radiance. A model considering this would be more complex and require not only the simple measurement of the irradiance on the tilted plane but many more variables that are not commonly measured, such as the spectral radiance. Thus, physical models must rely on assumptions that can result in high transient errors [6]. Some of the most referenced analytical physical models in the literature are reviewed in this work, such as the Klucher [7], Hay–Davies [8], and Modified Bugler [9] models. On the other hand, abundant research on the comparison of the different models at different locations and positions of the surfaces has been published [3,4,10,11]. From these, no single model emerges as the best for all studied locations/positions, but it has been widely demonstrated that anisotropic models tend to outperform isotropic models when compared to measured data because they consider the angular dependence of sky radiance, thus reflecting real-world conditions more closely compared to isotropic models.

Semi-empirical and empirical models, including machine learning approaches which are data-driven, tend to be site-dependent, which means that they are biased towards specific locations and/or tilt and azimuth angles [12–16], and will not be considered here. While these models are gaining relevance and can provide substantial benefits in some situations, their applicability is often limited because extensive local experimental data are required for the model calibration. This increases the challenge and difficulty of generalizing them to different geographic locations or environmental conditions. Given the focus of this study on developing a transposition model that is applicable across various conditions without the need for site-specific parameter adjustments, these models were excluded to maintain a broader applicability and robustness of proposed approach.

Most transposition models have been developed for a tilted surface in an open field. However, a solar energy power plant, such as photovoltaic power plants, comprises several rows of modules, where rows other than the first have a different view of the sky dome and the ground compared to the first row. Therefore, when computing the global tilted irradiance (GTI) on a panel not located in the front row, the transposition model should be adjusted to account for the fraction of the sky dome obscured by the rows at the front, as well as the possible blocking of the direct beam and circumsolar irradiance from the sun disk and the reflected irradiance coming from the front row and the ground between them. Since in large solar power plants the number of panels in the first row can be much smaller than the number of panels in the remaining rows, having a more accurate transposition model for these panels can improve the estimation of power output. Few authors have focused on this aspect, with the most notable and recent works being those by Applebaum et al. [17], Varga and Mayer [18], and Tschopp et al. [19].

As mentioned above, one aspect of higher complexity in transposition models is the parametrization of diffuse irradiance. This component of the GTI is strongly related to the diffuse horizontal irradiance (DIF) and the global horizontal irradiance (GHI), which are typically measured in radiometric stations. However, if only GHI is measured, a separation method is needed to estimate DNI and DIF. If there are no measurements at the location of interest, estimations or forecast values need to be used instead. This increases the uncertainty associated with the estimation of GTI and, consequently, of power output. Of particular interest in this work is the usage of forecast values of solar radiation components, with the goal of predicting power output in a time horizon up to three days with an acceptable accuracy. In this regard, when utilizing forecast DNI and DIF as inputs, a transposition model allows for the forecasting of GTI and power output of photovoltaic systems if coupled with thermal and electric models of photovoltaic systems.

These forecasts can be obtained from a global operational numerical weather prediction model (NWP) such as the Integrated Forecasting System (IFS) of the European Centre for Medium-range Forecasts (ECMWF) [20]. By incorporating constitutive and state equations describing physical phenomena in the atmosphere, subject to boundary and initial conditions, NWP models provide the evolution of the atmospheric state, which include surface-level GHI. In the case of IFS/ECMWFS, DNI is also included, allowing for the study and assessment of solar resources and operational forecasts [21,22].

This work presents a comprehensive review of transposition models, categorizing models that do not include shading or irradiance masking as "first-row models", and those that include these aspects as "inner-row models". The development of a transposition model for inner rows is also presented, which can be applied to a desired first-row model The proposed model is validated using experimental values, which are also used for the assessment and comparison of all models. The use of operational forecasts of solar irradiance is assessed to estimate the impact of the accuracy of the transposition model on GTI prediction at different time horizons. The primary aim is to develop a transposition model for inner rows of solar power plants, validate the proposed model against experimental measurements, and evaluate its performance using operational solar radiation forecasts. This study addresses gaps in the existing literature, which mainly focuses on the first row of panels, neglecting the impact of direct and anisotropic diffuse shading on inner rows. The hypotheses focus on the expected improvement in the accuracy of the developed model's results over existing models for the first row, as well as its robustness when used with forecasted irradiance data as input.

This paper is organized as follows: An initial review of the most used analytical models and recent models developed for rows that are not the first is presented in Section 2. Section 3 details the development of a transposition model based on the works of Tschopp et al. and Varga and Mayer [18,19] for surfaces not located in the first row. Section 4 presents the experimental setup and procedure for testing the different models, and Section 5 discusses the evaluation and results. Section 6 establishes a connection with solar irradiance forecasting by applying the developed model to forecasted values of direct normal and diffuse horizontal irradiance, integrating it as an essential part of an operational algorithm for forecasting solar irradiance on a tilted surface. Finally, Section 7 presents the conclusions of this work.

2. Solar Irradiance Transposition Models for Tilted Surfaces

This section provides a review of analytical transposition models, ranging from the most widely used and commonly referenced transposition models that do not consider shading and obscuration to the few and most recently proposed models designed for inner rows in solar power plants.

2.1. Transposition Models for the First Row of Collectors in Solar Power Plants

For modeling solar irradiance on a tilted surface (*GTI*), nine of the most commonly used analytical transposition models are reviewed, where one is isotropic and the remaining are anisotropic models. In all the following equations, α is the solar height, Φ is the solar zenith angle, γ_s is the solar azimuth, θ is the incidence angle on the surface, β is the tilt angle of the surface, and γ_p is the azimuth of the surface, taking the horizontal plane and the local meridian as references.

Analytical transposition models include the computation of tilt factors for determining the beam (R_b), diffuse (R_d), and reflected (R_r) irradiances on the tilted surface. These factors are used in Equation (1) for determining GTI, where DHI is the direct horizontal irradiance as defined in Equation (2):

$$GTI = R_b DHI + R_d DIF + R_r GHI \tag{1}$$

$$DHI = DNI \times \cos(\Phi) \tag{2}$$

For the first row of photovoltaic panels, assuming no obstruction of the sun's rays, the direct irradiance on the tilted plane can be calculated at any given instant by multiplying DHI with the beam radiation tilt factor described in Equation (3):

$$R_b = \frac{\cos(\theta)}{\cos(\Phi)} \tag{3}$$

The irradiance reflected onto the tilted surface by the surrounding surfaces is assumed isotropic and is modeled by multiplying GHI with the reflected irradiance tilt factor R_r (Equation (4)), which consists of the product of the albedo of the surrounding ground (ρ_g) and the view factor F_g (Equation (5)).

$$R_r = \rho_g F_g, \tag{4}$$

$$F_g = \frac{1 - \cos(\beta)}{2} \tag{5}$$

Numerous models have been proposed for determining the diffuse tilt factor R_d since the diffuse radiation is highly variable due to its dependency on atmospheric conditions. Models that consider the diffuse irradiance on a tilted surface coming from the entire sky dome to have the same intensity are isotropic, while others that divide the sky dome into regions with different intensities of diffuse radiance are called anisotropic models. These regions are usually the circumsolar region, which constitutes an annular region surrounding the sun, i.e., the horizon-brightening region, which is the region of the sky dome near the horizon, and the remaining sky dome region is termed the isotropic region.

In the following, several models are presented, including the widely used isotropic model that was developed by Liu and Jordan [23]. It assumes that the diffuse solar radiation is uniformly spread across the sky, which simplifies the calculation of solar radiation on inclined surfaces by considering the sky as a homogenous light source. This model only considers the tilt angle of the surface as input, as presented in Equations (6) and (7), where the diffuse tilt factor is simply a view factor between the surface and the sky. Its advantages lie in its simplicity and widespread adoption for preliminary solar energy designs and studies. However, the model's accuracy can be compromised due to its isotropic assumption, leading to potential inaccuracies in non-homogenous sky conditions and environments with significant obstructions. Despite these limitations, it remains a significant tool in solar energy research and engineering, often serving as a benchmark for more complex anisotropic models that account for the directional distribution of diffuse radiation.

Liu and Jordan model (LJ):

$$R_d = F_s \tag{6}$$

$$F_s = \frac{1 + \cos(\beta)}{2} \tag{7}$$

The remaining models analyzed are anisotropic models, which are often more representative of real sky conditions.

Bugler model (B):

The Bugler model [24] includes a circumsolar region in which its contribution to the diffuse irradiance is assumed to be 5% of DNI and can be computed using Equation (8) where F_s is the sky view factor given by Equation (7). Bugler derived this model by analyzing solar radiation data and developing equations that incorporate these anisotropic factors resulting in more accurate predictions of solar irradiance on tilted surfaces. However, this model overlooks the contribution of the circumsolar region to the isotropic irradiance and was later modified.

$$R_d = F_s + 0.05\frac{DHI \times R_b}{DIF} \tag{8}$$

Temps and Coulson model (TC):

The Temps and Coulson [25] model, defined by Equation (9), considers the circumsolar region through the factor $\left[1 + cos^2(\theta)sin^3(\Phi)\right]$ and the horizon brightening through $\left[1 + sin^3\left(\frac{\beta}{2}\right)\right]$. This model was obtained through measurements of direct and diffuse solar flux incident on slopes of various orientations for clear-sky conditions through a pyranometer.

$$R_d = F_s\left[1 + cos^2(\theta)sin^3(\Phi)\right]\left[1 + sin^3\left(\frac{\beta}{2}\right)\right] \tag{9}$$

Klucher model (K):

The Klucher model [7], given by Equation (10), is similar to the Temps and Coulson model, except for the inclusion of the clearness index f, defined by Equation (11). Under overcast conditions, this model reduces to the isotropic model (where f tends to 0). It was derived from a 6-month dataset of measured hourly diffuse and total solar radiation on a horizontal plane and total radiation on surfaces with 0° azimuth and tilts of 37° and 60°.

$$R_d = F_s\left[1 + fsin^3\left(\frac{\beta}{2}\right)\right]\left[1 + fcos^2(\theta)sin^3(\Phi)\right] \tag{10}$$

$$f = 1 - \left(\frac{DIF}{DHI + DIF}\right)^2 \tag{11}$$

Hay–Davies model (HD):

The Hay–Davies model [8], which does not consider horizon brightening, is given by Equation (12). Here, A is named the anisotropy index, also termed direct or beam irradiance index, and is defined in Equation (13), where ENI is the extraterrestrial normal irradiance This index represents the total transmittance of the atmosphere for beam radiation and is used to define the portion of circumsolar ($A \times R_b$) and isotropic ($F_s(1 - A)$) diffuse irradiance on the tilted surface.

$$R_d = A \times R_b + F_s(1 - A) \tag{12}$$

$$A = \frac{DNI}{ENI} \tag{13}$$

Ma Iqbal model (MI):

The Ma Iqbal model [26], described through Equation (14), is similar to the Hay–Davies model but uses the clearness index k_T (Equation (15)) to define the circumsolar portion of diffuse irradiance instead of the direct irradiance index. The clearness index is computed as the ratio of global horizontal irradiance (GHI) to extraterrestrial horizontal irradiance (EHI) This formulation allows the Ma Iqbal model to account for varying sky conditions and solar geometry, enhancing its applicability in solar energy studies and irradiance forecasting.

$$R_d = k_T R_b + F_s(1 - k_T) \tag{14}$$

$$k_T = \frac{GHI}{EHI} \tag{15}$$

Modified Bugler model (MB):

The Modified Bugler model [9] improves the Bugler model by considering the contribution of the circumsolar region to the background isotropic diffuse radiance, computed through Equation (16). This model was validated using extensive datasets from various locations around the world. These datasets included measured solar radiation data collected under different sky conditions, such as clear sky, cloudy sky, and overcast sky conditions. Additionally, the model was validated against experimental data that accounted for various surface orientations, including horizontal, tilted, and vertical surfaces.

$$R_d = \left(1 - 0.05\frac{DHI}{DIF}\right)F_s + 0.05\frac{DHI \times R_b}{DIF} \tag{16}$$

Modified Ma Iqbal model (MMI):

The modified Ma Iqbal model (Equation (17)) [3] is similar to the original model proposed by this author but uses a clearness index (Equation (18)) adjusted with an optical air mass as defined in Equation (19) [27].

$$R_d = k\prime_T R_b + (1 - k\prime_T)F_s \text{ a} = 1, \tag{17}$$

$$k\prime_T = \frac{k_T}{1.031exp\left(\frac{-1.4}{0.9+\frac{9.4}{M}}\right) + 0.1} \tag{18}$$

$$M = \frac{1}{\cos(\Phi) + 0.15(93.885 - \Phi)^{-1.253}} \tag{19}$$

Reindl model (R):

The Reindl model [28], based on the Hay–Davies model, uses the same anisotropic index, A (Equation (13)), but includes an additional term to account for horizon brightening (Equation (20)). This model was developed based on datasets from locations across the USA.

$$R_d = A \times R_b + F_s(1 - A)\left[1 + \sqrt{\frac{DHI}{DHI + DIF}}sin^3\left(\frac{\beta}{2}\right)\right] \tag{20}$$

Perez model (P):

The Perez model, one of the most widely used transposition models in the literature, is included in this work even though it cannot be classified as an analytical model since it is built upon measured data and is subject to continuous revisions for this reason. The version of the Perez model [29] used in this work, often referred to as Perez3, is considered the most widely accepted version of this model [4]. This model categorizes the sky in three zones: the circumsolar zone, horizon band, and isotropic zone. The diffuse irradiance tilt factor is computed using Equations (21)–(27), with the assumption that the circumsolar irradiance originates from a point source and irradiance from the horizon from an infinitesimally thin region. The zenith angle is exceptionally used in radians for Equation (26). The coefficients F_{ij} were determined using data from 10 American and 3 European locations and can be consulted in [4,29].

$$R_d = F_1\frac{a}{b} + (1 - F_1)F_s + F_2\sin\beta \tag{21}$$

$$a = \max(0, \cos\theta) \tag{22}$$

$$b = \max[\cos 85^\circ, \sin(90^\circ - \theta_z)] \tag{23}$$

$$F_1 = max[0, F_{11}(\varepsilon) + F_{12}(\varepsilon)\Delta + F_{13}(\varepsilon)\theta_z] \quad (24)$$

$$F_2 = F_{21}(\varepsilon) + F_{22}(\varepsilon)\Delta + F_{23}(\varepsilon)\theta_z \quad (25)$$

$$\varepsilon = \frac{GHI/DIF + 1.041\theta_z^3}{1 + 1.041\theta_z^3} \quad (26)$$

$$\Delta = \frac{DIF}{ENI\cos\theta_z} \quad (27)$$

2.2. Transposition Models for Surfaces Not Located in the Front Row of a Solar Power Plant

The models presented in Section 2.1 were developed for a single row. In the case of photovoltaic power plants, for instance, which are composed of various parallel rows, there is a partial obscuring of the sky radiance and, eventually, the blocking of direct sun rays by the front rows over the second and subsequent rows, thus affecting the global tilted irradiance on these surfaces. An obscuring of the ground between rows can also occur, which affects the reflected irradiance.

Some authors have considered this, namely Appelbaum et al. [17], who adjusted the Klucher transposition model [7] by computing the sky view factor for the second row and adjusting the indices of the circumsolar and horizon-brightening diffuse regions. For this, the authors considered the circumsolar irradiance coming from a point source (the sun) which would be completely obscured when the obscuring angle from the sun to the base of the second row caused by the front row is higher than the sun elevation angle The correction for the horizon-brightening region was considered near sunrise and sunset This adjusted model was validated with data obtained in Tel Aviv, showing that for rows other than the first, the effect of the anisotropic region defined as horizon brightening is negligible.

Varga and Mayer [18] modified the Hay–Davies model [8], which does not include a horizon-brightening region, to calculate the distribution of solar irradiance along a tilted surface for rows behind the first. Besides the adjustment of the view factors, the circumsolar irradiance was corrected through the modeling of the fraction of the region obscured by the front row by considering a circumsolar region with an apparent angular radius of 15°, and the ground was divided into two sections: a shaded section that reflects only isotropic diffuse irradiance and an unshaded section that reflects global irradiance. The model was validated against power generation data from Hungary showing a clear improvement when compared to a model that only considers shading of the direct irradiance.

Tschopp et al. [19] also adjusted the Hay–Davies model for surfaces that are not on the front row by dividing the length of the surface of the row being evaluated, the back surface of the row at its front, and the ground between them in segments and then computing diffuse and direct irradiance in each segment. For this, the view factors between all segments and between each segment and the sky are computed, while the obscuring of the direct and circumsolar irradiances are determined based on a simple formula for the shadow height on the surface. The model showed a significant improvement in the estimation of GTI for rows other than the first compared to the original transposition model.

Considering this, it is expected that combining the higher detail in the modeling of shadow and circumsolar irradiance obscuration presented in [18] and the inclusion of the back surface of the row at the front and the ground between rows presented in [19] will provide better estimations of GTI for rows that are not at the front of a solar power plant.

3. Development of Transposition Model for Surfaces Not Located in the Front Row of a Solar Power Plant

The model proposed in this work for determining solar irradiance on tilted rows adjacent to the first row builds upon the models presented in [18,19].Similar to these models, it assumes rows of panels with lengths much greater than their heights, resulting in a 2D representation as commonly used in the literature [17–19,30,31].

While the aforementioned models only consider cases when the sun is positioned at the front of the rows, here, the modeling of the direct and circumsolar irradiance shading for all involved surfaces is also included, considering any position of the sun, resulting in a more realistic model. Additionally, the developed model was made generic and can be applied to any first-row transposition model as long as it clearly considers direct, circumsolar, and isotropic irradiance, instead of relying on a predefined first-row model. Extra detail was also included in the modeling of ground shading by considering rows beyond the two main rows being modeled.

The three surfaces considered in this model, namely the front of the panel being evaluated, the back of the panel at its front, and the ground between rows, are divided into segments, as shown in Figure 1, where L, D, h_0, and β are the length of the panels, the horizontal distance between rows, the vertical distance between the ground and the base of the rows, and the tilt angle, respectively. The value of GTI is computed for each segment c of the panel being evaluated and considers the reflected solar irradiance from each segment of the ground u and the back of the front panel v. The number of segments into which these surfaces are divided is defined by the user and can be adjusted, namely the number of segments of the panel being evaluated $i = 1 : n$, the number of segments of the ground $j = 1 : p$, and the number of segments of the back of the front panel $k = 1 : q$.

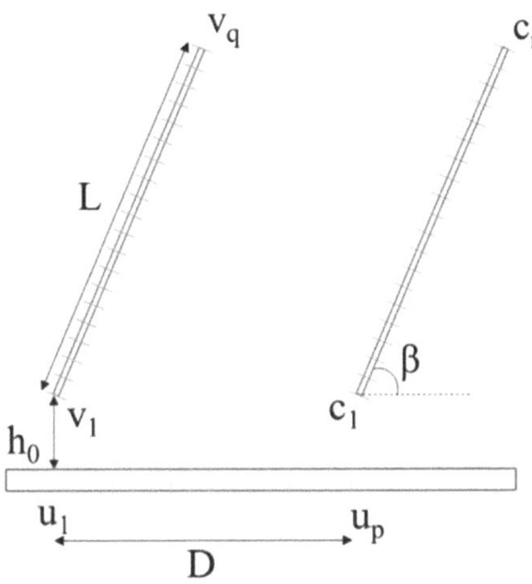

Figure 1. Schematic for modeling GTI in rows that are not the front row.

For each segment of both panels and ground, the GTI is computed as in [19] using Equation (28), where $\vec{GTI}$, $\vec{I}_r$, and $\vec{D}$ are vectors with the values of the global tilted irradiance, the direct normal irradiance, and the diffuse horizontal irradiance, respectively, on each segment of the panel, ground, and front panel. $\boldsymbol{I}$ is the identity matrix while $\boldsymbol{F}$ is the view factor matrix between all segments and $\boldsymbol{R}$ is the reflectivity matrix for each segment. The view factors between all segments are computed using the Hottel crossed string rule [32], and the reflectivity of the front of the panel is assumed to be 0 as in [19].

$$\vec{GTI} = (\boldsymbol{I} - \boldsymbol{FR})^{-1}\left(\vec{I}_r + \vec{D}\right), \tag{28}$$

$$\vec{I}_r = DHI \times \vec{R}_b \tag{29}$$

$$\vec{D} = DIF \times \vec{R}_d \quad (30)$$

$$\vec{R}_b = max\left(0, \frac{\cos\vec{\theta}}{\cos\vec{\Phi}}\right)\left(1 - \vec{S}\right) \quad (31)$$

$$\vec{R}_d = \vec{F}_s X_i + X_{cs}\left(1 - \vec{S}_{cs}\right) \quad (32)$$

For the computation of the transposed direct normal (Equation (29)) and diffuse (Equation (30)) irradiance on each segment, the vectors $\vec{R}_b$ and $\vec{R}_d$ are obtained from Equations (31) and (32) which consist of the tilt factors for the direct (beam) and diffuse irradiances. The shading of the direct irradiance in each segment is taken into consideration in the direct tilt factor through the vector $\vec{S}$, whose values are either 0, when direct irradiance is not obscured, or 1, when the direct irradiance is obscured, depending on the geometrical characteristics of the installation and the apparent position of the sun in the sky. Regarding the diffuse tilt factor, an isotropic component, X_i, and a circumsolar component, X_{cs}, are modeled as in a first-row transposition model, which, as mentioned above, can be any first-row model as long as it includes direct, circumsolar, and isotropic irradiance components. In the present work, a set of analytical transposition models for first rows are firstly assessed against experimental values and then one is selected, which will then be used in this model evaluation, as reported in the following sections. The sky view factors are computed for each segment considering the summation and reciprocity rules. The fraction of circumsolar irradiance that is obscured is modeled by the vector $\vec{S}_{cs}$ ranging from 0, when there is no obscuration, to 1, when all irradiance from the considered circumsolar region is obscured.

The vectors $\vec{S}$ and $\vec{S}_{cs}$ (Equations (33) and (34), respectively) are determined based on the model proposed by Varga and Mayer [18] with various modifications for the inclusion of cases in which the sun is positioned at the back of the row and the computations for the different segments of the back surface of the row in front of the one being evaluated and the ground. Depending on the surface of each segment, namely the surface of the panel being evaluated, c, the back surface of the panel of the row at its front, v, or the surface of the ground between them, u, the way the shading vectors are modeled differs. Since this is a two-dimensional model, firstly, the projection of the solar elevation angle to the azimuth of the panels, $\alpha\prime$, is needed, which is obtained through Equation (35), where γ_s is the solar azimuth and γ_p is the azimuth of the surfaces.

$$\vec{S} = \begin{bmatrix} \vec{S}_c \\ \vec{S}_u \\ \vec{S}_v \end{bmatrix} \quad (33)$$

$$\vec{S}_{cs} = \begin{bmatrix} \vec{S}_{cs,c} \\ \vec{S}_{cs,u} \\ \vec{S}_{cs,v} \end{bmatrix} \quad (34)$$

$$\alpha\prime = \tan^{-1}\left(\frac{\tan\alpha}{\cos\left(\gamma_s - \gamma_p\right)}\right) \quad (35)$$

The shading of direct beam and the fraction of obscured circumsolar irradiance are determined for each segment of the panel, ground, and back of the front panel according to the projected solar elevation angle relative to each of the angles shown in Figure 2. These angles are computed using Equations (36)–(43), where i, j, and k are the indices of the segments in the panel being evaluated, at the back of the front panel, and on the ground between the rows of panels, respectively. The angles with subscript c are obtained for each segment of the panel being evaluated, those with subscript v are obtained for each segment of the back of the front panel, and those with subscript u are obtained for each segment of ground between the two rows. The angles δ and ε result from the view of the top of an adjacent panel from a segment, the angles ζ and ξ result from the view of the bottom of an adjacent panel from a ground segment, and the angles λ and σ result from the view of the top of a subsequent panel from a ground segment.

$$\delta_c(i) = \tan^{-1}\left[\frac{\left(L - \frac{c(i+1)-c(i)}{2}\right)\sin\beta}{D + \left(\frac{c(i+1)-c(i)}{2} - L\right)\cos\beta}\right]\Pi\mathrm{a} = 1, \tag{36}$$

$$\delta_v(k) = \tan^{-1}\left[\left(\frac{\left(\mathrm{L} - \frac{v(k+1)-v(k)}{2}\right)\sin\beta}{\mathrm{D}}\right)\right] \tag{37}$$

$$\delta_u(j) = \tan^{-1}\left[\frac{\mathrm{L}\sin\beta + h_0}{\frac{u(j+1)-u(j)}{2} - \mathrm{L}\cos\beta}\right] \tag{38}$$

$$\varepsilon_u(j) = \tan^{-1}\left[\frac{\mathrm{L}\sin\beta + h_0}{D - \frac{u(j+1)-u(j)}{2} + \mathrm{L}\cos\beta}\right] \tag{39}$$

$$\zeta_u(j) = \tan^{-1}\left[\frac{h_0}{\frac{u(j+1)-u(j)}{2}}\right] \tag{40}$$

$$\lambda_u(j) = \tan^{-1}\left[\frac{\mathrm{L}\sin\beta + h_0}{D + \frac{u(j+1)-u(j)}{2} - \mathrm{L}\cos\beta}\right] \tag{41}$$

$$\xi_u(j) = \tan^{-1}\left[\frac{h_0}{D - \frac{u(j+1)-u(j)}{2}}\right] \tag{42}$$

$$\sigma_u(j) = \tan^{-1}\left[\frac{\mathrm{L}\sin\beta + h_0}{2D - \frac{u(j+1)-u(j)}{2} + \mathrm{L}\cos\beta}\right] \tag{43}$$

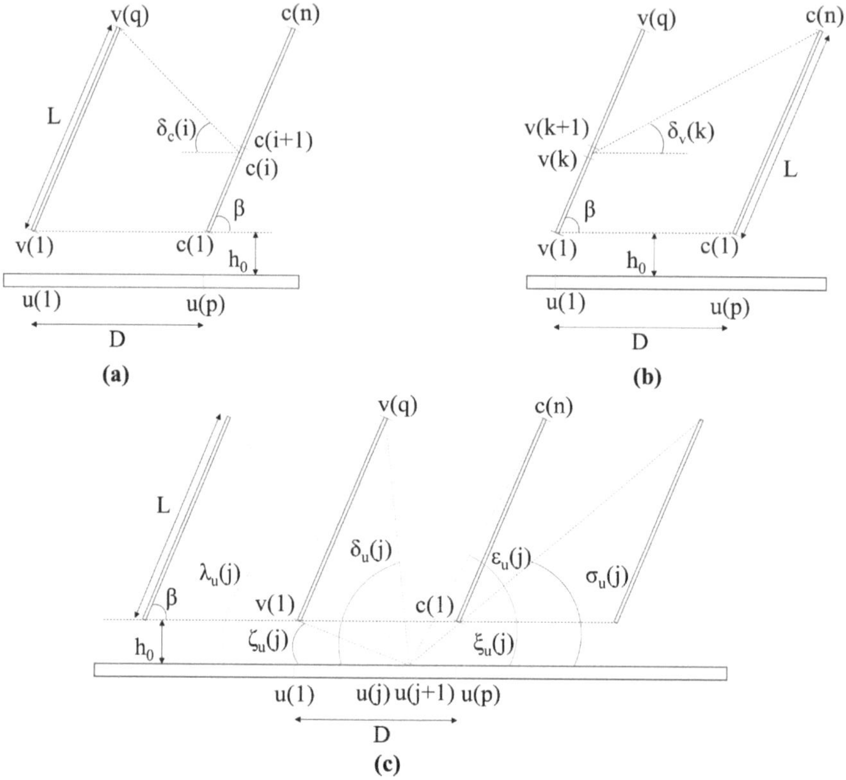

Figure 2. Schematic of the various angles for the computation of shadows and obscuring of circumsolar radiation for (**a**) a segment of the panel being evaluated, (**b**) the back of the front panel, and (**c**) the ground between the rows of panels.

The shading of direct irradiance for each segment of the panel surface, i, is given by Equation (44). Shading can occur ($S_c(i) = 1$) when the sun is positioned either in front of or behind the rows. Specifically, shading occurs when the projected solar elevation angle is positive and lower than the angle from the middle of the segment to the top of the front row (indicating that the sun is behind the front row), or when the projected solar elevation angle is negative and lower than the tilt angle of the surfaces (indicating that the sun is behind the panel). For the remaining cases, the segments are not shaded.

$$S_c(i) = \begin{cases} 1, & \left[\delta_c(i) > \alpha' \wedge \left|\gamma_s - \gamma_p\right| < 90^\circ\right] \vee \left(\beta > |\alpha'| \wedge \left|\gamma_s - \gamma_p\right| \geq 90^\circ\right) \\ 0, & \left[\delta_c(i) \leq \alpha' \wedge \left|\gamma_s - \gamma_p\right| < 90^\circ\right] \vee \left(\beta \leq |\alpha'| \wedge \left|\gamma_s - \gamma_p\right| \geq 90^\circ\right) \end{cases} \quad (44)$$

The shading of direct irradiance for each segment of the back surface of the front row, denoted by index k, is computed through Equation (45). Each segment is always shaded if the sun is positioned in front of the rows. Additionally, if the sun is not in front of the rows, shading occurs when the projected solar elevation angle exceeds the tilt angle of the surfaces or when it is lower than the angle from the middle of the segment to the top of the panel being evaluated.

$$S_v(k) = \begin{cases} 1, & \left|\gamma_s - \gamma_p\right| < 90^\circ \vee \left(\left|\gamma_s - \gamma_p\right| \geq 90^\circ \wedge [\delta_v(k) > |\alpha'| \vee \beta < |\alpha'|]\right) \\ 0, & \left|\gamma_s - \gamma_p\right| \geq 90^\circ \wedge \delta_v(k) \leq |\alpha'| \wedge \beta \geq |\alpha'| \end{cases} \quad (45)$$

The computation of the shading of direct irradiance on the ground between the rows is performed through Equation (46), which involves comparing the projected solar elevation angle with six other angles as depicted in Figure 2 and calculated through Equations (38) to (43).

Each ground segment, denoted by index j, is shaded under the following conditions: when the sun is positioned in front of the rows and behind the row in front, or when its projected solar elevation angle is lower than the angle of the middle of the segment to the top of the subsequent row in front. Additionally, ground segments are shaded if the sun is behind the row being evaluated or when the projected solar elevation angle is lower than the angle from the middle of the segment to the top of the subsequent row behind. Moreover, for ground segments positioned below the front row, shading occurs when the projected solar elevation angle exceeds the tilt of the surfaces.

$$S_u(j) = \begin{cases} 1, & \begin{array}{l} \left|\gamma_s - \gamma_p\right| < 90^\circ \wedge ([\zeta_u(j) < \alpha' < \delta_u(j) \wedge \delta_u(j) > 0] \vee \lambda_u(j) > \alpha' \vee [\delta_u(j) \leq 0 \wedge \alpha' > \zeta_u(j)]) \vee \\ \left|\gamma_s - \gamma_p\right| \geq 90^\circ \wedge \left(\begin{array}{l} [\varepsilon_u(j) > \xi_u(j) \wedge \varepsilon_u(j) > |\alpha'| > \xi_u(j)] \vee [\varepsilon_u(j) \leq \xi_u(j) \wedge \varepsilon_u(j) < |\alpha'| < \xi_u(j)] \vee \\ [\delta_u(j) < 0 \wedge |\delta_u(j)| < |\alpha'|] \vee |\alpha'| \leq \sigma_u(j) \end{array} \right) \end{array} \\ 0, & \begin{array}{l} \left|\gamma_s - \gamma_p\right| < 90^\circ \wedge ([\delta_u(j) \geq 0 \wedge \delta_u(j) \leq \alpha'] \vee [\lambda_u(j) \leq \alpha' \leq \zeta_u(j)]) \vee \\ \left|\gamma_s - \gamma_p\right| \geq 90^\circ \wedge \left(\begin{array}{l} [\delta_u(j) \geq 0 \wedge \varepsilon_u(j) \leq |\alpha'| \wedge \xi_u(j) \leq |\alpha'|] \vee [\sigma_u(j) \leq |\alpha'| \leq \xi_u(j) \wedge |\alpha'| \leq \varepsilon_u(j)] \vee \\ [\delta_u(j) < 0 \wedge \varepsilon_u(j) \leq |\alpha'| \wedge \xi_u(j) \leq |\alpha'| \wedge |\delta_u(j)| \geq |\alpha'|] \end{array} \right) \end{array} \end{cases} \quad (46)$$

The modeling of the circumsolar irradiance obscuring follows a similar approach, albeit with the utilization of the vector $\vec{S}_{cs}$ to represent the fraction of circumsolar irradiance that is obscured. Circumsolar irradiance is assumed as being uniformly distributed within the annular region surrounding the sun disk, with an apparent external angular radius, r, of 15°. The obscured area of the circumsolar region is given by Equation (47) as in [18]:

$$C(x) = \frac{sin^{-1}\frac{\sqrt{r^2-(|\alpha'|-x)^2}}{r}}{180} - \frac{\sqrt{r^2-(|\alpha'|-x)^2}}{2\pi r} \quad (47)$$

In the case of the segments of the panel being evaluated (Equation (48)), several conditions dictate the obscuring of circumsolar irradiance. When the sun is positioned in front of the rows, total obscuration ($S_{cs,c}(i) = 1$) occurs if the entire circumsolar region is below the angle defined by the middle of the segment to the top of the front row, denoted as $\delta_c(i)$. Substantial obscuration (more than 50%, $S_{cs,c}(i) = |1 - C(\delta_c(i))|$) occurs if this angle is higher than the projected solar elevation angle and the difference between these two angles is smaller than the angular radius. It is less obscured (less or equal to 50%, $S_{cs,c}(i) = C(\delta_c(i))$) if the projected solar elevation angle is higher than $\delta_c(i)$ and the difference between these two angles is smaller than the angular radius. Finally, no obscuration ($S_{cs,c}(i) = 0$) occurs if the projected solar elevation angle exceeds $\delta_c(i)$ and the difference between these two angles is higher than the angular radius. A similar principle applies when the sun is behind the rows, with the difference that instead of using the angle $\delta_c(i)$, the comparison is made with the tilt angle β.

$$S_{cs,c}(i) = \begin{cases} 1, & \left[\left|\gamma_s - \gamma_p\right| < 90^\circ \wedge \alpha' - \delta_c(k) \leq -r\right] \vee \left(\left|\gamma_s - \gamma_p\right| \geq 90^\circ \wedge |\alpha'| - \beta \leq -r\right) \\ |1 - C(\delta_c(i))|, & \left|\gamma_s - \gamma_p\right| < 90^\circ \wedge -r < \alpha' - \delta_c(i) < 0 \\ |1 - C(\beta)|, & \left|\gamma_s - \gamma_p\right| \geq 90^\circ \wedge -r < |\alpha'| - \beta < 0 \\ C(\delta_c(i)) & \left|\gamma_s - \gamma_p\right| < 90^\circ \wedge 0 \leq \alpha' - \delta_c(i) < r \\ C(\beta), & \left|\gamma_s - \gamma_p\right| \geq 90^\circ \wedge 0 \leq |\alpha'| - \beta < r \\ 0, & \left[\left|\gamma_s - \gamma_p\right| < 90^\circ \wedge \alpha' - \delta_c(i) \geq r\right] \vee \left(\left|\gamma_s - \gamma_p\right| \geq 90^\circ \wedge |\alpha'| - \beta \geq r\right) \end{cases} \quad (48)$$

The fraction of circumsolar irradiance obscured for each segment of the back surface of the front panels is given by Equation (49). It is assumed that all circumsolar irradiance is obscured when the sun is positioned in front of the rows. When the sun is positioned behind the rows, the projected solar elevation angles plus or minus the circumsolar angular radius are compared in a similar manner to Equation (48), but with reference to the tilt angle of the surface and the angle of the middle of the back surface segment to the top of the panel being evaluated.

$$S_{cs,v}(k) = \begin{cases} 1, & \left(\left|\gamma_s-\gamma_p\right| < 90^\circ\right) \vee \left(\left|\gamma_s-\gamma_p\right| \geq 90^\circ \wedge [|\alpha'| - \delta_v(k) \leq -r \vee |\alpha'| - \beta \geq r]\right) \\ |1 - C(\delta_v(k))|, & \left|\gamma_s-\gamma_p\right| \geq 90^\circ \wedge -r < |\alpha'| - \delta_v(k) < 0 \\ |1 - C(\beta)|, & \left(\left|\gamma_s-\gamma_p\right| \geq 90^\circ \wedge 0 < |\alpha'| - \beta < r\right) \\ C(\delta_v(k)) & \left|\gamma_s-\gamma_p\right| \geq 90^\circ \wedge 0 \leq |\alpha'| - \delta_v(k) < r \\ C(\beta), & \left(\left|\gamma_s-\gamma_p\right| \geq 90^\circ \wedge -r < |\alpha'| - \beta \leq 0\right) \\ 0, & \left[\left|\gamma_s-\gamma_p\right| \geq 90^\circ \wedge |\alpha'| - \delta_v(k) \geq r \wedge |\alpha'| - \beta \leq -r\right] \end{cases} \quad (49)$$

The computation of circumsolar irradiance obscuration for each segment of the ground between rows is more complex (Equation (50)). Complete obscuration is assumed when the entire circumsolar region is positioned either behind the front row panel or the panel being evaluated, considering the middle of the ground segment. Additionally, complete obscuration occurs when the angle of the top of the circumsolar region is lower than the angle of the top of the subsequent rows of panels, whether in front or behind. Similarly to the equations above (Equations (48) and (49)), the projected solar elevation angles plus or minus the circumsolar angular radius are compared with the six different angles shown in the scheme at the bottom of Figure 2 for the computation of the fraction of circumsolar irradiance that is obscured.

$$S_{cs,u}(j) = \begin{cases} 1, & \left[\left|\gamma_s-\gamma_p\right| < 90^\circ \wedge \begin{pmatrix} [\delta_u(j) \geq 0 \wedge \alpha' - \zeta_u(j) \geq r \wedge \alpha' - \delta_u(j) \leq -r] \vee \\ \alpha' - \lambda_u(j) \leq -r \vee \\ [\delta_u(j) < 0 \wedge \alpha' - \zeta_u(j) \geq r] \vee \\ [\alpha' - r \leq \lambda_u(j) \wedge \alpha' + r \geq \zeta_u(j)] \end{pmatrix}\right] \vee \\ & \left[\left|\gamma_s-\gamma_p\right| \geq 90^\circ \wedge \begin{pmatrix} [|\alpha'| - \xi_u(j) \geq r \wedge |\alpha'| - \varepsilon_u(j) \leq -r] \vee \\ |\alpha'| - \sigma_u(j) \leq -r \vee \\ [\delta_u(j) \leq 0 \wedge |\alpha'| - |\delta_u(j)| \geq r] \vee \\ [|\alpha'| - r \leq \sigma_u(j) \wedge |\alpha'| + r \geq \varepsilon_u(j)] \vee \\ [|\alpha'| - r \leq \sigma_u(j) \wedge |\alpha'| + r \geq \xi_u(j)] \end{pmatrix}\right] \\ |1 - C(|\delta_u(j)|)|, & \left[\left|\gamma_s-\gamma_p\right| < 90^\circ \wedge \delta_u(j) \geq 0 \wedge -r < \alpha' - \delta_u(j) < 0\right] \vee \\ & \left[\left|\gamma_s-\gamma_p\right| \geq 90^\circ \wedge \delta_u(j) < 0 \wedge 0 \leq |\alpha'| - |\delta_u(j)| < r\right] \\ |1 - C(\varepsilon_u(j))|, & \left|\gamma_s-\gamma_p\right| \geq 90^\circ \wedge -r < |\alpha'| - \varepsilon_u(j) < 0 \\ |1 - C(\zeta_u(j))|, & \left|\gamma_s-\gamma_p\right| < 90^\circ \wedge \alpha' - r > \lambda_u(j) \wedge 0 < \alpha' - \zeta_u(j) < r \\ |1 - C(\lambda_u(j))|, & \left|\gamma_s-\gamma_p\right| < 90^\circ \wedge \alpha' + r < \zeta_u(j) \wedge -r < \alpha' - \lambda_u(j) < 0 \\ |1 - C(\xi_u(j))| & \left|\gamma_s-\gamma_p\right| \geq 90^\circ \wedge |\alpha'| - r > \sigma_u(j) \wedge 0 < |\alpha'| - \xi_u(j) < r \\ |1 - C(\sigma_u(j))|, & \left|\gamma_s-\gamma_p\right| \geq 90^\circ \wedge |\alpha'| + r < \xi_u(j) \wedge -r < |\alpha'| - \sigma_u(j) < 0 \\ C(|\delta_u(j)|) & \left[\left|\gamma_s-\gamma_p\right| < 90^\circ \wedge \delta_u(j) \geq 0 \wedge 0 \leq \alpha' - \delta_u(j) < r\right] \vee \\ & \left[\left|\gamma_s-\gamma_p\right| \geq 90^\circ \wedge \delta_u(j) < 0 \wedge -r \leq |\alpha'| - |\delta_u(j)| < 0\right] \\ C(\varepsilon_u(j)), & \left|\gamma_s-\gamma_p\right| \geq 90^\circ \wedge 0 \leq |\alpha'| - \varepsilon_u(j) < r \\ C(\zeta_u(j)), & \left|\gamma_s-\gamma_p\right| < 90^\circ \wedge \alpha' - r > \lambda_u(j) \wedge -r < \alpha' - \zeta_u(j) < 0 \\ C(\lambda_u(j)), & \left|\gamma_s-\gamma_p\right| < 90^\circ \wedge \alpha' + r < \zeta_u(j) \wedge 0 < \alpha' - \lambda_u(j) < r \\ C(\xi_u(j)), & \left|\gamma_s-\gamma_p\right| \geq 90^\circ \wedge |\alpha'| - r > \sigma_u(j) \wedge -r < |\alpha'| - \xi_u(j) < 0 \\ C(\sigma_u(j)), & \left|\gamma_s-\gamma_p\right| \geq 90^\circ \wedge |\alpha'| + r < \xi_u(j) \wedge 0 < |\alpha'| - \sigma_u(j) < r \\ 0, & \left[\left|\gamma_s-\gamma_p\right| < 90^\circ \wedge \begin{pmatrix} [\delta_u(j) \geq 0 \wedge \alpha' - \delta_u(j) \geq r] \vee \\ [\alpha' - \zeta_u(j) \leq -r \wedge \alpha' - \lambda_u(j) \geq r] \end{pmatrix}\right] \vee \\ & \left[\left|\gamma_s-\gamma_p\right| \geq 90^\circ \wedge \begin{pmatrix} |\alpha'| - \varepsilon_u(j) \geq r \vee \\ [|\alpha'| - \xi_u(j) \leq -r \wedge |\alpha'| - \sigma_u(j) \geq r \wedge |\alpha'| - \varepsilon_u(j) \leq -r] \end{pmatrix}\right] \end{cases} \quad (50)$$

4. Experimental Setup and Procedure

In order to obtain observational data for both a first row and subsequent rows of panels with varying tilt angles and inter-row distances, a structure featuring a pyranometer for measuring GTI was constructed in an open field near Évora, Portugal (38.5306°, −8.0112°), as shown in Figures 3–5.

The experimental setup consists of three frames: a base, a front frame, and a rear frame, with a pyranometer installed on the rear frame. The base was leveled, and two transversal bars on the sides ensured that both front and rear frames maintained the same tilt angle. The apparatus allows for adjustment with three degrees of freedom: tilt angle of the front and rear frames (β, from 20° to 90°) through the solidary adjustment of the inclination of both frames; distance between frames (D, from 0.80 m to 1.10 m) through three positions of where the front frame can be fixed to the base; and position of the pyranometer along the length of the rear frame (c_s, from 0.08 m to 1.08 m) by sliding the instrument along its supporting bar. The uncertainty on the measurements of each of these variables is 1 cm in the cases of distances and 1° in the case of the tilt angles.

To represent the adjacent row, three Alveopan bilaminate white polypropylene boards, with a total width W of 3.03 m and length L of 1.08 m, were installed in the front frame. The reflectivity of these boards was measured using a FieldSpec HandHeld 2 spectroradiometer (ASD, Inc., Boulder, CO, USA) [33], yielding an average reflectivity of 0.921. Although potential edge effects were acknowledged due to board sizes, these were not factored into the general model. For data collection purposes relevant to the assessment of transposition models applied to the first rows, these boards were removed.

Figure 3. Experimental setup for measuring global tilt irradiance for different positions.

Figure 4. Overview of the experimental setup including (**a**) the Évora–PECS station and (**b**) the pyranometers used for albedo computations.

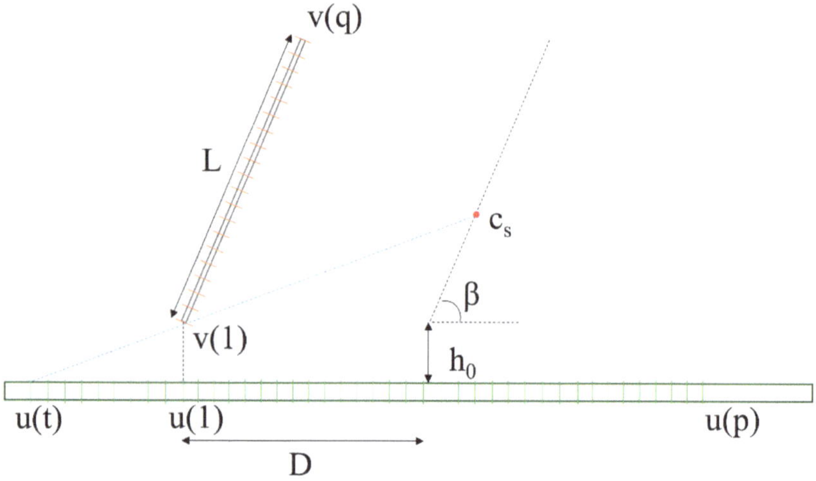

Figure 5. Schematic for modeling GTI on the sensor in the experimental apparatus.

Global tilted irradiance was measured using a Kipp & Zonen CMP11 pyranometer (Kipp & Zonen, Delft, The Netherlands), while global horizontal irradiance and reflected irradiances (for the computation of ground albedo) were measured using a Kipp & Zonen CM7B albedometer (Kipp & Zonen, Delft, The Netherlands). Both sensors were connected to a CR300 datalogger from Campbell Scientific (Shepshed, Loughborough, UK). Additionally, DNI, DIF, and GHI observation data were obtained from the Évora–PECS station of the DNI-ALENTEJO project network [34], located 5 m from the experimental setup.

The internal clock of the CR300 data logger used in the apparatus was synchronized with the data logger of the Évora–PECS station, both set to UTC time. Sensor outputs were sampled at 1 Hz and mean, maximum, minimum, and standard deviation values were recorded every minute. Observations were corrected following the best practices in the field, namely the WMO recommendations and the BSRN (Baseline Solar Radiation Network) guide, including corrections for sensor zero offset and filters according to the BSRN quality control procedure, considering the extremely rare limits [35] and removing measurements for zenith angles equal or above 85°.

Prior to the field measurements, a calibration procedure was conducted specifically for the CMP11 pyranometer and CM7B albedometer using a reference CMP21 pyranometer (Kipp & Zonen, Delft, The Netherlands), according to the ISO 9847:1992 standard [36].

For the accurate application of transposition models, the ground albedo value is needed. While a standard value of 0.2 is often used, this study estimated the ground albedo using Equation (51) [37], where BSA and WSA represent the black-sky albedo and white-sky albedo, respectively. BSA is defined as the albedo in the absence of the diffuse component and is a function of the solar zenith angle, while the WSA is the albedo considering the diffuse component as isotropic and in the absence of the direct component. The mean values of BSA and WSA were obtained by fitting Equation (44) to experimental data, where the albedo, ρ, was computed by applying the ratio of reflected (GRI) to global horizontal (GHI) irradiance observations from the albedometer. Following data treatment, including filtering and removal of the records for solar zenith angles lower than 70°, BSA and WSA were found to be 0.206 and 0.208, respectively, across all recorded data periods.

$$\rho = BSA + (WSA - BSA)\frac{DIF}{GHI} \tag{51}$$

Observations were conducted between 14 April and 1 June 2022. For first-row tests, 5 datasets or periods were generated, each corresponding to a specific tilt angle, as shown in Table 1, with the pyranometer positioned at c_s = 0.50 m. Testing of the developed model for other rows resulted in 19 periods with various tilt angles, distance between rows, and pyranometer positions, as shown in Table 2. Periods 4, 9, and 19 include instances in which the pyranometer is shaded.

It should be noted that measurements were conducted during a period of relatively high solar elevation, minimizing shading effects. To better capture shading effects, the inter-row distance during experimental tests were shorter than typical photovoltaic power plant configurations. Nonetheless, the developed model is designed to encompass diverse real-world conditions in the field.

Table 1. Structure position and number of 1 min data points for each testing period of the first-row tests.

Period	β (°)	Number of Data Points
1	20	551
2	38	582
3	50	620
4	70	545
5	90	508
All	-	2806

Table 2. Structure position and number of 1 min data points for each testing period of the proposed model for rows other than the first.

Period	D (m)	β (°)	c_s (m)	Number of Data Points
1	1.10	38	0.08	585
2	1.10	38	0.58	604
3	1.10	38	1.08	1528
4	0.81	38	0.18	2345
5	0.81	38	0.38	605
6	0.81	38	0.58	615
7	0.81	38	0.78	608
8	0.81	38	0.98	585
9	0.80	50	0.15	581
10	0.80	50	0.35	2412
11	0.80	50	0.55	617
12	0.80	50	0.75	632
13	0.80	50	0.95	651
14	0.84	20	0.26	1807
15	0.84	20	0.46	365
16	0.84	20	0.66	648
17	0.84	20	0.86	612
18	0.84	20	1.06	633
19	0.81	38	0.18	718
All	-	-	-	17,151
Shaded	-	-	-	2163
Unshaded	-	-	-	14,988

5. Results and Discussion

The different transposition models, including the developed model for rows that are not the first, were applied to the observations of DNI and DIF from the Évora–PECS station. Subsequently, the model outputs were compared with GTI observations from the experimental setup using multiple evaluation metrics developed in the software MATLAB R2018b. These metrics comprised the coefficient of determination (R^2), mean bias error (MBE), and root-mean-squared error (RMSE) along with a global performance index (GPI) based on R^2, MBE, and RMSE, where a higher value represents the better accuracy of the model [22].

Given the significant influence of atmospheric conditions, particularly cloud cover, on global irradiance, the mean and standard deviation of the clearness index were computed for each period based on 1 min observations. The clearness index, typically ranging from 0 to 1, represents the ratio of global horizontal irradiance measured at ground level to its counterpart estimated at the top of the atmosphere [34] (Equation (15)). It serves as an indicator of the total transmittance of the atmosphere, reflecting higher values under clear-sky conditions and lower values under overcast conditions. The subsequent subsections provide detailed tables presenting these clearness index values for each period.

Some clearness index values exceeded unity, with a maximum value of 1.079, attributed to cloud enhancement events. These phenomena occur when partly cloudy skies lead to a temporary increase in local GHI above the extraterrestrial irradiance, facilitated by multiple scatterings and reflections by clouds [35]. These values were kept in the analysis, as they capture the transient nature of atmospheric conditions during the observation periods.

5.1. Results for the First Row

The transposition models presented in Section 2.1 were computed for the periods shown in Table 1, with the resulting mean and standard deviation of the clearness index and GTI, alongside various evaluation metrics, summarized in Table 3. Across all models, GTI values were generally underestimated, with the exception of the Bugler and Modified Bugler models, where GTI was overestimated. This discrepancy could stem from the treatment of the direct normal component within the factor R_b used for modeling the diffuse component. Nevertheless, the Modified Bugler model showed better results compared with the other models. The Modified Bugler model tends to perform best except for vertical surfaces (period 5), where the Klucher model seems to show better results. Given that the global performance index (GPI) for the overall data presented the Modified Bugler model as the best performing model, it was chosen as the primary model for first-row applications in the proposed model. Despite its tendency to overestimate GTI, with an overall MBE of 23.8 W/m^2 and RMSE of 30.5 W/m^2, it delivered optimal results for period 3, characterized by a tilt angle of 50° and small variation in sky conditions (standard deviation of clearness index of 0.087).

Table 3. Mean and standard deviation of the clearness index and GTI and metrics of GTI from the transposition models analyzed for each and all measuring periods of the first-row tests. The best performing model is represented in bold for each indicator and period and the clearness index data are repeated for each metric for better readability.

Metric	Period	k_t Mean	k_t Std	GTI Mean (W/m^2)	GTI Std (W/m^2)	Liu Jordan	Klucher	Hay Davies	Reindl	Ma Iqbal	Modified Ma Iqbal	Bugler	Modified Bugler	Temps Coulson	Perez
R^2	1	0.697	0.100	700.0	322.0	0.9995	0.9992	0.9995	0.9995	**0.9995**	0.9995	0.9994	0.9995	0.9987	0.9994
	2	0.720	0.081	702.1	330.0	0.9991	0.9988	0.9991	**0.9992**	0.9991	0.9991	0.9990	0.9991	0.9985	0.9991
	3	0.674	0.087	585.5	297.9	0.9955	0.9941	0.9965	0.9967	0.9959	0.9959	0.9946	0.9957	0.8341	**0.9975**
	4	0.686	0.141	474.8	288.4	**0.9946**	0.9936	0.9945	0.9938	0.9944	0.9944	0.9942	0.9944	0.9317	0.9934
	5	0.676	0.102	327.7	202.8	0.9982	0.9963	0.9978	0.9953	0.9979	0.9978	0.9991	0.9987	0.9837	**0.9993**
	All	0.690	0.105	558.1	323.6	0.9959	0.9969	0.9974	0.9975	0.9970	0.9970	0.9938	0.9947	0.9216	**0.9978**
MBE (W/m^2)	1	0.697	0.100	700.0	322.0	−44.4	**−28.8**	−36.6	−36.4	−38.6	−38.4	58.1	29.8	−43.7	−31.5
	2	0.720	0.081	702.1	330.0	−53.7	−40.7	−44.6	−43.9	−46.9	−46.7	51.2	**23.6**	−34.6	−36.5
	3	0.674	0.087	585.5	297.9	−66.5	−32.4	−48.6	−42.4	−53.9	−53.5	23.0	**5.3**	−157.4	−24.1
	4	0.686	0.141	474.8	288.4	−51.4	−28.4	−45.1	−38.4	−46.5	−46.5	37.2	**19.7**	−75.0	−24.0
	5	0.676	0.102	327.7	202.8	−23.2	**−2.0**	−27.3	−18.3	−25.8	−26.2	57.9	44.6	−19.6	−10.3
	All	0.690	0.105	558.1	323.6	−48.7	−27.1	−40.9	−36.4	−42.9	−42.8	44.8	**23.8**	−68.7	−25.6
RMSE (W/m^2)	1	0.697	0.100	700.0	322.0	49.7	36.1	41.3	41.2	43.5	43.3	59.1	**30.4**	45.8	37.5
	2	0.720	0.081	702.1	330.0	56.1	43.5	46.3	45.6	48.8	48.5	53.9	**25.8**	36.1	38.9
	3	0.674	0.087	585.5	297.9	71.9	38.0	51.7	45.1	57.6	57.2	27.4	**15.2**	181.8	26.9
	4	0.686	0.141	474.8	288.4	56.5	34.8	48.7	42.7	50.4	50.3	42.9	**26.7**	94.8	30.8
	5	0.676	0.102	327.7	202.8	24.9	11.8	28.3	21.7	27.0	27.3	62.6	48.2	41.9	11.3
	All	0.690	0.105	558.1	323.6	54.9	35.0	44.5	40.7	47.1	47.0	50.2	**30.5**	100.2	30.9
GPI	All	0.690	0.105	558.1	323.6	1.068	1.848	1.412	1.569	1.324	1.329	1.195	**1.959**	−1.000	1.953

5.2. Results for Other Rows

For the evaluation of the developed transposition model for rows other than the first, some adjustments were implemented to accommodate the experimental setup (refer to Figure 5). Given the absence of panels in the second row, this row was not considered in the modeling. Instead, GTI was computed for each segment of the back of the front panel and ground, followed by the calculation of GTI at a designated point which represents the sensor. In this case, the angles ε_u, λ_u, ξ_u, σ_u, and δ_v used for shadow computation and circumsolar irradiance obscuration were not applicable. Another modification involved the length of ground considered in the model. Since the setup comprised only one panel, reflections from the ground beyond the modeled rows could significantly impact the measured GTI and were thus incorporated into the model validation process (depicted in Figure 5).

GTI estimation was performed using both the developed model and the Modified Bugler model for reference and comparison, which is a common practice in the absence of a specific model for other rows. It is important to note that the configurations used in periods 4, 9, and 19 result in direct shading of the pyranometer for a certain time span of the day. As example, periods 1 and 12, when there was no shading, and period 19, when there was shading and obscuration of the pyranometer by the front row, are shown in Figures 6–8 (the small data gaps during the day are a result of the filtering procedure mentioned in Section 4). Despite the fact that slightly lesser improvements are observed for period 12 (Figure 6), which is attributed to partially cloudy conditions, the effectiveness of the developed model over the Modified Bugler is evident across the evaluated periods.

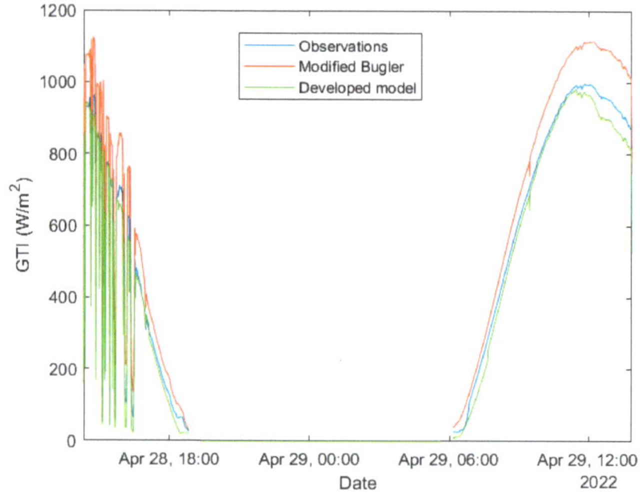

Figure 6. Global tilted irradiance observed and modeled by the Modified Bugler model (first-row model for reference and comparison) and the developed model for period 1 (1 min timestep).

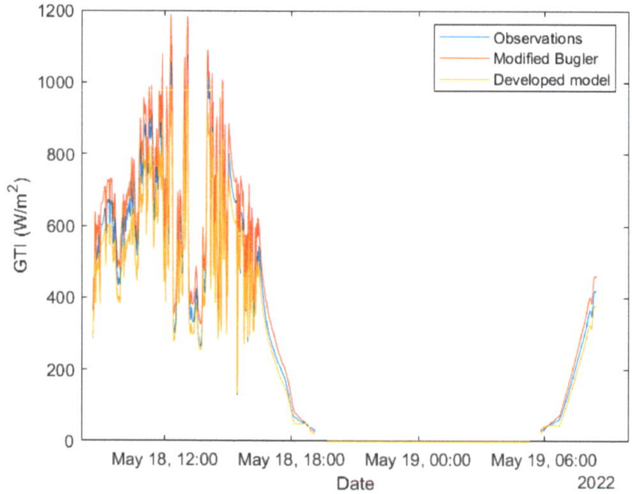

Figure 7. Global tilted irradiance observed and modeled by the Modified Bugler model (first-row model for reference and comparison) and the developed model for period 12 (1 min timestep).

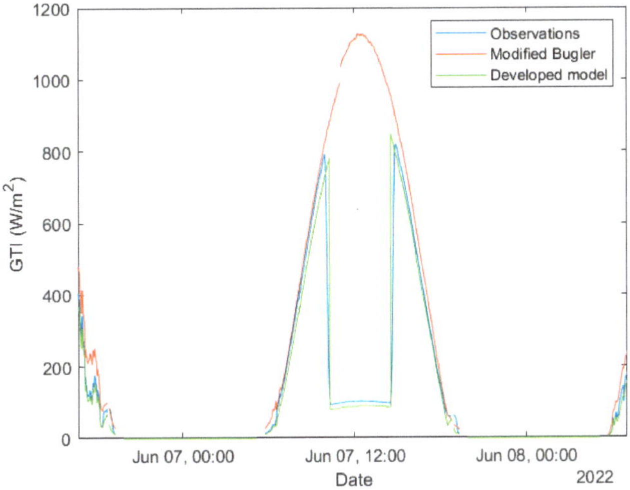

Figure 8. Global tilted irradiance observed and modeled by the Modified Bugler model (first-row model for reference and comparison) and the developed model for period 19 (1 min timestep).

The results for each period, all periods, and for the data when the pyranometer is shaded or unshaded are presented in Table 4. When compared with the original Modified Bugler model, which overestimates the GTI, the proposed model improves the MBE for most periods, albeit with a slight underestimation. Typically, the Modified Bugler performs better in periods characterized by higher pyranometer positioning and greater frame-to-frame distances, resembling first-row irradiance conditions. During periods of direct irradiance shading (periods 4, 9, and 19), the Modified Bugler model, which does not consider shading, shows significantly higher errors. Another aspect to highlight is the impact of clouds in the performance of the model. In periods 9 and 14, for example, when the mean clearness index is lower and its standard deviation is higher (indicating cloudier skies) the metrics show lower performance of both models.

Table 4. Mean and standard deviation of the clearness index and metrics of GTI from the Modified Bugler model and proposed model for other rows.

Period	k_t		GTI		Modified Bugler Model			Developed Model for Other Rows		
	Mean	Std	Mean (W/m^2)	Std (W/m^2)	R^2	MBE (W/m^2)	RMSE (W/m^2)	R^2	MBE (W/m^2)	RMSE (W/m^2)
1	0.687	0.077	566.1	342.1	0.9905	100.0	106.1	0.9921	−31.1	39.5
2	0.675	0.108	646.2	324.4	0.9988	75.6	76.6	0.9987	−38.5	44.8
3	0.584	0.204	456.2	338.0	0.9413	57.9	96.2	0.9410	−24.7	82.0
4	0.641	0.191	176.2	126.3	0.0873	521.5	651.9	0.9844	−15.7	22.7
5	0.704	0.095	627.8	335.9	0.9984	76.4	78.5	0.9990	−38.6	42.3
6	0.690	0.103	599.6	356.2	0.9995	68.4	70.8	0.9991	−39.4	42.3
7	0.672	0.098	582.9	345.2	0.9996	69.3	71.0	0.9996	−39.4	43.8
8	0.684	0.104	634.2	335.2	0.9990	36.6	37.6	0.9988	−69.5	77.3
9	0.540	0.206	248.4	187.0	0.1385	290.2	376.0	0.3225	−81.1	161.5
10	0.590	0.190	420.5	307.1	0.9769	103.4	112.0	0.9944	−42.3	50.6
11	0.674	0.145	548.9	307.6	0.9986	61.2	62.1	0.9962	51.6	57.8
12	0.622	0.157	432.0	270.4	0.9949	52.4	55.2	0.9946	−54.0	62.1
13	0.561	0.134	436.4	272.7	0.9962	37.9	40.6	0.9961	−44.7	58.7
14	0.351	0.192	124.5	79.55	0.4669	250.6	304.9	0.4610	87.0	141.2
15	0.636	0.154	440.7	300.0	0.9980	28.3	30.3	0.9987	−62.7	69.6
16	0.524	0.275	481.7	370.5	0.9792	35.1	61.7	0.9793	−50.3	78.2
17	0.695	0.172	679.9	360.3	0.9907	41.5	50.6	0.9909	−65.9	78.0
18	0.677	0.147	643.9	334.1	0.9966	37.4	42.4	0.9964	−63.4	69.0
19	0.644	0.177	248.6	232.2	0.0372	373.2	570.2	0.9063	−12.4	75.5
All	0.638	0.188	323.8	304.6	0.4698	381.2	301.2	0.9543	−12.9	76.8
Shaded	0.682	0.197	203.7	169.2	0.0028	717.3	775.0	0.2585	−28.9	84.2
Unshaded	0.632	0.186	421.6	336.7	0.9047	88.9	131.0	0.9552	−26.5	75.7

Due to the variation in sky conditions along the different periods, the comparison between the different positioning of the setup proved challenging and thus, more importance is given to the overall results instead of each period. In this regard, the developed model for rows affected by the presence of rows in front showed an MBE of −12.9 W/m^2 and a root-mean-squared error of 76.8 W/m^2. As expected, this model outperforms the first-row model when the pyranometer is shaded. Even under unshaded conditions, the developed model is better than the model for the first row, showing the impact of the obscuring of the sky dome due to the other rows, considering the sky radiance anisotropy, namely the circumsolar region, and of the reflections from the front row and ground on the GTI.

To quantify the impact of the proposed model on reducing the error for each irradiance component on a tilted surface compared with the Modified Bugler model, a weight (w_i) for each component i of GTI was computed through Equation (52):

$$w_i = \frac{1}{n}\sum \frac{i_D - i_{MB}}{GTI_D - GTI_{MB}} \quad (52)$$

Here, D stands for developed model and MB for the Modified Bugler model. The results are presented in Table 5 for the direct, I_b, diffuse circumsolar, D_{cs}, diffuse isotropic D_{iso}, and reflected, I_{refl}, components and for both unshaded and shaded conditions. For the proposed model, the reflected component includes reflections in the ground and in the back side of the front panel. The mean bias error of each model is also included in the table for reference.

Table 5. MBE and weight of each component of GTI on the reduction in bias error when comparing the developed model with the Modified Bugler model.

Period	MBE Modified Bugler (W/m^2)	MBE Developed Model (W/m^2)	w_{Ib} (%)	w_{Dcs} (%)	w_{Diso} (%)	w_{Irefl} (%)
Unshaded	88.9	−26.5	0.0	0.5	56.6	42.9
Shaded	717.3	−28.9	66.5	2.6	28.7	2.2
All	381.2	−12.9	8.4	0.8	53.1	37.7

Overall, the masking of the isotropic diffuse irradiance has the highest weight in the difference between the Modified Bugler and the developed model followed by the modeling of reflected irradiance. As expected, when the direct irradiance is shaded, this becomes the most impactful component, while for unshaded conditions, it has no impact.

6. Operational Algorithm for Forecasting of Solar Irradiance on Tilted Surfaces

This section presents the application of the developed model to predict irradiance on tilted surfaces using operational direct normal and diffuse irradiance forecasts as inputs.

6.1. Forecast Input Data

The transposition models presented in this work, in particular the developed model for rows that are not the first, require several geometrical details on the solar panels of the power plant as input. However, only two meteorological variables, specifically DNI and DIF, are necessary for the determination of the GTI. The European Centre for Medium-range Weather Forecasts (ECMWF) developed the Integrated Forecasting System (IFS), which issues global hourly forecasts every day at 00 UTC up to 10 days ahead. These forecasts include the GHI and DNI variables, while DIF can be computed through the closure function $GHI = BHI + DIF$ (where BHI is given by Equation (2)).

In the work by Pereira et al. [22], a method based on artificial neural networks (ANNs) was developed to generate improved spatial and temporal downscaled DNI forecasts using operational forecast outputs from the ECMWF/IFS and the Copernicus Atmospheric Monitoring Service (CAMS). This work employs the same models and procedure, while also including the analysis and usage of GHI as compared with previous work. Figures 9–11 provide a direct comparison between ECMWF/IFS forecasts and observations made through calibrated sensors at Évora–Verney station in Portugal (38.5678, −7.9114). These figures show data for DNI, GHI, and DIF for each forecast day of the forecast horizon (day 0: 1 h to 24 h; day 1: 25 h to 48 h; day 2: 49 h to 72 h) over the period from 1 December 2016 to 31 May 2021, with a timestep of 10 min.

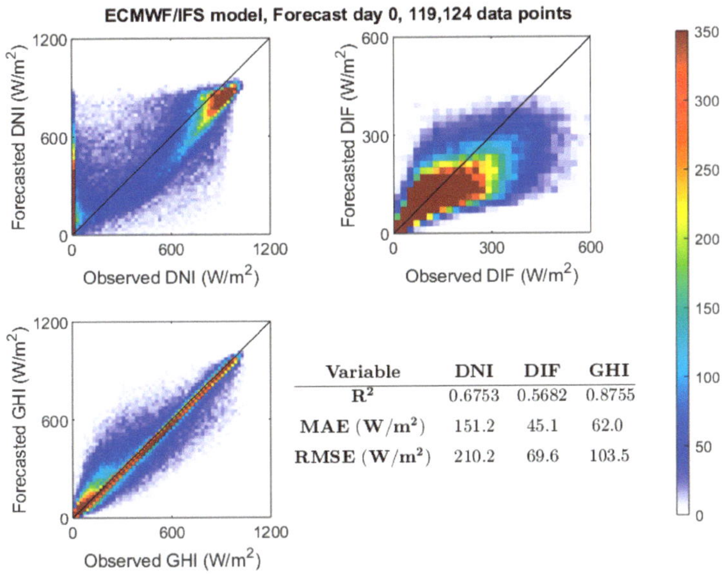

Variable	DNI	DIF	GHI
R^2	0.6753	0.5682	0.8755
MAE (W/m^2)	151.2	45.1	62.0
RMSE (W/m^2)	210.2	69.6	103.5

Figure 9. Comparison between downscaled 10 min forecasts of the ECMWF/IFS and observations made at Évora–Verney of DNI, DIF, and GHI for forecast day 0 (the colormap represents the number of data points in each bin; bin size: 20 × 20 W/m^2).

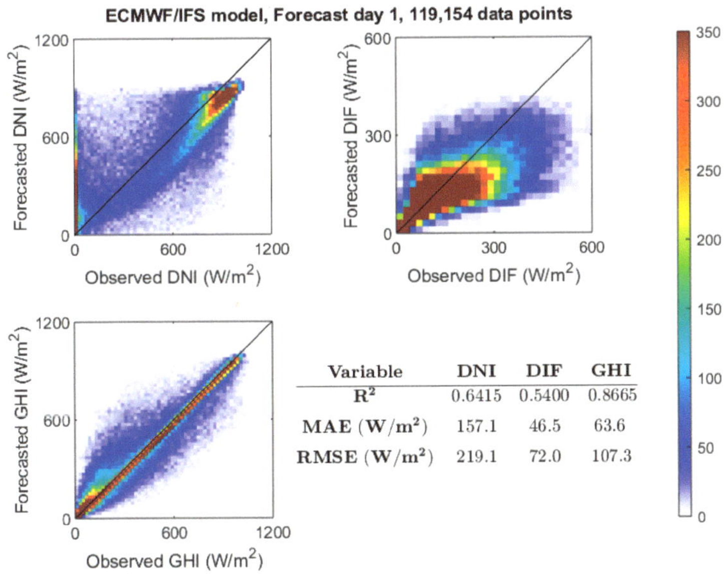

Variable	DNI	DIF	GHI
R^2	0.6415	0.5400	0.8665
MAE (W/m^2)	157.1	46.5	63.6
RMSE (W/m^2)	219.1	72.0	107.3

Figure 10. Comparison between 10 min downscaled forecasts of the ECMWF/IFS and observations made at Évora–Verney of DNI, DIF, and GHI for forecast day 1 (the colormap represents the number of data points in each bin; bin size: 20 × 20 W/m^2).

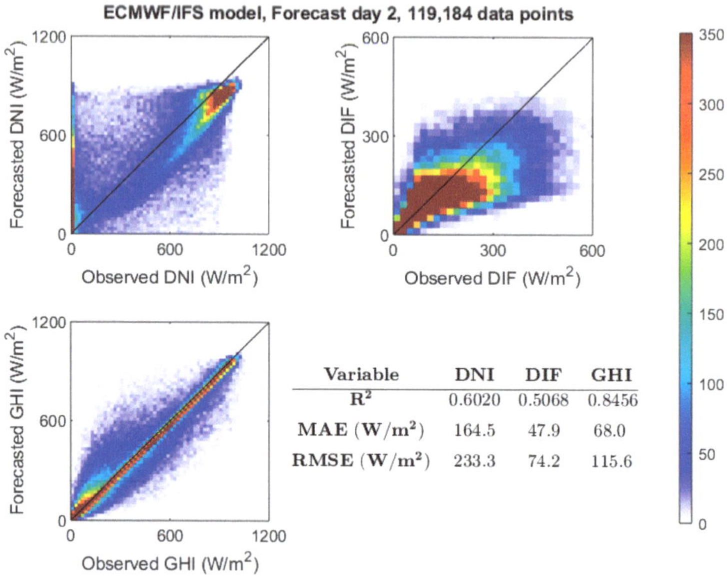

Variable	DNI	DIF	GHI
R^2	0.6020	0.5068	0.8456
MAE (W/m^2)	164.5	47.9	68.0
RMSE (W/m^2)	233.3	74.2	115.6

Figure 11. Comparison between 10 min downscaled forecasts of the ECMWF/IFS and observations made at Évora–Verney of DNI, DIF, and GHI for forecast day 2 (the colormap represents the number of data points in each bin; bin size: 20 × 20 W/m^2).

As expected, shorter forecast horizons correspond to smaller errors in the forecasts. Improved DNI forecasts were obtained using artificial neural network models, accounting for the nonlinear relationships between the DNI and various meteorological forecasted variables, such as GHI, cloud cover, and aerosol optical depth, as well as the variation in DNI over a given period before the forecasted instant. A brief explanation of this model and procedure can be found in Appendix A. The application of this procedure results in the metrics shown in Figure 12, where the diffuse component was computed using the GHI forecasted by the ECMWF/IFS and the improved DNI forecasts generated by the ANN models.

The application of the ANN model shows improvements in the forecasting of both DNI and DIF over a dataset that encompasses different sky conditions. The same procedure was applied by retrieving weather forecasts for the measurement periods used in this work to evaluate the different transposition models (Section 4). The resulting 10 min improved DNI and computed DIF forecasts were then evaluated against the 10 min mean observations obtained from the experimental data presented in Section 4.

Table 6 shows the results for first-row and inner-row tests. Detailed results for each testing period can be found in Appendix B. Similar to the larger dataset, the longer the forecast horizon, the higher the errors. However, the errors obtained tend to be greater than the ones aforementioned. This discrepancy is due to the smaller dataset used for this analysis, where even minor differences between forecasts and observations can lead to significant errors. Forecasts of DNI and DIF tend to have larger errors compared to global variables such as GHI or GTI, since accurate forecasting of clouds, specifically their location, is critical for this temporal resolution. The GTI results from the transposition models, based on the output of these forecast models, are presented in the following.

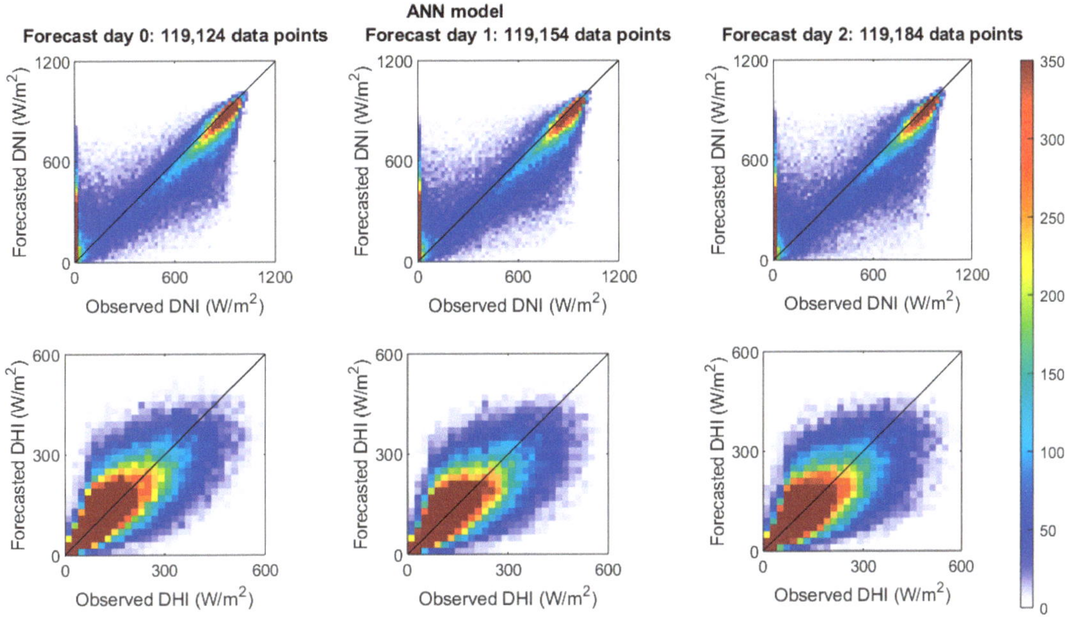

Variable	DNI	DHI
R^2	0.7498	0.6087
MAE (W/m^2)	125.5	42.2
RMSE (W/m^2)	184.4	66.6

Variable	DNI	DHI
R^2	0.7234	0.5883
MAE (W/m^2)	130.9	43.0
RMSE (W/m^2)	193	68.1

Variable	DNI	DHI
R^2	0.6789	0.5475
MAE (W/m^2)	139.2	45.4
RMSE (W/m^2)	209.0	71.8

Figure 12. Comparison between improved 10 min forecasts and observations made at Évora–Verney of DNI and DIF for forecast days 0, 1, and 2 (the colormap represents the number of data points in each bin; bin size: 20 × 20 W/m^2).

Table 6. DNI and DIF forecast results for all data of first-row tests and for all shaded and unshaded data of the developed model testing (MBE, MAE, and RMSE in W/m^2).

Data and Variable		Day 0 R^2	Day 0 MBE	Day 0 MAE	Day 0 RMSE	Day 1 R^2	Day 1 MBE	Day 1 MAE	Day 1 RMSE	Day 2 R^2	Day 2 MBE	Day 2 MAE	Day 2 RMSE
First-row tests													
All	DNI	0.4859	−38.5	87.2	132.1	0.2568	−41.5	107.4	169.2	0.2373	−99.8	151.2	225.7
	DIF	0.5258	−1.6	44.2	62.1	0.5540	4.0	41.3	60.3	0.5382	7.7	48.6	62.4
Inner-row tests													
All	DNI	0.5386	34.0	146.3	208.7	0.5458	39.4	145.1	208.6	0.4930	31.8	156.0	224.6
	DIF	0.4252	66.4	66.5	94.6	0.4720	64.0	64.0	91.5	0.4316	68.3	68.3	95.2
Shaded	DNI	0.6166	126.4	185.6	267.4	0.6488	134.8	183.6	263.8	0.5840	121.6	196.2	272.8
	DIF	0.4670	−49.7	92.8	126.8	0.5025	−53.1	97.1	126.9	0.4255	−44.4	101.0	131.0
Unshaded	DNI	0.5351	16.9	139.0	196.0	0.5389	21.8	138.0	196.7	0.4876	15.2	148.5	214.6
	DIF	0.3884	−8.9	61.6	87.4	0.4437	−17.6	57.8	83.3	0.4038	−15.6	62.3	87.1

6.2. Operational Analysis of Transposition Models

To understand how the selected transposition models perform in an operational forecast context, the transposition models were applied using the resulting forecasts of DNI and DIF for the testing location with a 10 min timestep. The results were then compared with the observations.

Table 7 presents the results for the first row using the Modified Bugler model and for inner rows using the developed model for each day of the forecast horizon.

Table 7. GTI results of the developed model for rows that are not the front row for each day o forecast horizon using DNI and DIF forecasts (MAE and RMSE in W/m^2; MAPE and RMSPE in %)

Data	Day 0				Day 1				Day 2			
	R^2	MBE	MAE	RMSE	R^2	MBE	MAE	RMSE	R^2	MBE	MAE	RMSE
First-row tests												
All	0.9646	6.7	34.5	48.7	0.9607	15.5	39.4	55.0	0.8001	71.5	67.3	128.5
Inner-row tests												
All	0.7197	−31.5	100.6	164.0	0.7162	−32.1	101.1	165.6	0.6727	−37.2	106.8	179.4
Shaded	0.0074	−115.3	152.4	285.1	0.0242	−116.7	152.4	281.8	0.0078	−113.9	151.3	284.5
Unshaded	0.8031	−16.0	91.0	129.8	0.7907	−16.5	91.6	133.6	0.7318	−23.0	98.6	152.2

Regarding first-row results, when compared with the results using DNI and DIF measurements instead of forecasts (Table 3), it is evident that for the analyzed periods, the use of forecasts shows better results for days 0 and 1 regarding MBE. However, there is a deterioration for day 2 and across all forecast horizons regarding the MAE and RMSE.

As for the inner-row test results, when compared with GTI based on DNI and DIF measurements instead of forecasts (Table 4), there is a general deterioration in the results except for the MBE in unshaded conditions for forecast days 0 and 1. Detailed results for each testing period can be found in Appendix B. The use of the developed transposition model resulted in overall MBE and RMSE values of 33.6 W/m^2 and 169.7 W/m^2, respectively, for the entire forecast horizon. These values show that the model is eneficial for linking irradiance forecast models with energy generation modeling in solar power plants aiding in the production of power output forecasts. These forecasts are essential for better decision-making by operators of such energy systems due to the variability of resource and energy demand.

To understand how the mean bias error of DNI and DIF forecasts affects the bias of the computed GTI values, Figure 13 was created. This figure shows the difference between the MBE of GTI from the developed model using forecasts and experimental data as input as a function of the MBE of DNI and DIF predictions for the forecast at day 0. Positive values of this difference are represented with blue circles, while negative values are represented with white circles. For better readability, only periods with an MBE difference of less than 100 W/m^2 are represented, which resulted in excluding only period 1.

Since the MBE of the proposed model is typically negative for the tests carried out (Tables A5 and 7), and considering the relationship between the MBE values for the forecasted DNI and DIF components (when one tends to higher positive values, the other tends to more negative values (Tables A3 and 6) as a consequence of the closure equation and knowing that the MBE of the GHI forecasts is lower), the bias of the GTI using prediction values may decrease as shown in Figure 13 due to a favorable combination of forecast and model errors. This is the reason why the MBE for unshaded conditions using forecast values (Table 7) is lower than the bias error for the same conditions using experimental values (Table 4). However, this trend is not observed for the shaded conditions, where the mean bias error is still lower when using ground-based measurements. A more detailed analysis of these aspects is needed in the future, which extends beyond the scope of the present work. Also, due to spatial resolution limitations, the forecasted variables from ECMWF (current operational horizontal resolution: ~9 km) are identical for the first and second rows in this experiment. In future work, this algorithm should be validated for a larger power plant that encompasses multiple grid points of a high-resolution forecast model at a hectometric scale.

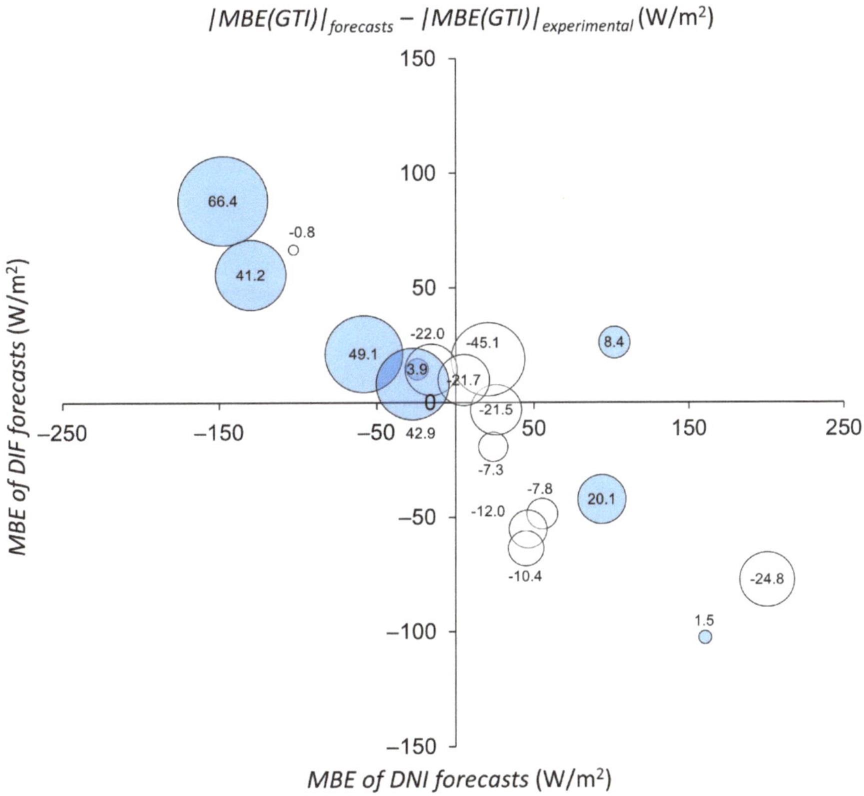

Figure 13. Difference between mean bias errors of GTI from the developed model results using forecasted or experimental data as input for day 0 (negative differences in white and positive differences in blue).

7. Conclusions

This work presented a comprehensive analysis of nine analytical transposition models based on physics alongside the Perez transposition model to compute the global tilted irradiance on photovoltaic module surfaces. Additionally, it presented a model for the computation of this variable in rows of modules other than the first, which usually comprises most rows of solar power plants. The developed model can be applied to any first-row transposition model, provided it considers direct, circumsolar, and isotropic diffuse irradiance. This model computes the GTI for different longitudinal segments of the surfaces of the row of modules, the back of the row in front, and the ground between the rows. It takes into consideration the different view factors and the obscuring of direct and circumsolar irradiance for each of the segments for any apparent solar position and includes the shading effect of the succeeding rows on the ground segments.

The evaluation of these models utilized data collected in Évora, Portugal, for different tilt angles for first-row tests and also for different inter-row distances, including shading conditions, to assess the performance of the developed model. The clearness index helps address potential confounding variables by providing a baseline for sunshine conditions. However, we acknowledge that factors such as wind, relative humidity, precipitation, and aerosols also affect the experiment to some extent. In future research, these factors should be considered. Also, our dataset is limited to our experimental setup. While expanding to diverse locales would enhance generalizability, the focus of this study was on addressing challenges specific to inner rows of solar panels, rather than aiming for global applicability. Thus, while our findings offer valuable insights, they may not directly apply to all regions.

Results showed that the best analytical transposition model for the first row is the Modified Bugler model, showing an overall MBE of 23.8 W/m^2 and RMSE of 30.5 W/m^2. Conversely, for other rows, the developed model showed an MBE of -12.9 W/m^2 and RMSE of 76.8 W/m^2, resulting in an improvement of 368.3 W/m^2 and 224.4 W/m^2, respectively, compared to using the selected reference transposition model for first rows. This shows the importance of considering the direct shading and obscuring of the sky dome when computing GTI for surfaces in rows that are not the first.

Furthermore, the operational performance of transposition models was evaluated for GTI forecasting, using improved irradiance forecast values instead of measurements of DNI and DIF. These forecast values were obtained from artificial neural network models using numerical weather prediction and aerosol forecast data. Results of the first-row tests showed a MBE and RMSE for all data of 6.7 W/m^2 and 48.7 W/m^2 for forecast day 0, 15.5 W/m^2 and 55.0 W/m^2 for forecast day 1, and 71.5 W/m^2 and 128.5 W/m^2 for forecast day 2. This shows an increased error compared to results using observations which are a mean MBE and RMSE increase across the three days of forecast of 7.4 W/m^2 and 46.9 W/m^2. It also shows how the forecast performance tends to deteriorate with time. The same is visible for the tests performed for other rows, which show an overall MBE and RMSE of -31.5 W/m^2 and 164.0 W/m^2 for forecast day 0, -32.1 W/m^2 and 165.6 W/m^2 for forecast day 1, and -37.2 W/m^2 and 179.4 W/m^2 for forecast day 2.

This work demonstrated that transposition models that neglect shading and irradiance obscuration are not suitable for the accurate estimation of GTI in surfaces that are not in the front row of a solar power plant. The use of a dedicated model for these conditions, such as the one presented in this work, is of great importance, given that GTI is the main factor influencing the energy generation of solar photovoltaic systems.

Author Contributions: Conceptualization, S.P. and P.C.; methodology, S.P. and P.C.; validation, S.P. and P.C.; formal analysis, S.P. and P.C.; investigation, S.P., P.C. and R.S.; data curation, S.P. and P.C.; writing—original draft preparation, S.P.; writing—review and editing, S.P., P.C. and R.S.; visualization, S.P. and P.C.; supervision, P.C. and R.S. All authors have read and agreed to the published version of the manuscript.

Funding: This work was funded by National funds through FCT—Fundação para a Ciência e Tecnologia, I.P. (projects UIDB/04683/2020 and UIDP/04683/2020). S. Pereira acknowledges the support of FCT (the Portuguese Science and Technology Foundation) through the grant with reference SFRH/BD/145378/2019.

Data Availability Statement: The raw data supporting the conclusions of this article will be made available by the authors on request. Restrictions apply to the availability of these data. Data were obtained from ECMWF.

Acknowledgments: The authors are thankful to Afonso Cavaco, Josué Figueira, and Samuel Ramos Bárias for their technical support.

Conflicts of Interest: The authors declare no conflicts of interest.

Glossary

Nomenclature

A	Anisotropy index
BHI	Beam horizontal irradiance (W/m^2)
C	Fraction of circumsolar area obscured by the adjacent row
c	Length along the panel (m)
c_s	Position of the pyranometer along the length of the panel (m)
D	Distance between rows in the horizontal plane (m)
$\vec{D}$	Vector of diffuse horizontal irradiance vector (W/m^2)
D_{cs}	Circumsolar diffuse component of GTI (W/m^2)
DIF	Diffuse horizontal irradiance (W/m^2)
D_{iso}	Isotropic diffuse component of GTI (W/m^2)
DNI	Direct normal irradiance (W/m^2)
EHI	Extraterrestrial horizontal irradiance (W/m^2)
ENI	Extraterrestrial normal irradiance (W/m^2)
F	View factor matrix
F_g	Ground view factor
F_s	Sky view factor
f	Clearness index as defined in the Klucher model
GTI	Global tilted irradiance (W/m^2)
GRI	Global reflected irradiance (W/m^2)
h_0	Vertical distance between the ground and the panel base (m)
I	Identity matrix
I_b	Direct component of GTI (W/m^2)
$\vec{I}_r$	Vector of direct horizontal irradiance vector (W/m^2)
I_{refl}	Reflection component of GTI (W/m^2)
k_T	Clearness index
L	Length of the panel (m)
MAE	Mean absolute error (W/m^2)
MBE	Mean bias error (W/m^2)
MAPE	Mean absolute percentage error (%)
R	Reflectivity matrix
R_b	Beam irradiance tilt factor
R_d	Diffuse irradiance tilt factor
RMSE	Root-mean-squared error (W/m^2)
RMSPE	Root-mean-squared percentage error (%)
R^2	Coefficient of determination
R_r	Reflected irradiance tilt factor
r	Apparent angular radius of circumsolar irradiance (°)
S	Shading of direct component coefficient
S_{cs}	Shading of circumsolar component coefficient
u	Length along the ground (m)
v	Length along the back of the front panel (m)
W	Row width (m)
w_i	Weight of irradiance component i to the bias reduction (%)
X_i	Isotropic component of first-row model
X_{cs}	Circumsolar component of first-row model

Acronyms

ANN	Artificial neural networks
CAMS	Copernicus Atmospheric Monitoring Service
ECMWF	European Centre for Medium-range Weather Forecasts
IFS	Integrated forecasting system
NWP	Numerical weather prediction

Greek symbols

Φ	Solar zenith angle (°)
α	Solar elevation angle (°)
α'	Projection of α in the vertical plane of the local meridian (°)
β	Surface tilt angle (°)
γ_p	Surface azimuth (°)
γ_s	Solar azimuth (°)
δ	Angle between the horizontal and the top of the front panel (°)
ε	Angle between the horizontal and the top of the panel being considered (°)
ζ	Angle between the horizontal and the bottom of the front panel (°)
θ	Angle of incidence (°)
λ	Angle between the horizontal and the bottom of the panel being considered (°)
ξ	Angle between the horizontal and the top of the panel in front of the front panel (°)
ρ	Ground albedo
σ	Angle between the horizontal and the top of the panel behind the panel being considered (°)

Appendix A

The flowchart of the model used to generate forecast data of direct normal irradiance (DNI) [22] used in this work is shown in Figure A1.

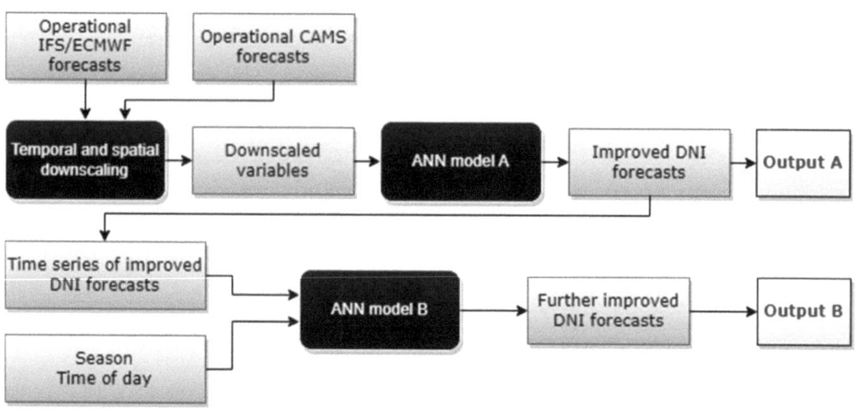

Figure A1. Flowchart of the model used to generate DNI forecasts [22].

The model takes as input data the variables shown in Table A1 from the operational Integrated Forecasting System (IFS) of the European Centre for Medium-range Weather Forecasts (ECMWF) and the Copernicus Atmospheric Monitoring Service (CAMS). These are run every day at 00 UTC, providing hourly forecast values up to 90 h ahead at discrete points of a global grid with a horizontal spatial resolution of 0.125° × 0.125°.

A temporal and spatial downscaling is performed on these variables, which results in forecasts for a specific location and higher temporal resolution. This downscaling is obtained through a bi-linear interpolation of the values in the four surrounding grid points of the desired location for the spatial downscaling and a piecewise cubic hermite interpolation of the hourly values into smaller timesteps (in this work, it was 10 min values) for the temporal downscaling.

Table A1. Input variables obtained from numerical weather prediction models.

Variables Obtained from IFS/ECMWF	Variables Obtained from CAMS
Direct normal irradiance	Total aerosol optical depth at 670 nm
Global horizontal irradiance	Total aerosol optical depth at 865 nm
Low cloud cover	Total aerosol optical depth at 1240 nm
Medium cloud cover	Sea salt aerosol optical depth at 550 nm
High cloud cover	
Total cloud cover	
Wind speed	
Air temperature	
Solar zenith angle	

The downscaled variables are then fed to an artificial neural network (ANN model A), which is a feed-forward network with one hidden layer comprising seven neurons and uses (i) a backpropagation learning function, namely Bayesian regularization backpropagation; (ii) a linear layer output with an initialization function that initializes the weights and biases of the layers according to the Nguyen–Widrow initialization algorithm; (iii) the hyperbolic tangent sigmoid transfer function; and (iv) the mean-squared error as a performance function. The input and output data are processed by removing rows with constant values and scaling the mean of each row to 0 and deviations to 1. This ANN takes into consideration the nonlinear relationships between the different atmospheric and aerosol variables and DNI resulting in improved DNI forecasts for the specified location and temporal resolution.

A second artificial neural network (ANN model B) is similar to ANN model A, but uses the Levenberg–Marquardt backpropagation algorithm and has eight neurons in the hidden layer, taking as input a time series of 12 timesteps of the improved DNI forecasts prior to the forecast moment (from the output of the ANN model A) along with the season and time of day. This model takes into consideration the temporal variation in DNI and further improves the DNI forecasts from ANN model A.

Appendix B

Table A2. Metrics of DNI and DIF forecasts for the measuring periods used for first-row tests (MBE, MAE, and RMSE in W/m^2).

Period and Variable		Day 0				Day 1				Day 2			
		R^2	MBE	MAE	RMSE	R^2	MBE	MAE	RMSE	R^2	MBE	MAE	RMSE
1	DNI	0.5114	−33.1	57.0	87.6	0.6120	−14.9	38.3	67.9	0.4664	−15.3	72.2	99.5
	DIF	0.1228	18.5	26.5	39.1	0.2874	6.0	11.0	13.7	0.0912	−2.8	32.0	37.9
2	DNI	0.9082	−68.7	68.9	72.7	0.7794	−30.4	39.5	58.6	0.6228	−36.1	61.6	85.1
	DIF	0.8119	40.8	41.1	46.5	0.0063	6.8	15.9	21.1	0.0019	7.0	31.7	36.5
3	DNI	0.3542	−50.9	92.0	115.0	0.0803	−17.7	175.5	204.5	0.0014	−13.2	131.0	164.0
	DIF	0.6937	−8.7	59.7	70.4	0.5290	0.6	78.9	86.8	0.6843	−9.4	70.4	77.7
4	DNI	0.2470	47.4	74.0	127.7	0.2273	−33.2	79.1	128.7	0.1022	−322.1	322.1	391.3
	DIF	0.2873	−57.5	58.4	94.0	0.4425	−24.5	40.2	69.2	0.4884	11.7	43.2	69.3
5	DNI	0.4456	−86.6	149.2	220.4	0.1410	−121.6	206.9	283.7	0.2980	−132.1	180.7	256.1
	DIF	0.5380	−3.2	32.5	39.9	0.0513	33.5	57.9	70.8	0.0195	36.9	65.1	76.6

Table A3. DNI and DIF forecast results for the measuring periods used for testing of the developed model (MBE, MAE, and RMSE in W/m^2).

Period and	Variable	Day 0				Day 1				Day 2			
		R^2	MBE	MAE	RMSE	R^2	MBE	MAE	RMSE	R^2	MBE	MAE	RMSE
1	DNI	0.1957	26.3	101.8	141.4	0.1127	14.1	92.3	146.1	0.0151	−28.4	128.6	162.6
	DIF	0.1687	−59.2	64.2	76.0	0.4663	−36.4	36.9	52.9	0.0006	4.7	59.5	67.6
2	DNI	0.9552	55.4	55.6	66.8	0.4768	28.5	59.8	80.9	0.9153	9.4	21.7	33.9
	DIF	0.5153	−48.5	48.5	52.3	0.0240	−20.4	31.3	40.1	0.2740	−23.8	30.7	39.8
3	DNI	0.6753	24.0	121.2	190.3	0.6811	−16.6	134.1	187.7	0.6930	2.9	134.2	184.3
	DIF	0.6857	−19.4	48.7	72.4	0.6798	−15.5	51.3	71.1	0.5439	−15.4	60.7	85.3
4	DNI	0.0879	46.2	178.6	273.2	0.1934	59.5	162.1	253.6	0.2277	51.8	162.0	251.3
	DIF	0.2560	−55.3	63.1	97.3	0.5052	−60.9	65.2	90.1	0.4056	−55.6	64.6	91.0
5	DNI	0.1872	−58.2	106.8	140.4	0.0197	−45.0	133.4	184.8	0.0065	0.4	83.2	149.6
	DIF	0.1999	21.2	40.3	46.0	0.0189	−11.8	38.5	58.5	0.0143	−4.1	35.6	56.4
6	DNI	0.3289	−147.5	169.6	210.3	0.3673	46.18	47.2	89.7	0.3013	38.0	46.6	89.3
	DIF	0.0152	88.0	96.9	129.0	0.0550	−33.5	33.5	46.3	0.0019	−25.4	28.5	44.2
7	DNI	0.8260	−24.4	36.7	44.5	0.8981	11.9	31.0	37.3	0.8394	37.6	49.0	65.9
	DIF	0.6205	14.6	22.6	28.4	0.4374	−3.9	23.9	26.6	0.0125	−25.9	38.1	49.2
8	DNI	0.8672	25.8	53.5	60.2	0.8169	30.5	74.9	85.3	0.8656	57.5	61.8	69.2
	DIF	0.5586	−3.2	11.8	15.8	0.3760	−13.7	26.8	30.6	0.0822	−26.5	26.5	30.9
9	DNI	0.1415	160.6	213.8	270.0	0.1635	99.9	184.8	223.0	0.0117	190.2	290.4	341.6
	DIF	0.3964	−102.6	115.7	143.9	0.0298	−96.5	126.6	157.6	0.0196	−102.5	131.1	161.4
10	DNI	0.3396	93.7	196.6	266.0	0.2661	33.3	204.1	268.5	0.4108	−20.3	201.3	235.9
	DIF	0.2564	−42.3	90.2	122.9	0.2894	−20.0	85.2	113.8	0.3719	−17.3	87.1	105.8
11	DNI	0.2717	−129.9	170.9	207.2	0.6058	−42.8	83.0	111.2	0.6257	22.0	69.2	103.2
	DIF	0.0832	55.7	70.8	96.5	0.0929	22.9	37.4	45.3	0.3765	−7.9	23.0	30.9
12	DNI	0.0420	45.0	197.2	228.6	0.1069	95.6	258.8	310.0	0.3188	197.5	295.7	363.3
	DIF	0.5633	−63.8	97.1	113.2	0.6490	−114.3	128.1	152.1	0.7377	−108.5	121.3	150.0
13	DNI	0.2652	200.3	217.9	265.3	0.2650	231.0	252.1	300.2	0.2762	262.2	299.9	341.5
	DIF	0.1240	−77.5	104.1	127.2	0.1153	−99.3	119.9	141.8	0.1656	−119.1	135.3	154.5
14	DNI	0.3747	101.8	137.5	163.5	0.4259	156.6	162.3	210.6	0.2201	107.0	145.8	186.7
	DIF	0.4555	26.2	89.6	111.4	0.4369	14.8	85.0	112.2	0.4314	55.2	95.9	122.8
15	DNI	0.1945	−102.7	125.5	161.5	0.2203	−148.3	173.5	190.0	0.2292	−116.1	139.4	170.9
	DIF	0.3872	66.8	68.7	93.0	0.4400	81.7	81.7	97.4	0.4533	73.0	76.7	102.6
16	DNI	0.0805	20.7	230.0	265.3	0.1606	−34.3	209.9	246.7	0.0197	−197.8	326.9	405.4
	DIF	0.5675	18.9	53.0	65.6	0.4792	31.2	64.2	77.2	0.7111	−15.6	46.0	52.8
17	DNI	0.3496	−15.5	92.7	146.0	0.4274	−37.6	119.2	158.3	0.4238	−96.0	151.4	205.5
	DIF	0.1479	14.3	29.4	38.5	0.2745	11.2	31.1	39.7	0.3782	31.2	41.1	50.6
18	DNI	0.0262	5.2	145.8	211.6	0.0983	19.4	139.8	198.7	0.0154	39.0	142.2	210.7
	DIF	0.0994	9.8	76.4	102.3	0.1921	3.7	72.6	95.8	0.0518	−4.0	81.5	111.6
19	DNI	0.7739	−27.3	65.6	99.7	0.8940	0.1	42.2	67.5	0.7421	−33.4	89.0	121.4
	DIF	0.4384	8.1	20.8	27.4	0.8150	−2.1	12.9	18.2	0.0654	34.8	45.4	60.6

Table A4. GTI metrics of first-row tests for each day of forecast horizon and for each period using the Modified Bugler transposition model with DNI and DIF forecasts (MBE, MAE, and RMSE in W/m^2).

Period	Day 0				Day 1				Day 2			
	R^2	MBE	MAE	RMSE	R^2	MBE	MAE	RMSE	R^2	MBE	MAE	RMSE
1	0.9795	29.2	33.5	46.4	0.9798	29.5	35.6	46.4	0.9719	23.0	41.4	48.7
2	0.9980	8.9	13.8	16.6	0.9941	12.7	22.6	25.6	0.9933	11.0	23.8	27.7
3	0.9224	−29.4	45.4	63.9	0.9511	31.6	45.7	65.9	0.9345	16.3	36.4	56.3
4	0.9177	9.8	44.7	59.4	0.8715	19.8	51.8	74.6	0.0893	−203.9	210.1	277.7
5	0.9633	21.6	34.9	40.6	0.9329	21.8	41.7	48.4	0.9742	47.0	29.9	34.6

Table A5. GTI results of the developed model for rows that are not the front row for each day of forecast horizon and for each period using DNI and DIF forecasts (MAE and RMSE in W/m2, MAPE and RMSPE in %).

Period	Day 0				Day 1				Day 2			
	R^2	MBE	MAE	RMSE	R^2	MBE	MAE	RMSE	R^2	MBE	MAE	RMSE
1	0.0043	−163.0	207.1	364.9	0.0059	−163.5	195.8	364.0	0.0048	−179.4	215.5	365.9
2	0.9913	−30.7	34.9	38.8	0.9835	−41.0	42.7	57.2	0.9768	−50.3	53.2	70.5
3	0.6449	−17.4	117.5	175.0	0.6302	−36.0	129.9	181.5	0.5980	−25.8	128.2	187.2
4	0.3640	−3.7	81.8	137.8	0.3764	−3.0	84.0	138.3	0.3630	1.1	85.3	142.8
5	0.4550	−87.7	108.9	214.5	0.4017	−99.6	131.8	234.1	0.4038	−72.3	105.8	217.9
6	0.9196	−105.8	113.3	141.0	0.9651	−29.2	45.2	57.8	0.9636	−30.8	46.8	59.7
7	0.9938	−43.3	43.8	48.8	0.9981	−31.1	31.6	35.0	0.9971	−26.5	27.0	30.9
8	0.9958	−48.0	50.0	52.5	0.9963	−50.2	51.5	53.8	0.9949	−49.7	51.7	54.6
9	0.1442	−82.6	112.5	166.3	0.1479	−93.7	114.7	166.2	0.0815	−84.5	127.3	179.2
10	0.1607	−62.4	142.8	242.5	0.1907	−78.1	143.9	239.9	0.2246	−105.8	147.0	242.8
11	0.8884	−92.8	97.4	125.5	0.9728	−62.1	67.4	78.0	0.9823	−37.5	44.1	51.3
12	0.5099	−43.6	108.8	150.6	0.4232	−36.7	136.7	181.2	0.5488	44.0	111.4	167.7
13	0.9111	19.9	57.1	71.5	0.9154	28.6	54.7	71.8	0.9025	41.8	64.9	87.6
14	0.1142	95.4	105.1	135.2	0.1097	107.8	113.6	155.5	0.0710	111.3	121.3	155.7
15	0.8651	−61.9	74.8	91.6	0.8613	−68.5	87.4	96.9	0.8723	−64.5	78.0	91.8
16	0.4764	5.2	166.3	204.4	0.4233	−30.8	172.6	212.8	0.0174	−227.0	295.8	380.3
17	0.9180	−43.9	70.8	85.8	0.9211	−59.2	76.7	91.8	0.9147	−79.1	90.5	107.8
18	0.8772	−41.7	81.4	99.7	0.8803	−39.9	78.1	97.9	0.8741	−41.9	83.2	100.8
19	0.6672	−55.3	65.9	114.3	0.6605	−51.8	62.1	114.6	0.6384	−39.4	51.7	109.7

References

1. SolarPower Europe. *Global Market Outlook for Solar Power 2023–2027*; SolarPower Europe: Brussels, Belgium, 2023. Available online: https://www.solarpowereurope.org/insights/market-outlooks/global-market-outlook-for-solar-power-2023-2027-1 (accessed on 5 June 2024).
2. *International Energy Agency Renewables 2023*; Analysis and Forecast to 2028; International Energy Agency: Paris, France, 2023.
3. Nassar, Y.F.; Hafez, A.A.; Alsadi, S.Y. Multi-Factorial Comparison for 24 Distinct Transposition Models for Inclined Surface Solar Irradiance Computation in the State of Palestine: A Case Study. *Front. Energy Res.* **2020**, *7*, 1–19. [CrossRef]
4. Yang, D. Solar Radiation on Inclined Surfaces: Corrections and Benchmarks. *Sol. Energy* **2016**, *136*, 288–302. [CrossRef]
5. Gueymard, C.A. Direct and Indirect Uncertainties in the Prediction of Tilted Irradiance for Solar Engineering Applications. *Sol. Energy* **2009**, *83*, 432–444. [CrossRef]
6. Gueymard, C. An Anisotropic Solar Irradiance Model for Tilted Surfaces and Its Comparison with Selected Engineering Algorithms. *Sol. Energy* **1987**, *38*, 367–386. [CrossRef]
7. Klucher, T.M. Evaluation of Models to Predict Insolation on Tilted Surfaces. *Sol. Energy* **1979**, *23*, 111–114. [CrossRef]
8. Hay, J.E. Calculating Solar Radiation for Inclined Surfaces: Practical Approaches. *Renew. Energy* **1993**, *3*, 373–380. [CrossRef]
9. Hay, J.; McKay, D. *Final Report IEA Task IX-Calculation of Solar Irradiances for Inclined Surfaces: Verification of Models Which Use Hourly and Daily Data*; International Energy Agency: Paris, France, 1988.
10. Nassar, Y.F.; Abuhamoud, N.M.; Miskeen, G.M.; El-Khozondar, H.J.; Alsadi, S.Y.; Ahwidi, O.M. Investigating the Applicability of Horizontal to Tilted Sky-Diffuse Solar Irradiation Transposition Models for Key Libyan Cities. In Proceedings of the 2022 IEEE 2nd International Maghreb Meeting of the Conference on Sciences and Techniques of Automatic Control and Computer Engineering (MI-STA 2022), Sabratha, Libya, 23–25 May 2022; pp. 9–14. [CrossRef]

11. Mayer, M.J.; Gróf, G. Extensive Comparison of Physical Models for Photovoltaic Power Forecasting. *Appl. Energy* **2021**, *283*, 116239. [CrossRef]
12. Khan, M.M.; Ahmad, M.J. Estimation of Global Solar Radiation Using Clear Sky Radiation in Yemen. *J. Eng. Sci. Technol. Rev.* **2012**, *5*, 12–19. [CrossRef]
13. Padovan, A.; Del Col, D. Measurement and Modeling of Solar Irradiance Components on Horizontal and Tilted Planes. *Sol. Energy* **2010**, *84*, 2068–2084. [CrossRef]
14. Muneer, T.; Gueymard, C.; Kambezidis, H. Hourly Slope Irradiation and Illuminance. In *Solar Radiation and Daylight Models*; Butterworth-Heinemann: Oxford, UK, 2004; p. 143. [CrossRef]
15. Badescu, V. 3D Isotropic Approximation for Solar Diffuse Irradiance on Tilted Surfaces. *Renew. Energy* **2002**, *26*, 221–233. [CrossRef]
16. Tian, Y.Q.; Davies-Colley, R.J.; Gong, P.; Thorrold, B.W. Estimating Solar Radiation on Slopes of Arbitrary Aspect. *Agric. Meteorol.* **2001**, *109*, 67–74. [CrossRef]
17. Appelbaum, J.; Massalha, Y.; Aronescu, A. Corrections to Anisotropic Diffuse Radiation Model. *Sol. Energy* **2019**, *193*, 523–528. [CrossRef]
18. Varga, N.; Mayer, M.J. Model-Based Analysis of Shading Losses in Ground-Mounted Photovoltaic Power Plants. *Sol. Energy* **2021**, *216*, 428–438. [CrossRef]
19. Tschopp, D.; Jensen, A.R.; Dragsted, J.; Ohnewein, P.; Furbo, S. Measurement and Modeling of Diffuse Irradiance Masking on Tilted Planes for Solar Engineering Applications. *Sol. Energy* **2022**, *231*, 365–378. [CrossRef]
20. ECMWF IFS Documentation. Available online: https://www.ecmwf.int/en/publications/ifs-documentation (accessed on 10 December 2021).
21. Pereira, S.; Abreu, E.F.M.; Iakunin, M.; Cavaco, A.; Salgado, R.; Canhoto, P. Method for Solar Resource Assessment Using Numerical Weather Prediction and Artificial Neural Network Models Based on Typical Meteorological Data: Application to the South of Portugal. *Sol. Energy* **2022**, *236*, 225–238. [CrossRef]
22. Pereira, S.; Canhoto, P.; Salgado, R. Development and Assessment of Artificial Neural Network Models for Direct Normal Solar Irradiance Forecasting Using Operational Numerical Weather Prediction Data. *Energy AI* **2023**, *15*, 100314. [CrossRef]
23. Liu, B.Y.H.; Jordan, R.C. The Long-Term Average Performance of Flat-Plate Solar-Energy Collectors. *Sol. Energy* **1963**, *7*, 53–74. [CrossRef]
24. Bugler, J.W. The Determination of Hourly Insolation on an Inclined Plane Using a Diffuse Irradiance Model Based on Hourly Measured Global Horizontal Insolation. *Sol. Energy* **1977**, *19*, 477–491. [CrossRef]
25. Temps, R.C.; Coulson, K.L. Solar Radiation Incident upon Slopes of Different Orientations. *Sol. Energy* **1977**, *19*, 179–184. [CrossRef]
26. Iqbal, M. *An Introduction to Solar Radiation*; Elsevier: Vancouver, BC, Canada, 1983. ISBN 9780123737502.
27. Kasten, F. A New Table and Approximation Formula for the Relative Optial Air Mass. *Arch. Für. Meteorol. Geophys. Und Bioklimatol. Ser. B* **1965**, *14*, 206–223. [CrossRef]
28. Reindl, D.T.; Beckman, W.A.; Duffie, J.A. Evaluation of Hourly Tilted Surface Radiation Models. *Sol. Energy* **1990**, *45*, 9–17. [CrossRef]
29. Perez, R.; Ineichen, P.; Seals, R.; Michalsky, J.; Stewart, R. Modeling Daylight Availability and Irradiance Components from Direct and Global Irradiance. *Sol. Energy* **1990**, *44*, 271–289. [CrossRef]
30. Saint-Drenan, Y.-M.; Barbier, T. Data-Analysis and Modelling of the Effect of Inter-Row Shading on the Power Production of Photovoltaic Plants. *Sol. Energy* **2019**, *184*, 127–147. [CrossRef]
31. Maor, T.; Appelbaum, J. View Factors of Photovoltaic Collector Systems. *Sol. Energy* **2012**, *86*, 1701–1708. [CrossRef]
32. Eckert, E.R.G. Radiative Transfer, H. C. Hottel and A. F. Sarofim, McGraw-Hill Book Company, New York, 1967. 52 Pages. *AIChE J.* **1969**, *15*, 794–796. [CrossRef]
33. ASD FieldSpec® HandHeld 2 User Manual 2010, 1–140. Available online: https://www.geo-informatie.nl/courses/grs60312/material2017/manuals/600860-dHH2Manual.pdf (accessed on 5 June 2024).
34. Cavaco, A.; Canhoto, P.; Collares Pereira, M. Procedures for Solar Radiation Data Gathering and Processing and Their Application to DNI Assessment in Southern Portugal. *Renew. Energy* **2021**, *163*, 2208–2219. [CrossRef]
35. Long, C.N.; Dutton, E.G. BSRN Global Network Recommended QC Tests, V2.0. Available online: https://epic.awi.de/id/eprint/30083/1/BSRN_recommended_QC_tests_V2.pdf (accessed on 28 May 2024).
36. *ISO 9847:1992*; Solar Energy—Calibration of Field Pyranometers by Comparison to a Reference Pyranometer. International Standard Organization: Geneva, Switzerland, 1992.
37. Betti, A.; Blanc, P.; David, M.; Saint-Drenan, Y.-M.; Driesse, A.; Freeman, J.; Fritz, R.; Gueymard, C.; Habte, A.; Höller, R.; et al. *Best Practices Handbook for the Collection and Use of Solar Resource Data for Solar Energy Applications, Report IEA-PVPS 16-04:2021*, 3rd ed. International Energy Agency: Paris, France, 2021; ISBN 978-3-907281-19-2.

technologies

MDPI

Article

A New LCL Filter Design Method for Single-Phase Photovoltaic Systems Connected to the Grid via Micro-Inverters

Heriberto Adamas-Pérez [1], Mario Ponce-Silva [1,*], Jesús Darío Mina-Antonio [1], Abraham Claudio-Sánchez [1], Omar Rodríguez-Benítez [1] and Oscar Miguel Rodríguez-Benítez [2]

1 Tecnológico Nacional de México-CENIDET, Cuernavaca 62490, Mexico; heriberto.adamas17ee@cenidet.edu.mx (H.A.-P.); jesus.ma@cenidet.tecnm.mx (J.D.M.-A.); abraham.cs@cenidet.tecnm.mx (A.C.-S.); d19ce044@cenidet.tecnm.mx (O.R.-B.)

2 Facultad de Ingeniería, Universidad Nacional Autónoma de México, Mexico City 04510, Mexico; oscar.rodriguezb@fi.unam.edu

* Correspondence: mario.ps@cenidet.tecnm.mx

Abstract: This paper aims to propose a new sizing approach to reduce the footprint and optimize the performance of an LCL filter implemented in photovoltaic systems using grid-connected single-phase microinverters. In particular, the analysis is carried out on a single-phase full-bridge inverter, assuming the following two conditions: (1) a unit power factor at the connection point between the AC grid and the LCL filter; (2) a control circuit based on unipolar sinusoidal pulse width modulation (SPWM). In particular, the ripple and harmonics of the LCL filter input current and the current injected into the grid are analyzed. The results of the Simulink simulation and the experimental tests carried out confirm that it is possible to considerably reduce filter volume by optimizing each passive component compared with what is already available in the literature while guaranteeing excellent filtering performance. Specifically, the inductance values were reduced by almost 40% and the capacitor value by almost 100%. The main applications of this new design methodology are for use in single-phase microinverters connected to the grid and for research purposes in power electronics and optimization.

Keywords: passive filters; DC-AC power converters; photovoltaic systems; SPWM

Citation: Adamas-Pérez, H.; Ponce-Silva, M.; Mina-Antonio, J.D.; Claudio-Sánchez, A.; Rodríguez-Benítez, O.; Rodríguez-Benítez, O.M. A New LCL Filter Design Method for Single-Phase Photovoltaic Systems Connected to the Grid via Micro-Inverters. *Technologies* **2024**, *12*, 89. https://doi.org/10.3390/technologies12060089

Academic Editors: Jayanta Deb Mondol, Annamaria Buonomano and Biplab Das

Received: 6 May 2024
Revised: 4 June 2024
Accepted: 7 June 2024
Published: 12 June 2024

1. Introduction

Solar energy is transformed into electrical energy using photovoltaic panels (PVs). This electrical energy is then connected to the grid or isolated loads through power converters. This collection of components is known as a photovoltaic system [1–6]. The basic block diagram of a typical single-phase PV is shown in Figure 1. This system consists of a photovoltaic panel, a DC–DC converter, a link capacitor between the DC–DC converter and the inverter, an output filter, and the grid [4,7]. The blocks shown are those present in a commercial PV system. This work is focused on the design of the output LCL filter.

Prior studies have offered mathematical formulations for computing the LCL filter connected to the grid in single-phase systems [8–11]. Recently, optimization methods for three-phase LCL filters based on complex algorithms have been proposed [12–19]. Few optimization methods have been reported for single-phase systems [11,20–22], and the proposed methods are based on the analyses performed in [8–10]. Methods based on cost minimization have also been reported for three-phase systems [23,24].

This paper analyzes the LCL filter using phasor analysis for the fundamental harmonic and an nth harmonic. The analysis considers the ripple percentage at the input and output of the LCL filter. The value of the inductors and capacitors are based on two parameters called Alpha (α), where alpha is the ratio between the reactance of the inductor L_1 and the reactance of the filter capacitor LCL (C_f) and Beta (β), where Beta is the ratio between the values of L_1 and L_2. These parameters are the ones that are optimized, resulting in the minimum and optimum values of the capacitor and inductors. These analysis and optimization methods are not found in the literature at present; therefore, they constitute the main contribution of this paper. The main applications of this research are photovoltaic systems connected to the grid that can be used in microgrids [25]. The main limitations of this work are that the method can only be used in single-phase systems connected to the grid and that it cannot be used for three-phase systems and isolated systems.

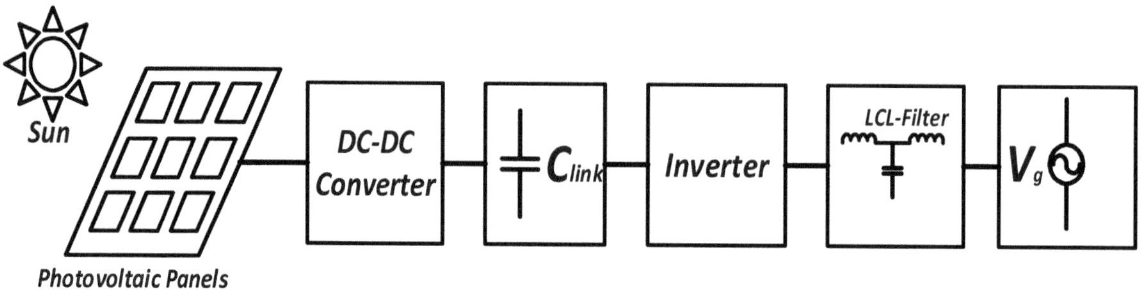

Figure 1. Photovoltaic system with LCL filter.

The theoretical calculations were validated with simulations in SIMULINK and experimentally. The paper is organized as follows: Section 2 presents the proposed mathematical analysis of the LCL filter. Section 3 presents the proposed optimization method. The simulation results are presented in Section 4. Section 5 presents the experimental validation. Section 6 presents a discussion of the traditional methods for calculating the LCL filter [8,11,26–29] and the methodology proposed in this paper. Finally, the conclusions are shown in Section 7.

2. LCL Filter Mathematical Analysis

The analyzed system uses an LCL filter, connected to a single-phase full bridge inverter and the grid as shown in Figure 2. The SPWM modulation activates and deactivates the gates that make up the full bridge inverter. The generated signal passes through the LCL filter, which is used to reduce the harmonics of the current to be injected into the grid.

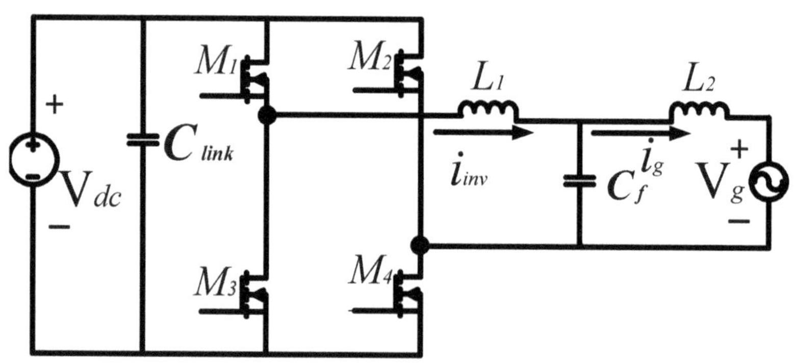

Figure 2. Grid-connected full bridge inverter with an LCL filter.

2.1. Mathematical Analysis of the LCL Filter for the Fundamental Component

One previous analysis of this type was performed for an L-filter connected to the grid and was published in [4]. In this paper, to analyze the performance of the LCL filter, a Fourier analysis was performed to identify each of the harmonics in the inverter output voltage. Figure 3 shows a flow chart describing step by step how the mathematical analysis with the fundamental and n-harmonics is organized.

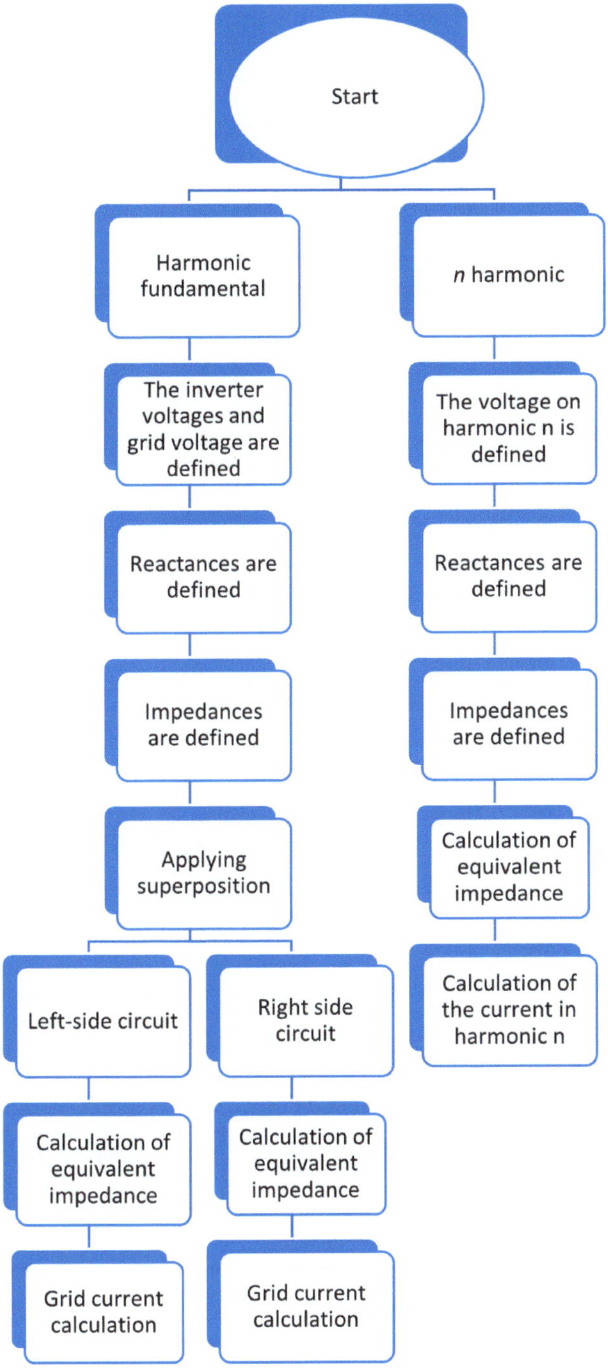

Figure 3. General diagram of mathematical analysis using harmonics.

Figure 4 shows a step-by-step diagram of the mathematical analysis of the current ripple percentage from which the proposed values and parameters for the LCL filter are obtained.

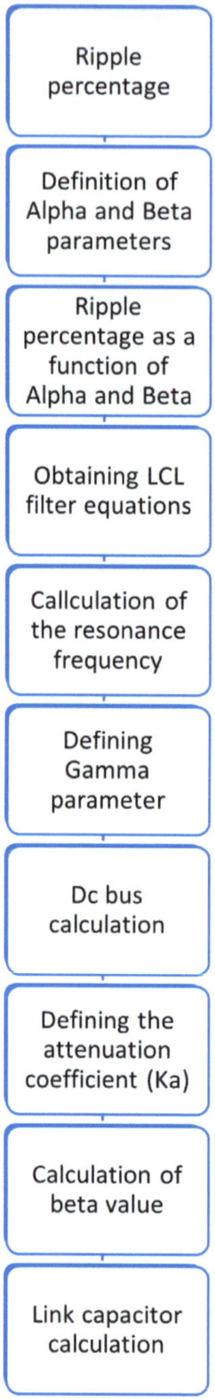

Figure 4. Step-by-step diagram for LCL filter calculation and proposed parameters.

First, the circuit is analyzed for the fundamental harmonic at the frequency of the grid and after for the next harmonic that appears according to the modulation technique of the output inverter. For the fundamental harmonic, the simplified circuit to be analyzed is shown in Figure 5.

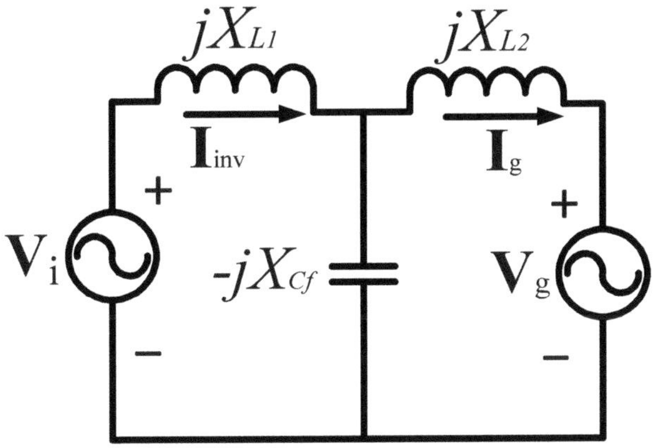

Figure 5. LCL filter connected to the grid for the fundamental harmonic.

$\mathbf{V}_\mathrm{i}$ is the phasor of the fundamental harmonic at the LCL filter input, $\mathbf{I}_\mathrm{inv}$ is the phasor of the LCL filter input current, $\mathbf{I}_\mathrm{g}$ is the phasor of the grid current, $\mathbf{V}_\mathrm{g}$ is the phasor of the grid voltage, and X_{L1} and X_{L2} are the inductive reactances of the inverter-side inductor (L_1) and the grid-side inductor (L_2). X_{Cf} is the capacitive reactance of the filter capacitor (C_f). Filter input voltage and grid voltage are defined as shown in Equations (1) and (2).

$$\mathbf{V}_\mathrm{i} = V_i \angle \phi_i \tag{1}$$

$$\mathbf{V}_\mathrm{g} = V_g \angle 0^\circ \tag{2}$$

where V_i is the peak voltage magnitude of the phasor $\mathbf{V}_\mathrm{i}$ at LCL filter input, ϕ_i is the phase angle caused by the LCL filter in the inverter voltage, V_g is the peak grid voltage magnitude, and 0° is the grid reference angle for a unity power factor connection. The reactances of each element of the LCL filter are defined as shown in Equations (3)–(5).

$$X_{L_1} = \omega L_1 \tag{3}$$

$$X_{C_f} = \frac{1}{\omega C_f} \tag{4}$$

$$X_{L_2} = \omega L_2 \tag{5}$$

where ω is the grid angular frequency. The impedances of each element of the LCL filter are defined as shown in Equations (6)–(8).

$$Z_{L_1} = jX_{L_1} \tag{6}$$

$$Z_{C_f} = -jX_{C_f} \tag{7}$$

$$Z_{L_2} = jX_{L_2} \tag{8}$$

where Z_{L1} is the impedance of inductor L_1, j is the imaginary number, Z_{Cf} is the LCL filter capacitor impedance, and Z_{L2} is the impedance of the grid side inductor. Applying the superposition theorem to the circuit shown in Figure 3 results in the circuit shown in Figure 6.

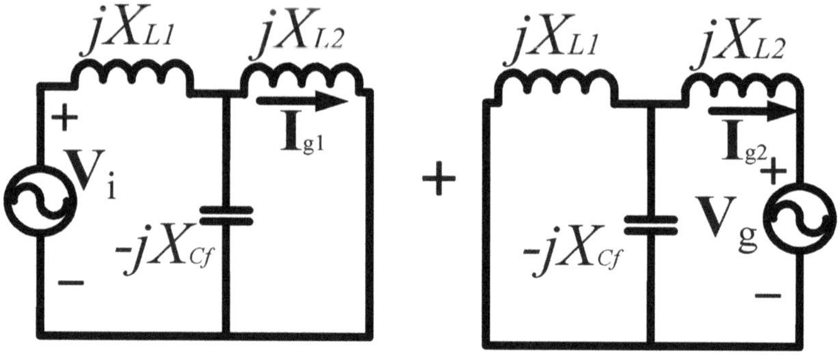

Figure 6. LCL filter divided by superposition theorem.

Solving the circuit on the left, which is fed by the peak voltage of the fundamental harmonic of the output in unipolar SPWM modulation, X_{L2} and X_{Cf} are in parallel. Solving the parallel is shown in Equation (9).

$$Z_{eq} = \frac{Z_{L_2} Z_{C_f}}{Z_{L_2} + Z_{C_f}} = \frac{jX_{C_f} X_{L_2}}{X_{C_f} - X_{L_2}} \tag{9}$$

Applying voltage divider on the Z_{eq}. This is shown in Equation (10).

$$\mathbf{V}_{\mathrm{C_f}} = \frac{\mathbf{V}_{\mathrm{i}} Z_{eq}}{Z_{eq} + Z_{L_1}} = \frac{V_i \angle \phi_i X_{C_f} X_{L_2}}{X_{C_f} X_{L_1} + X_{C_f} X_{L_2} - X_{L_1} X_{L_2}} \tag{10}$$

where $\mathbf{V}_{\mathrm{Cf}}$ is the voltage phasor in the parallel formed by X_{L2} and X_{Cf}. The current to be injected into the grid of the left-hand side ($\mathbf{I}_{\mathrm{g1}}$) circuit is solved using Equation (11).

$$\mathbf{I}_{\mathrm{g_1}} = \frac{\mathbf{V}_{\mathrm{C_f}}}{Z_{L_2}} = -\frac{V_i \angle \phi_i j X_{C_f}}{X_{C_f} X_{L_1} + X_{C_f} X_{L_2} - X_{L_1} X_{L_2}} \tag{11}$$

Solving the LCL filter for the right side of Figure 4. Z_{L1} and Z_{Cf} are in parallel, solving the parallel. This is shown in Equation (12).

$$Z_{eq2} = \frac{Z_{L_1} Z_{C_f}}{Z_{L_1} + Z_{C_f}} = \frac{jX_{C_f} X_{L_1}}{X_{C_f} - X_{L_1}} \tag{12}$$

Solving for grid current from the right-hand side ($\mathbf{I}_{\mathrm{g2}}$) of the circuit. This is shown in Equation (13).

$$\mathbf{I}_{\mathrm{g_2}} = \frac{\mathbf{V}_{\mathrm{g}}}{Z_{L_2} + Z_{eq2}} = -\frac{V_g \angle 0^\circ \left(X_{C_f} - X_{L_1} \right) j}{X_{C_f} X_{L_1} + X_{C_f} X_{L_2} - X_{L_1} X_{L_2}} \tag{13}$$

Adding (11) and (13), Equation (14) is obtained, which is the total current injected into the grid at the fundamental harmonic ($\mathbf{I}_{\mathrm{g}}$).

$$\mathbf{I}_{\mathrm{g}} = -\frac{V_g \angle 0^\circ j \left(X_{C_f} - X_{L_1} \right) - V_i \angle \phi_i \left(j X_{C_f} \right)}{X_{C_f} X_{L_1} + X_{C_f} X_{L_2} - X_{L_1} X_{L_2}} \tag{14}$$

2.2. *Mathematical Analysis of the LCL Filter for a Harmonic n*

An LCL filter analysis was performed for any harmonic n. The circuit to be analyzed is shown in Figure 7. The analysis omits the grid voltage since the grid provides only the fundamental harmonic.

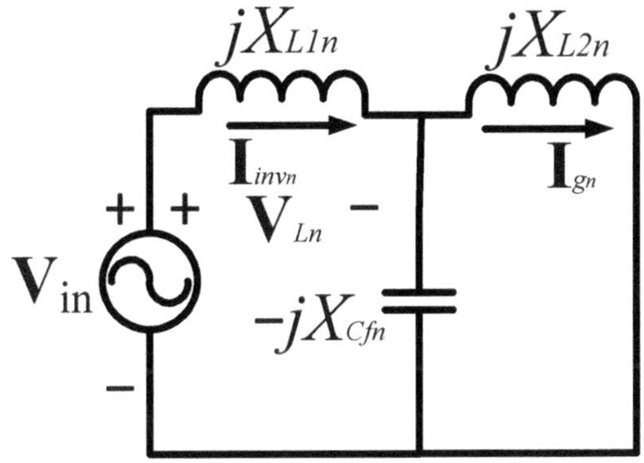

Figure 7. LCL filter for any harmonic n.

The reactances of each LCL filter element for Figure 5 are defined as shown in the following (Equations (15)–(17)).

$$X_{L_{1_n}} = \omega_n L_1 \tag{15}$$

$$X_{C_{fn}} = \frac{1}{\omega_n C_f} \tag{16}$$

$$X_{L_{2_n}} = \omega_n L_2 \tag{17}$$

where X_{L1n} is the inductive reactance for a harmonic n in the inductor L_1, ω_n is the angular frequency for a harmonic n, X_{Cfn} is the capacitive reactance for a harmonic n in the LCL filter capacitor, and X_{L2n} is the inductive reactance for a harmonic n in the grid side inductor. Defining the filter input voltage and the grid voltage, this is shown in Equation (18)

$$\mathbf{V}_{\mathrm{i}_n} = V_{i_n} \angle \phi_{i_n} \tag{18}$$

where $\mathbf{V}_{\mathrm{i}n}$ is the phasor of the LCL filter input voltage for one harmonic n, V_{in} is the Peak magnitude of the LCL filter input voltage for one harmonic n, ϕ_n is the phase angle of the LCL filter input voltage for one harmonic n.

The impedances of each element of the LCL filter for Figure 5 are defined as shown in Equations (19)–(21).

$$Z_{L_{1_n}} = jX_{L_{1_n}} \tag{19}$$

$$Z_{C_{fn}} = -jX_{C_{fn}} \tag{20}$$

$$Z_{L_{2_n}} = jX_{L_{2_n}} \tag{21}$$

Figure 5 shows that Z_{Cfn} and Z_{L2n} are in parallel, solving the given Equation (22).

$$Z_{eq3} = \frac{Z_{C_{fn}} Z_{L_2}}{Z_{C_{fn}} + Z_{L_2}} = \frac{jX_{L_{2_n}} X_{C_{fn}}}{X_{C_{fn}} - X_{L_{2_n}}} \tag{22}$$

where Z_{eq3} is the equivalent impedance of the parallel between Z_{Cfn} and Z_{L2n}. The inverter side current for one harmonic n in the LCL filter is calculated using Equation (23)

$$\mathbf{I}_{\mathrm{inv}_n} = \frac{V_{i_n} \angle \phi_{i_n}}{Z_{L_{1_n}} + Z_{eq3}} \tag{23}$$

Substituting (19) and (22) into Equation (23), it results in (24).

$$\mathbf{I}_{\text{inv}n} = \frac{V_{i_n}\angle\phi_{i_n} j\left(X_{C_{fn}} - X_{L_{2n}}\right)}{-X_{C_{fn}}X_{L_{1n}} - X_{C_{fn}}X_{L_{2n}} + X_{L_{1n}}X_{L_{2n}}} \tag{24}$$

where $\mathbf{I}_{\text{inv}n}$ is the phasor of the inverter-side current for one harmonic n. The magnitude of Equation (24) is shown in (25).

$$I_{inv_n} = |\mathbf{I}_{\text{inv}_n}| = \frac{V_{i_n}\left(X_{C_{fn}} - X_{L_{2n}}\right)}{X_{C_{fn}}X_{L_{1n}} + X_{C_{fn}}X_{L_{2n}} - X_{L_{1n}}X_{L_{2n}}} \tag{25}$$

where $I_{\text{inv}n}$ is the peak magnitude of the inverter-side current for harmonic n. The maximum grid current of the fundamental harmonic (I_g) is defined by Equation (26). For a unity power factor.

$$I_g = \frac{2P_{avg}}{V_g} \tag{26}$$

where P_{avg} is the average power in the connection point. The current ripple at the input of the LCL filter ($\%r_{inv}$) is approximately equal to Equation (27) as shown in [4]. The current ripple analyzed in this paper is the L_1 ripple, as most of the articles are the proposed current ripple [30–32]

$$\%r_{i_{inv}} \approx \frac{(2I_{inv_n})(100)}{I_g} \tag{27}$$

Substituting (25) and (26) into (27), Equation (28) results in the following:

$$\%r_{inv} = \frac{100V_g V_{i_n}\left(X_{C_{fn}} - X_{L_{2n}}\right)}{P_{avg}\left(-X_{C_{fn}}X_{L_{1n}} - X_{C_{fn}}X_{L_{2n}} + X_{L_{1n}}X_{L_{2n}}\right)} \tag{28}$$

2.3. Mathematical Analysis to Calculate LCL Filter Elements

Defining the Alpha and Beta parameters shown in Equations (29) and (30).

$$\alpha = \frac{X_{L_{1n}}}{X_{C_{fn}}} = \frac{\omega_n L_1}{\frac{1}{\omega_n C_f}} = \omega_n{}^2 L_1 C_f \tag{29}$$

$$\beta = \frac{X_{L_{1n}}}{X_{L_{2n}}} = \frac{\omega_n L_1}{\omega_n L_2} = \frac{L_1}{L_2} \tag{30}$$

where α is the ratio between X_{L1n} and X_{Cfn} and β is the ratio between inductors L_1 and L_2. The reactance of L_1 as a function of Alpha is shown in Equation (31)

$$X_{L_{1n}} = \alpha X_{C_{fn}} \tag{31}$$

By subtracting X_{L2n} from Equation (30), the result is (32).

$$X_{L_{2n}} = \frac{\alpha X_{C_{fn}}}{\beta} \tag{32}$$

Substituting (31) and (32) in Equation (28) and simplifying results in (33).

$$\%r_{i_{inv}} = \frac{100V_g V_{i_n}(\alpha - \beta)}{X_{C_{fn}} P_{avg}\alpha(-\beta + \alpha - 1)} \tag{33}$$

Substituting the capacitive reactance (16) in the current ripple, Equation (33) results in (34).

$$\%r_{i_{inv}} = \frac{100V_g V_{i_n}(\alpha - \beta)}{\frac{1}{\omega_n C_f} P_{avg}\alpha(-\beta + \alpha - 1)} \tag{34}$$

By subtracting C_f from Equation (34), the result is shown in Equation (35).

$$C_f = \frac{\%r_{i_{inv}} P_{avg}\alpha(-\beta + \alpha - 1)}{100V_g V_{i_n}\omega_n(\alpha - \beta)} \tag{35}$$

Combining (35) and (16), substituting in (31), Equation (36) results in the following:

$$X_{L_{1n}} = \omega_n L_1 = \frac{100V_g V_{i_n}(\alpha - \beta)}{\%r_{i_{inv}} P_{avg}(-\beta + \alpha - 1)} \tag{36}$$

Solving for L_1 of Equation (36), the result is (37).

$$L_1 = \frac{100V_g V_{i_n}(\alpha - \beta)}{\omega_n \%r_{i_{inv}} P_{avg}(-\beta + \alpha - 1)} \tag{37}$$

By subtracting L_2 from Equation (30) and substituting into (37), the result is (38).

$$L_2 = \frac{100V_g V_{i_n}(\alpha - \beta)}{\beta\omega_n \%r_{i_{inv}} P_{avg}(-\beta + \alpha - 1)} \tag{38}$$

2.4. Resonance Frequency

In most articles [8,28,33,34], the resonant frequency for the LCL filters is reported. Equation (39) shows how to calculate the resonant frequency using an LCL filter.

$$f_{res} = \frac{1}{2\pi}\sqrt{\frac{L_1 + L_2}{L_1 L_2 C_f}} \tag{39}$$

Substituting the design Equations (35), (37), and (38), into Equation (39). The result for the resonant frequency is the expression (40).

$$f_{res} = \frac{1}{2\pi}\sqrt{\frac{\omega_n{}^2(\beta + 1)}{\alpha}} \tag{40}$$

According to the literature [9,26,33], the resonant frequency must satisfy the condition shown in (41).

$$10f_g \leq f_{res} \leq \frac{f_{sw}}{2} \tag{41}$$

2.5. DC Bus Calculation

Solving for the V_i of Equation (14). This is shown in (42).

$$V_i\angle\phi_i = \frac{V_g\angle 0^\circ\left(jX_{C_f} - jX_{L_1}\right) - \mathbf{I}_g\left(X_{C_f}X_{L_1} + X_{C_f}X_{L_2} - X_{L_1}X_{L_2}\right)}{\left(jX_{C_f}\right)} \tag{42}$$

Substituting (3), (4) and (5) into Equation (42), the magnitude of V_i is obtained as follows in (43).

$$V_i = \sqrt{\left(V_g - \omega^2 L_1 C_f V_g\right)^2 + \left(\omega L_1 I_g + \omega L_2 I_g - \omega^3 L_1 L_2 C_f I_g\right)^2} \tag{43}$$

Substituting (35), (37), and (38) in Equation (43). The result is shown in Equation (44).

$$V_i = \sqrt{\left(V_g\left(1 - \frac{\alpha\omega^2}{\omega_n{}^2}\right)\right)^2 + \left(\frac{200V_{i_n}\omega(\alpha-\beta)(\beta\omega_n{}^2 - \alpha\omega^2 + \omega_n{}^2)}{\beta\%r_{i_{inv}}\omega_n{}^3(-\beta+\alpha-1)}\right)^2} \quad (44)$$

The gamma parameter (γ) is defined by Equation (45).

$$\gamma = \frac{\omega_n}{\omega} \quad (45)$$

where γ is the ratio of the angular frequency of a harmonic n to the angular frequency of the fundamental harmonic. Combining (45) into Equation (44) results in (46):

$$V_i = \sqrt{\left(V_g\left(1 - \frac{\alpha}{\gamma^2}\right)\right)^2 + \left(\frac{200V_{i_n}(\alpha-\beta)(\gamma^2\beta - \alpha + \gamma^2)}{\beta\%r_{i_{inv}}\gamma^3(-\beta+\alpha-1)}\right)^2} \quad (46)$$

The modulation index (m) relates the DC bus voltage (V_{dc}) and the input voltage (V_i) of the fundamental harmonic in the LCL filter [35,36]. This is shown in Equation (47).

$$V_i = mV_{dc} \quad (47)$$

Substituting V_i and clearing V_{dc} from Equation (47) results in Equation (48).

$$V_{dc} = \frac{\sqrt{\left(V_g\left(1 - \frac{\alpha}{\gamma^2}\right)\right)^2 + \left(\frac{200V_{i_n}(\alpha-\beta)(\gamma^2\beta-\alpha+\gamma^2)}{\beta\%r_{i_{inv}}\gamma^3(-\beta+\alpha-1)}\right)^2}}{m} \quad (48)$$

For a periodical waveform, the Fourier series expression is calculated with Equation (49) [37,38].

$$V_{in} = \frac{a_0}{2} + \sum_{n=1}^{\infty} a_n\cos(n\theta) + b_n\sin(n\theta) \quad (49)$$

where α_0 is the constant DC term, α_n is the Fourier cosine coefficient, and b_n is the Fourier sine coefficient. A unipolar SPWM modulation only has b_n components, and the harmonics after the fundamental appear at double the switching frequency and are odd [38–40]. This is shown in Equation (50).

$$b_n = \frac{4V_{dc}}{n\pi}\left[\sum_{i=1}^{N}(-1)^i\cos(n\alpha_i)\right] \quad (50)$$

where n is the harmonic order and α_i is the ith switching angle. With the Fourier analysis performed in [41–43], the harmonics of the highest magnitude after the fundamental are located at $n = 2f_{sw} - f_g$ and $n = 2f_{sw} + f_g$, where f_{sw} is the switching frequency and f_g is the grid frequency. Evaluating (49) and (50) for the n harmonic, the relationship shown in Equation (51) is established.

$$V_{i_n} = m_n V_{dc} \quad (51)$$

V_{in} is the input voltage of the LCL filter at harmonic n; m_n is a ratio of V_{dc} and V_{in} m_n has a direct relationship to the modulation index (m). This value can be obtained from Equation (50). The calculated values are shown in Table 1.

Substituting (51) in (48). Equation (52) results.

$$V_{dc} = \frac{\sqrt{\left(V_g\left(1 - \frac{\alpha}{\gamma^2}\right)\right)^2 + \left(\frac{200(m_nV_{dc})(\alpha-\beta)(\gamma^2\beta-\alpha+\gamma^2)}{\beta\%r_{i_{inv}}\gamma^3(-\beta+\alpha-1)}\right)^2}}{m} \quad (52)$$

Table 1. Relationship between modulation index and mn parameter.

Modulation Index (m)	m_n
1	0.2116
0.9	0.2824
0.8	0.3917
0.7	0.5061
0.6	0.6178
0.5	0.7220
0.4	0.8140

By dividing Equation (52) into two parts (A and B), Equations (53) and (54) result in the following:

$$A = \left(V_g\left(1 - \frac{\alpha}{\gamma^2}\right)\right)^2 \tag{53}$$

$$B = \left(\frac{200(m_n)(\alpha - \beta)\left(\gamma^2\beta - \alpha + \gamma^2\right)}{\beta\%r_{i_{inv}}\gamma^3(-\beta + \alpha - 1)}\right)^2 \tag{54}$$

After compacting Equation (52) and clearing V_{dc}, the result is shown in Equation (55).

$$V_{dc} = \frac{\sqrt{A + BV_{dc}{}^2}}{m} = \sqrt{\frac{A}{m^2 - B}} \tag{55}$$

Substituting (53) and (54) into (55). This is shown in Equation (56).

$$V_{dc} = \sqrt{\frac{\left(V_g\left(1 - \frac{\alpha}{\gamma^2}\right)\right)^2}{m^2 - \left(\frac{200(m_n)(\alpha-\beta)(\gamma^2\beta-\alpha+\gamma^2)}{\beta\%r_{i_{inv}}\gamma^3(-\beta+\alpha-1)}\right)^2}} \tag{56}$$

With Equation (56), it is possible to calculate the dc bus level as a function of the Alpha, Beta, and Gamma parameters, the grid voltage (V_g), the percentage of ripple current ($\%r_{inv}$) on the inverter side, the m_n parameter, and the modulation index (m). Substituting Equation (56) in Equation (51) results in (57).

$$V_{i_n} = m_n\sqrt{\frac{\left(V_g\left(1 - \frac{\alpha}{\gamma^2}\right)\right)^2}{m^2 - \left(\frac{200(m_n)(\alpha-\beta)(\gamma^2\beta-\alpha+\gamma^2)}{\beta\%r_{i_{inv}}\gamma^3(-\beta+\alpha-1)}\right)^2}} \tag{57}$$

Equation (57) is evaluated for different values of Alpha and Beta. The values obtained from Equation (57) are substituted into Equations (35)–(37), which correspond to the LCL filter elements.

2.6. Attenuation Coefficient to Determine Beta

To calculate the grid side inductor (L_1), some papers [10,44,45] use the parameter (K_a) attenuation coefficient, which is defined as the ratio of the grid side current to the inverter side current at the n-harmonic. This is shown in Equation (58).

$$\frac{I_{gn}}{I_{inv_n}} \cong K_a = \frac{1}{\left|1 + r\left[1 - L_1C_f\omega_{sw}{}^2\right]\right|} \tag{58}$$

where r is the relationship between L_2 and L_1, ω_{sw} is the angular frequency of commutation and K_a is the attenuation coefficient. Figure 6 shows the relation between K_a, y, and r.

Figure 8 shows that to have a lower attenuation coefficient, the beta should be close to 1. In this work, the beta chosen in this paper—β—is the ratio of L_1 and L_2. Therefore, β is the inverse of r. This is shown in Equation (59). In this paper, the parameter beta will be equal to 1.

$$\beta = \frac{1}{r} \quad (59)$$

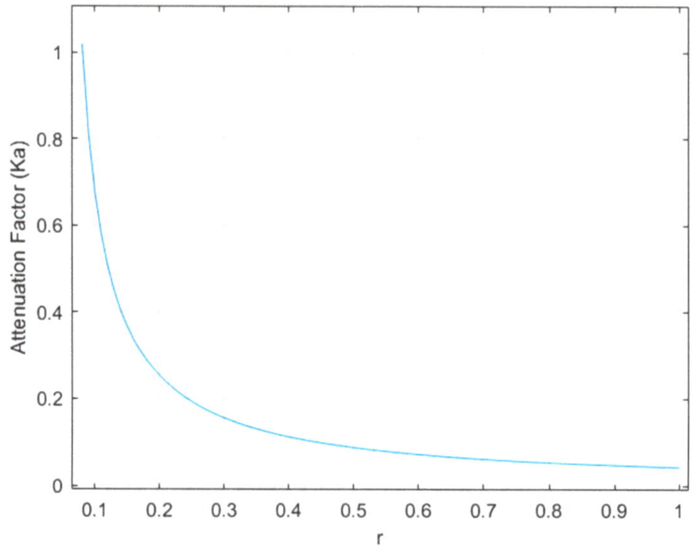

Figure 8. Attenuation factor (K_a) versus r.

3. Design and Optimization of an LCL Filter Connected to the Grid

To validate the above equations, a step-by-step design methodology is proposed in this work. The design specifications are based on the information shown in Table 2. Where it is shown that the frequency $f_n = 2f_{sw} - f_g$, as explained in Section 2.3 of this paper and in Equations (49) and (50). A design and optimization of the LCL filter are proposed under the specifications shown in the following table.

Table 2. General design specifications.

Parameter	Symbol	Value
Average power	P_{avg}	90 W
Peak grid voltage	V_g	180 V
Switching frequency	f_{sw}	10 KHz
Frequency of harmonic n	f_n	19.94 kHz
Angular frequency at harmonic n	ω_n	2π (19,940)
Gamma	γ	332.33
Peak grid current	I_g	1 A
Alpha	α	Variable
Betha	β	1
Modulation index	m	0.9
Relationship between V_{dc} and V_{in}	m_n	0.28242
Percentage of current ripple in L_1	$\%r_{i_{inv}}$	15%
Grid frequency	f_g	60 Hz

Table 3 shows the step-by-step design methodology for the inverter and LCL filter.

Table 3. Proposed design methodology for the inverter and LCL filter.

Step	Parameter	Symbol	Equation
1	DC bus voltage	V_{dc}	$V_{dc} = \sqrt{\frac{\left(V_g\left(1-\frac{\alpha}{\gamma^2}\right)\right)^2}{m^2-\left(\frac{200(m_n)(\alpha-\beta)(\gamma^2\beta-\alpha+\gamma^2)}{\beta\%r_{i_{inv}}\gamma^3(-\beta+\alpha-1)}\right)^2}}$
2	Inverter output voltage on harmonic n	V_{i_n}	$V_{i_n} = m_n\sqrt{\frac{\left(V_g\left(1-\frac{\alpha}{\gamma^2}\right)\right)^2}{m^2-\left(\frac{200(m_n)(\alpha-\beta)(\gamma^2\beta-\alpha+\gamma^2)}{\beta\%r_{i_{inv}}\gamma^3(-\beta+\alpha-1)}\right)^2}}$
3	Inductor L_1	L_1	$L_1 = \frac{100V_g V_{in}(\alpha-\beta)}{\omega_n\%r_{i_{inv}}P_{avg}(-\beta+\alpha-1)}$
4	Inductor L_2	L_2	$L_2 = \frac{100V_g V_{in}(\alpha-\beta)}{\beta\omega_n\%r_{i_{inv}}P_{avg}(-\beta+\alpha-1)}$
5	Filter capacitor	C_f	$C_f = \frac{\%r_{i_{inv}}P_{avg}\alpha(-\beta+\alpha-1)}{100V_g V_{in}\omega_n(\alpha-\beta)}$
6	Resonance frequency	f_r	$f_{res} = \frac{1}{2\pi}\sqrt{\frac{\omega_n{}^2(\beta+1)}{\alpha}}$

3.1. Step 1. V_{dc} Calculation

Equation (56) was evaluated using the data from Table 2 to obtain the value of the DC bus. This is shown in Figure 9.

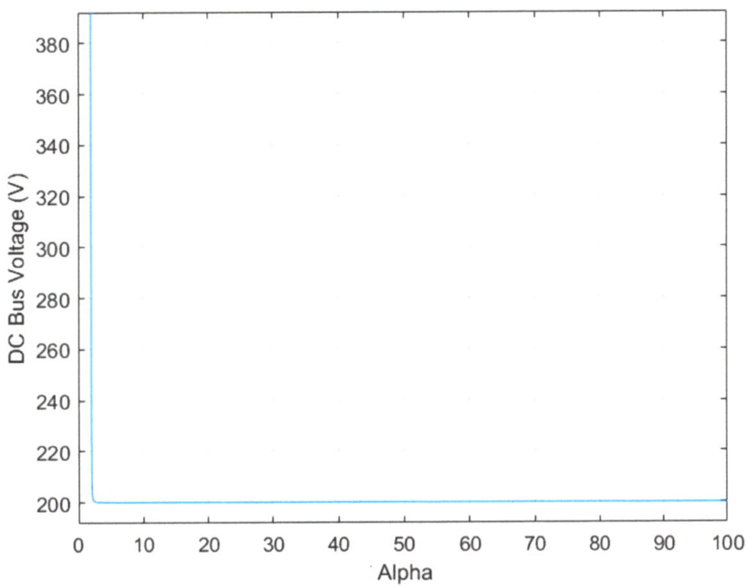

Figure 9. The voltage on the DC bus as a function of alpha.

Figure 9 shows a curve that is asymptotic at 200 V. Therefore, this would be the minimum value of the voltage on the DC bus.

3.2. Step 2. V_{in} Calculation

Equation (57) was evaluated using the data from Table 2 to obtain the value of the DC bus. This is shown in Figure 10.

Figure 10 shows that v_{in} as a function of alpha is asymptotic at 56.4 V. This can be shown by substituting the dc bus value (see Figure 7) into Equation (51) with a modulation index of 0.9, which results in (0.282) × (200 V) = 56.4 V.

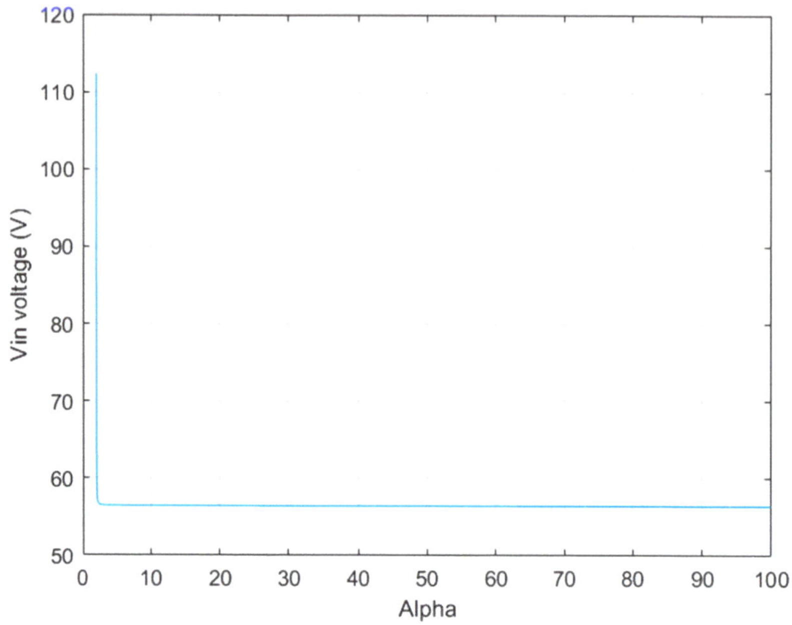

Figure 10. V_{in} as a function of alpha.

3.3. Step 3. Inductance L_1 Calculation

Referring to Figure 8, r is selected as equal to 1 to ensure the lowest attenuation factor (K_a). Equation (37) is evaluated for $\beta = 1$ and different values of alpha. The graph is shown in Figure 11. Using a Pareto optimal point at alpha = 3.29, a value of L_1 equal to 10.68 mH is obtained.

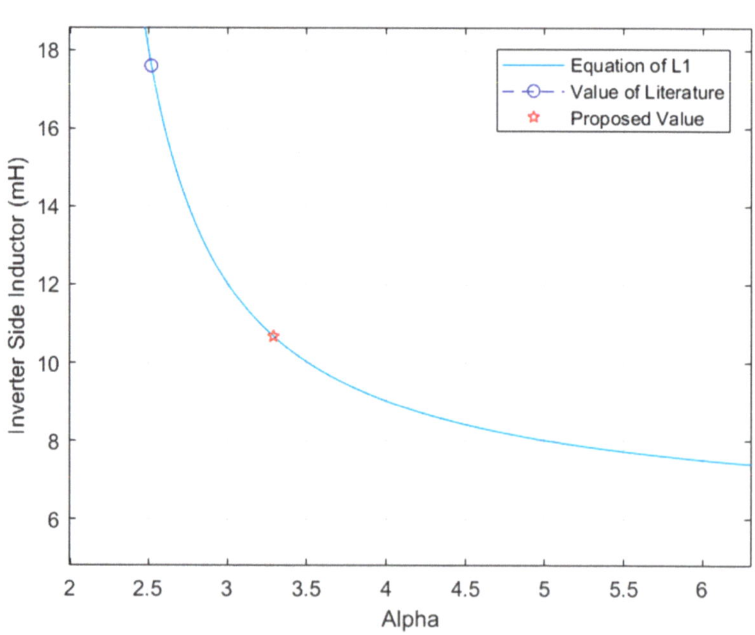

Figure 11. Proposed value for L_1.

3.4. Step 4. Inductance L_2 Calculation

Equation (38) is evaluated for $\beta = 1$ and different values of alpha. The same graph is shown in Figure 11. Using a Pareto optimal point at alpha = 3.29, a value of L_2 equal to 10.68 mH is obtained.

3.5. Step 5. LCL Filter Capacitor Calculation (C_f)

Equation (35) is evaluated for $\beta = 1$ and different values of alpha. Using the alpha obtained in L_1 and L_2 results in C_f equal to 0.0269 µF. This is shown in Figure 12.

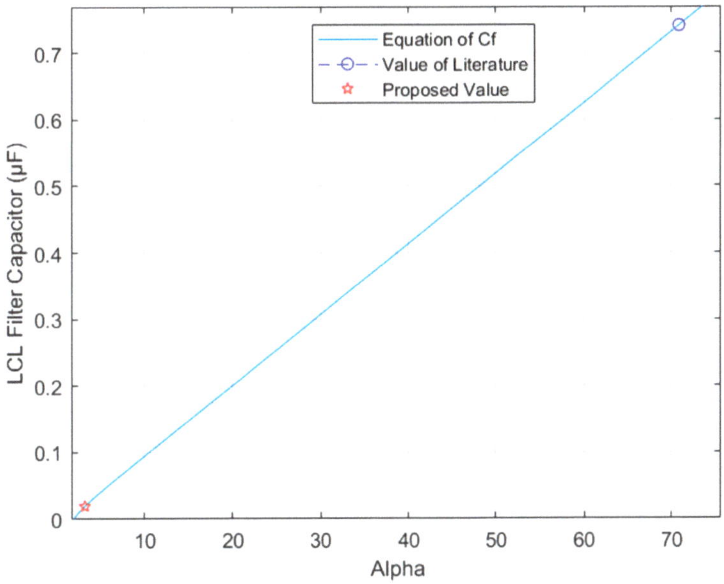

Figure 12. Proposed value for C_f.

Figure 10 shows that the higher the alpha, the higher the capacitor value of the LCL filter increases linearly. It also shows a comparison between the value calculated with the equation published in [12,13,17,26] and the value calculated with the equation proposed in this work. The values obtained for alpha, beta L_1, L_2, and C_f are shown in Table 4.

Table 4. Values calculated with the design methodology.

Parameter	Symbol	Value
DC Bus Voltage	V_{dc}	200.1 V
Inverter side voltage at harmonic n	V_{i_n}	56.4 V
Alpha	α	3.29
Inductor 1	L_1	10.68 mH
Inductor 2	L_2	10.68 mH
LCL filter capacitor	C_f	0.0269 µF

3.6. Step 6. Resonance Frequency Calculation (f_r)

Equation (40) was evaluated using the data obtained in Table 4. A resonance frequency of 15.54 KHz was obtained.

3.7. Step 7. Calculation of Link Capacitor (C_f)

The equation published in [4] is used to calculate the required link capacitor (C_{link}). This is shown in Equation (60).

$$C_{link} = \frac{P_{avg}(2 - \cos(\phi_i))}{V_g \omega_g \Delta V_{dc}} \quad (60)$$

where ϕ_i is the phase angle required in SPWM modulation to ensure connection to the grid with a unity power factor. ΔV_{dc} is the proposed DC bus voltage ripple. The angle ϕ_i is the phase obtained from Equation (42) as shown in (61).

$$\phi_i = \tan^{-1} \frac{\omega L_1 I_g + \omega L_2 I_g - \omega^3 L_1 L_2 C_f I_g}{V_g - \omega^2 L_1 C_f V_g} \quad (61)$$

Substituting Equation (61) into (60) gives the expression shown in Equation (62), which is used to calculate the required link capacitor as a function of the value of the LCL filter elements and the phase shift caused by the LCL filter. This phase shift is compensated by the closed-loop control to obtain the connection to the grid with a unity power factor.

$$C_{link} = \frac{P_{avg}\left(2 - \cos\left(\tan^{-1} \frac{\omega L_1 I_g + \omega L_2 I_g - \omega^3 L_1 L_2 C_f I_g}{V_g - \omega^2 L_1 C_f V_g}\right)\right)}{V_g \omega_g \Delta V_{dc}} \quad (62)$$

Regarding the data in Tables 2 and 4, assuming a voltage ripple of 29 V_{pp} or 14.5% results in a link capacitor equal to 45.78 µF.

4. Simulation Results

To verify the performance of the design, a closed-loop simulation was performed with Simulink software version 9.0. The block diagram is shown in Figure 13, where the LCL filter is connected to the inverter and the grid. The control is performed using a phase-locked-loop (PLL) with a proportional resonant (PR) current controller.

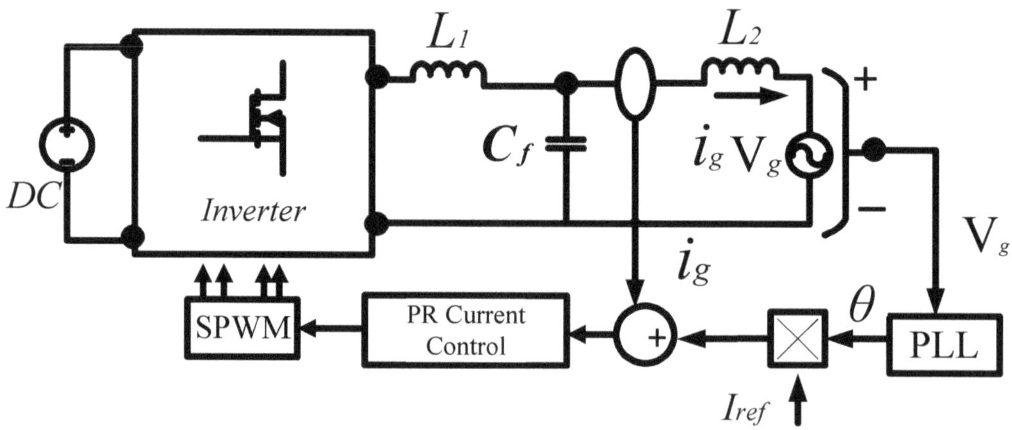

Figure 13. LCL filter control diagram connected to the grid.

Figure 11 shows the block diagram of the implemented control. The grid voltage is sensed by the PLL to calculate the angle to be implemented in the modulation. The reference current is 1 A, which is compared with the error current to obtain the error to be introduced to the controller. The schematic of the inverter is shown in Figure 14.

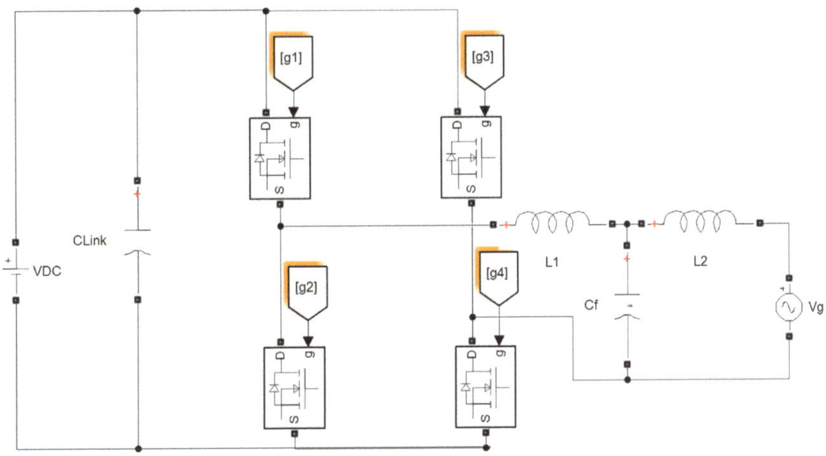

Figure 14. Full bridge inverter with LCL filter connected to the grid.

The control implemented in Simulink is shown in Figure 15.

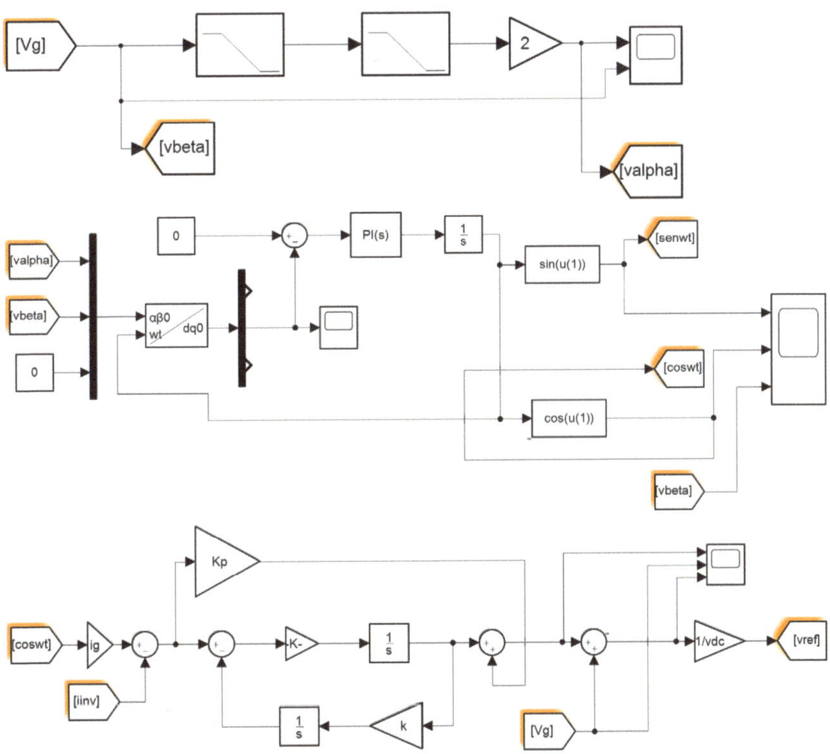

Figure 15. Schematic of control implemented in Simulink.

Figure 16 shows the simulation results for voltage, current, instantaneous power, and average power in the grid. The theoretical value of the grid current is 1 A, and the value obtained in the simulation is 1.06 A, resulting in an error of 5.6%. The theoretical value of the maximum instantaneous power is 182 W, and the measured value is 182.8 W, resulting in an error of 1.5%. The calculated average power is 90 W, and the measured value is 89.1

W, which is a 1.01% error. The current and voltage are at a fundamental frequency of 60 Hz and the instantaneous power is at a frequency of 120 Hz.

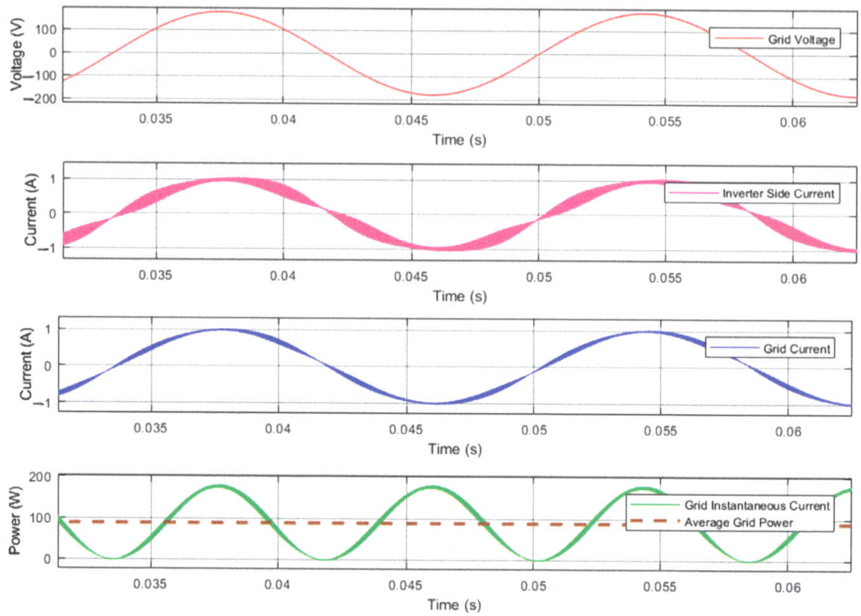

Figure 16. Results of simulation.

The FFT computed with MATLAB shows on the (y) axis the magnitude of the harmonics concerning the peak magnitude of the fundamental harmonic; however, in this work, it was calculated for the peak-to-peak magnitude; therefore, Equation (34) must be divided by two as shown in Equation (63). The current signal on the inverter side is shown in Figure 17.

$$\%I_{inv} = \frac{\%r_{inv}}{2} = 7.5\% \quad (63)$$

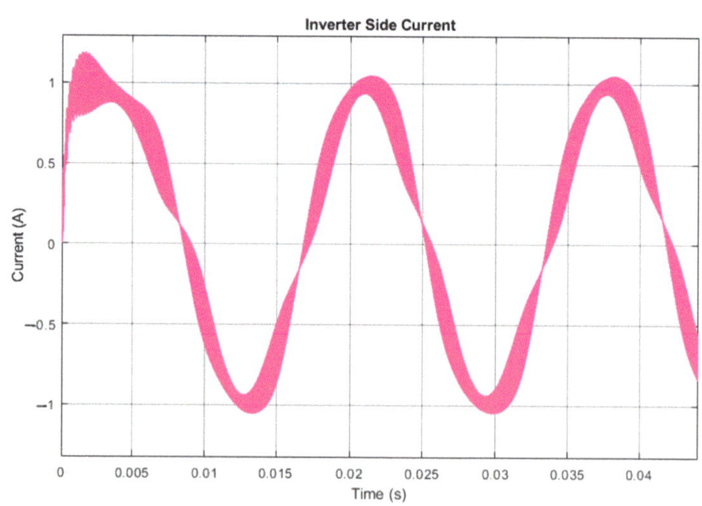

Figure 17. Current in L_1.

The percentage of current ripple on the inverter side was proposed to be 15%; therefore, by substituting this value in Equation (64), we obtain a percentage of 7.5%, which is what is shown in the FFT (Figure 18).

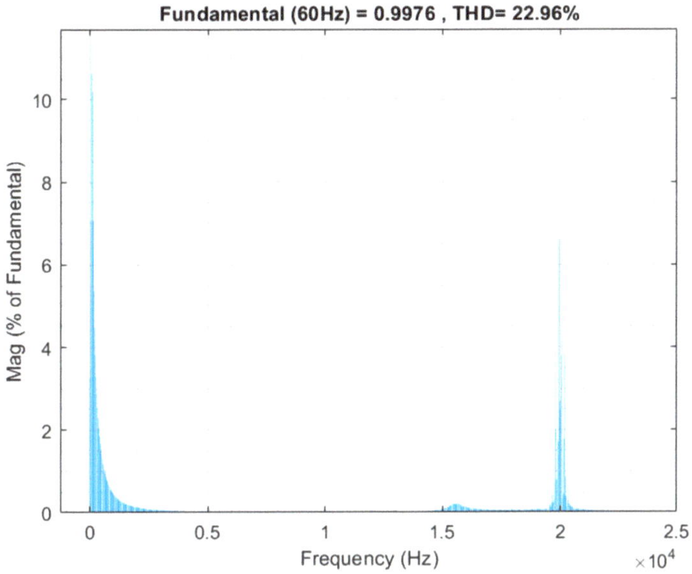

Figure 18. FFT of the inverter side current signal.

Zooming from Figure 18 to the harmonics near 20 kHz, these are shown in Figure 19. The ripple percentage of the proposed harmonic n was 7.5%, and the simulation result was 6.64%, resulting in an error of 12.95%.

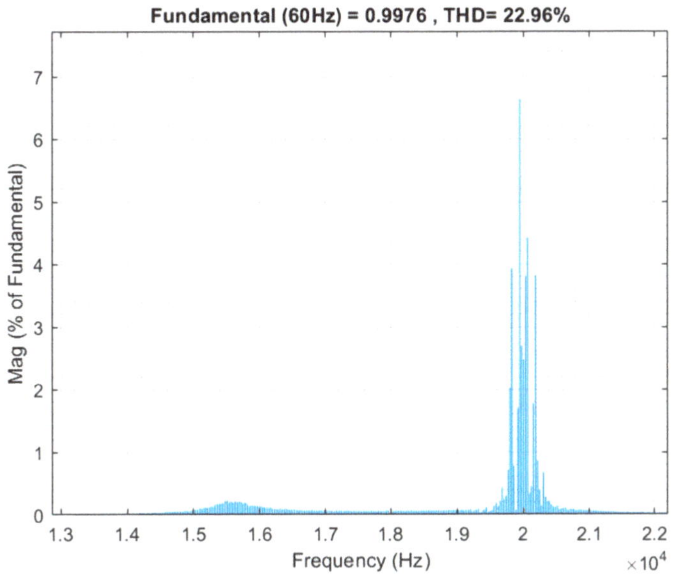

Figure 19. FFT of harmonics near harmonic n for the inverter side current.

The grid side current is shown in Figure 20.

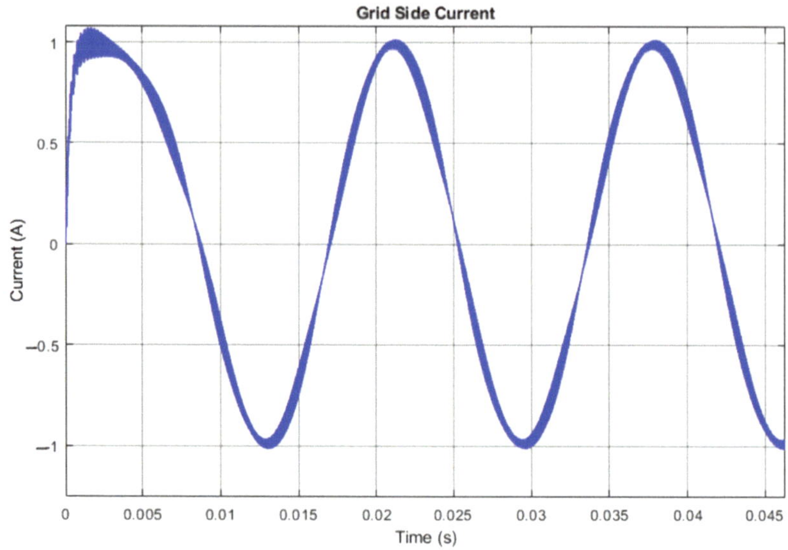

Figure 20. Grid side current.

The FFT of Figure 20 is shown in Figure 21. The THD of the grid current is 5.12% which complies with the IEEE 519-2022 standard [46].

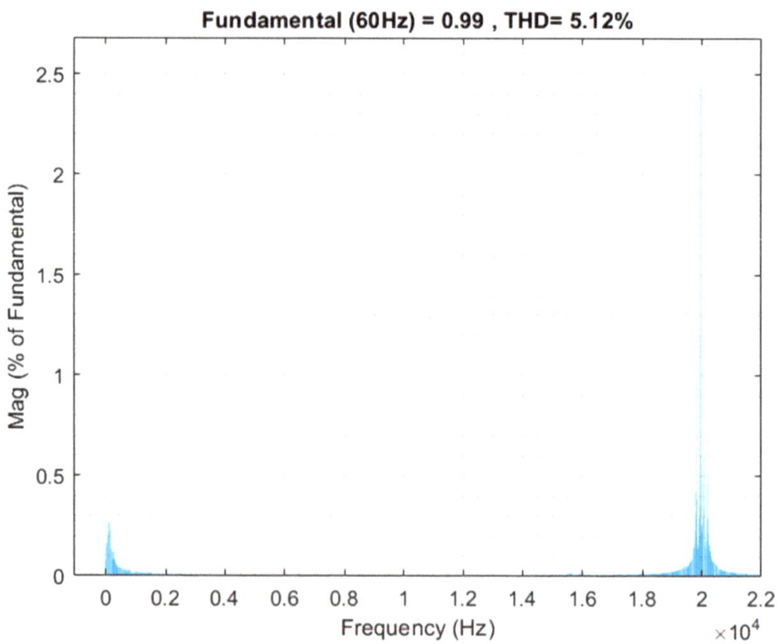

Figure 21. FFT of the grid side current signal.

If one zooms in from Figure 21 to the harmonics near harmonic n (see Figure 22), it becomes apparent that the harmonics associated with the resonant frequency are lower than the harmonics associated with harmonic *n*.

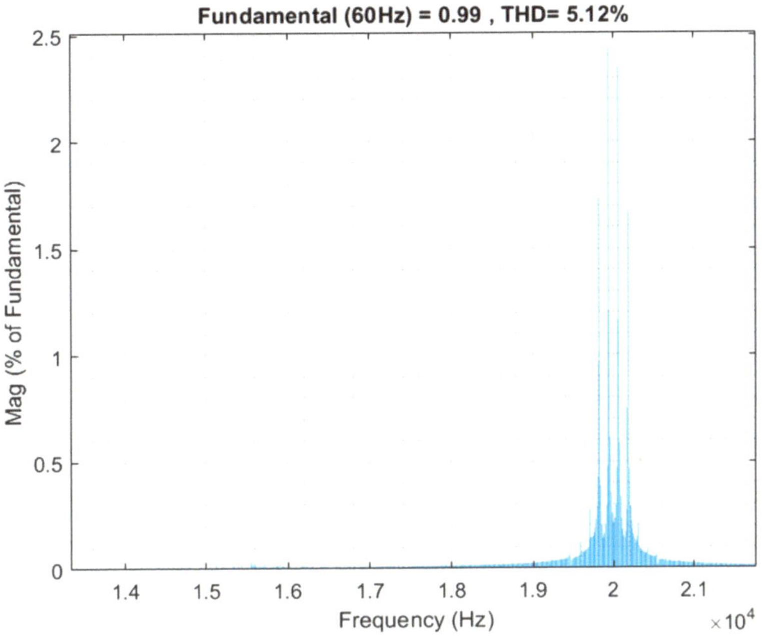

Figure 22. FFT of I_g (harmonics near harmonic n).

5. Experimental Results

A prototype was implemented experimentally to validate the design methodology and the calculations performed. The values obtained from Tables 2 and 4 were used. The prototype used is shown in Figure 23.

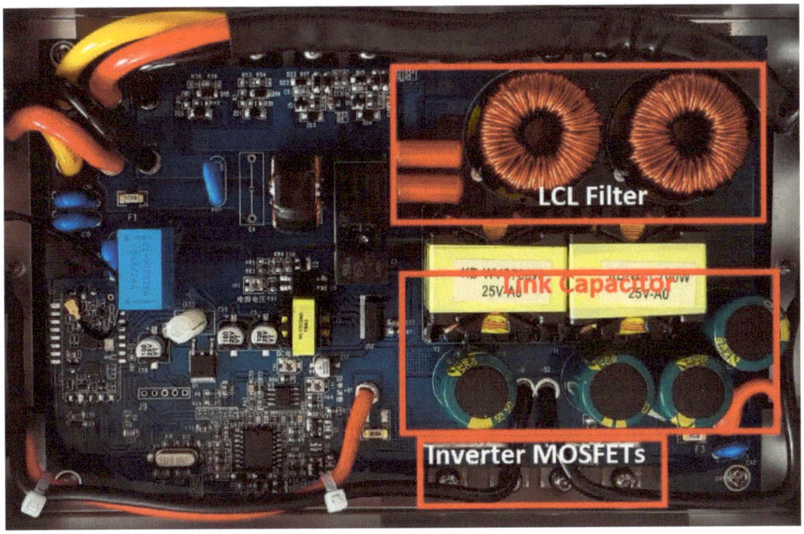

Figure 23. Prototype implemented.

Figure 24 shows the grid voltage, L_1 current, and the instantaneous power. The data obtained are shown in Table 5.

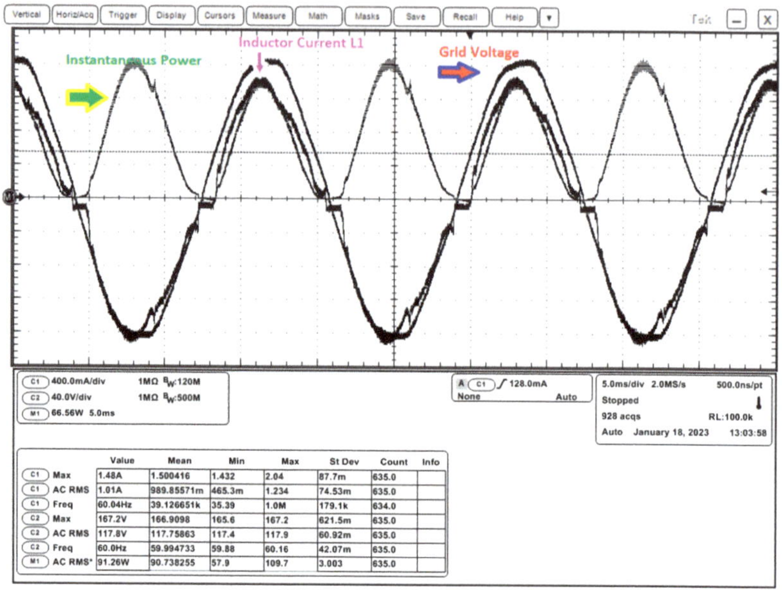

Figure 24. Measured inverter-side current and grid voltage.

Table 5. Experimental results of L_1 current, grid voltage, and instantaneous power.

Parameter	Symbol	Measured Value	% Error
Inductor Current L_1	I_{inv}	1.01 A	0.99%
Average Power	P_{avg}	91.26	1.38 W

With the data obtained from the oscilloscope, the FFT is computed in Simulink, and this is shown in Figure 25.

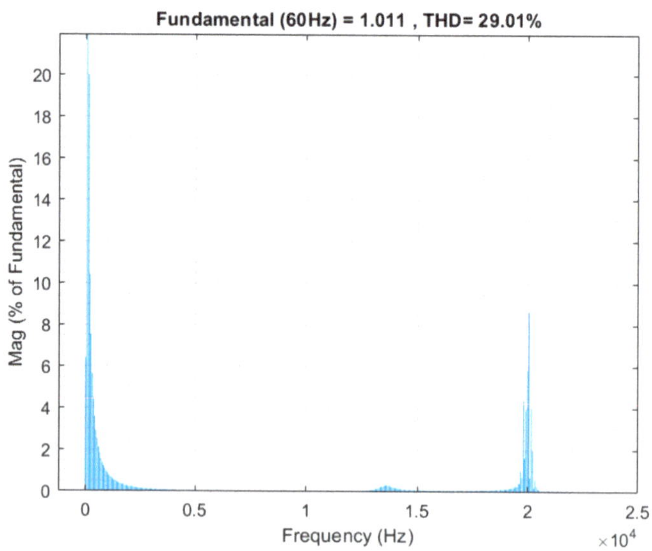

Figure 25. Experimental FFT of the L_1 current.

The magnitude of the harmonic n for the inverter side current (Figure 25) is 8.5%, and the proposed theoretical value was 7.5%. This results in an error of 11.76%. The grid current was measured with a spectrum analyzer. This is shown in Figure 26.

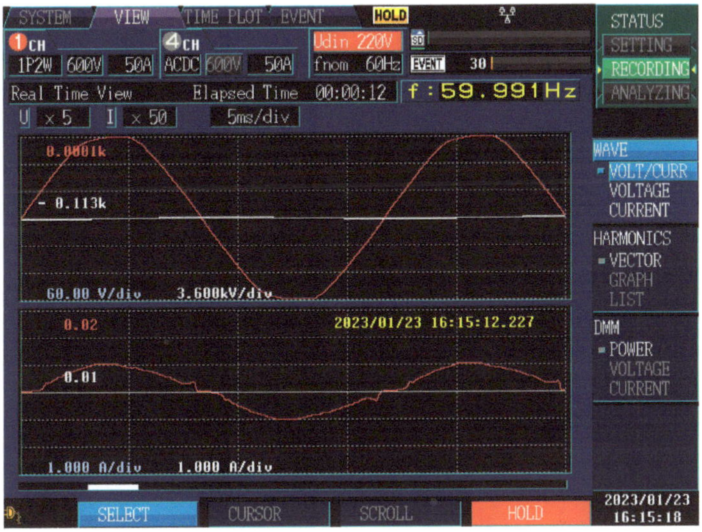

Figure 26. Measured grid current with a spectrum analyzer.

Figure 26 shows the grid voltage (upper signal) with a peak magnitude of 180 V on a scale of 60.0 V/div. Also shown is the grid current (lower signal) with a measured magnitude of 1 A on a scale of 1.00 A/div. The data obtained from the spectrum analyzer are exported to Simulink, and the FFT shown in Figure 25 is calculated.

Figure 27 shows the magnitude of the harmonics obtained with the data (Figure 24). The magnitude of harmonic n (19,940 Hz) is less than 2%. The magnitude of the harmonics close to the resonance frequency is smaller than the magnitude of the harmonics close to harmonic n. The THD calculated from the experimental data for the grid current was 4.4%, thus complying with the IEEE standard.

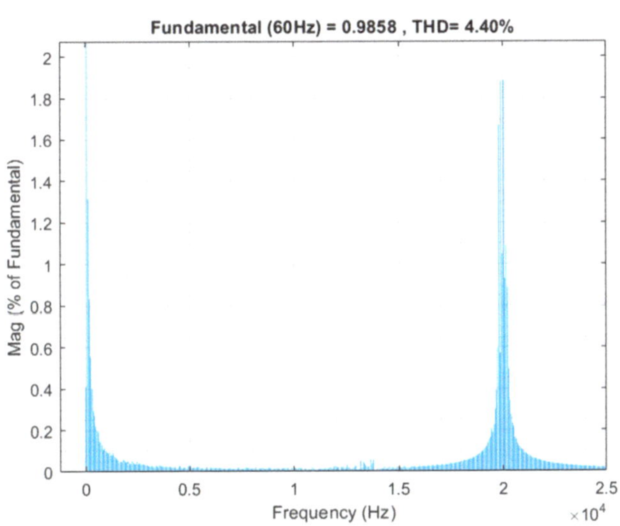

Figure 27. Experimental FFT of grid current.

6. Discussion

The LCL filter design for single-phase grid-connected systems published in [8–11,26,44,47] uses the following equations. For inductor L_1, Equation (64) is used.

$$L_1 = \frac{V_{dc}}{8 f_{sw} \Delta_I} \tag{64}$$

where Δ_I is the ripple of the current proposed. This current ripple is calculated using Equation (65).

$$\Delta_I = \frac{\% r_{inv} \sqrt{2} P_{avg}}{100 V_{g_{rms}}} \tag{65}$$

where V_{grms} is the rms grid voltage. For the inductor on the grid side, Equation (66) is used.

$$L_2 = r L_1 \tag{66}$$

where r is the relation between L_2 and L_1. This relation is shown in Figure 6. The relation r produces values from 0 to 1. For the LCL filter capacitor, the base impedance (Z_b) is shown in Equation (67).

$$Z_b = \frac{V_{g_{rms}}{}^2}{P_{avg}} \tag{67}$$

Z_b is used to define a base capacitance (C_b). This is shown in Equation (68).

$$C_b = \frac{1}{\omega_g Z_b} \tag{68}$$

where ω_g is the grid angular frequency. For the calculation of the LCL filter capacitor; it is proposed that the capacitor should handle a maximum of 5% reactive power; therefore, C_b is multiplied by 0.05, as shown in Equation (69).

$$C_f = 0.05 C_b \tag{69}$$

Equation (69) calculates the maximum value of the capacitor [10,45,48]. However, this formula is not used in this work since the aim is to minimize this value. Table 6 shows the values obtained with the equations. The values are calculated using Equations (35), (37) and (38) proposed in this work.

Table 6. Comparison of the values reported in the literature and those proposed in this work for the LCL filter.

Element	Equations from the Literature	Value Obtained	Proposed Equations	Value Obtained
L_1	$L_1 = \frac{V_{dc}}{8 f_{sw} \Delta_I}$	16.63 mH	$L_1 = \frac{100 V_g V_{in} (\alpha - \beta)}{\omega_n \% r_{i_{inv}} P_{avg} (-\beta + \alpha - 1)}$	10.68 mH
L_2	$L_2 = r L_1$	16.63 mH	$L_2 = \frac{100 V_g V_{in} (\alpha - \beta)}{\beta \omega_n \% r_{i_{inv}} P_{avg} (-\beta + \alpha - 1)}$	10.68 mH
C_f	$C_f = 0.05 C_b$	740 nF	$C_f = \frac{\% r_{i_{inv}} P_{avg} \alpha (-\beta + \alpha - 1)}{100 V_g V_{in} \omega_n (\alpha - \beta)}$	19.62 nF

The main objective of this work is to optimize the values of an LCL filter. This objective has been achieved and is shown in Table 6. The values of the inductors have been reduced by 39.1% while the capacitor is the element with the highest reduction of 97%.

7. Conclusions

In this paper, a mathematical analysis and a Pareto-based optimization method for the design of LCL filters in single-phase grid-connected photovoltaic systems have been presented. New design equations have been proposed for the calculation of the inductances and capacitors of an LCL filter. The errors presented using the current ripple at harmonic n are less than 10%. The THD of the grid current is lower than the IEEE standard. The inductor size has been reduced to 39.31% and the capacitor value by 97%. This reduction in capacitor value is because the equation reported in the literature is obtained from a maximum value of reactive power that will handle a capacitor with a maximum value of this capacitor. However, in this work, it was calculated based on the value of the inductors, the grid voltage, the current ripple in the inductors, and the average power, resulting in a minimum value of the capacitor. These are the main contributions of this work. The main applications of this work are as follows: use in microinverters for photovoltaic applications that are interconnected to the grid through an LCL filter, as well as use in the microgrids of the proposed LCL filter. A limitation of this work is that it is only applicable to single-phase connections connected to the grid. In three-phase and isolated systems, it is not possible to use the proposed methodology to design the filters.

Author Contributions: Conceptualization, H.A.-P., M.P.-S., J.D.M.-A. and A.C.-S.; methodology, M.P.-S., J.D.M.-A., A.C.-S. and H.A.-P.; software, H.A.-P., A.C.-S., O.R.-B. and H.A.-P.; validation, M.P.-S., J.D.M.-A., O.R.-B., O.M.R.-B. and A.C.-S.; formal analysis, M.P.-S., A.C.-S. and H.A.-P.; investigation, H.A.-P., M.P.-S., J.D.M.-A., A.C.-S. and O.M.R.-B.; resources, H.A.-P., O.R.-B. and O.M.R.-B.; data curation, J.D.M.-A., A.C.-S., O.R.-B., O.M.R.-B. and H.A.-P.; writing—original draft preparation, H.A.-P., J.D.M.-A., M.P.-S. and A.C.-S.; writing—review and editing, H.A.-P., M.P.-S., J.D.M.-A., A.C.-S. and O.R.-B.; visualization, A.C.-S., O.R.-B., O.M.R.-B. and H.A.-P.; supervision, M.P.-S., J.D.M.-A., O.R.-B. and H.A.-P.; project administration, M.P.-S. and O.R.-B.; funding acquisition, M.P.-S., H.A.-P. and O.R.-B. All authors have read and agreed to the published version of the manuscript.

Funding: This research received no external funding.

Data Availability Statement: The original contributions presented in the study are included in the article, further inquiries can be directed to the corresponding author.

Conflicts of Interest: The authors declare no conflicts of interest.

References

1. Lai, C.S.; Jia, Y.; Lai, L.L.; Xu, Z.; McCulloch, M.D.; Wong, K.P. A comprehensive review on large-scale photovoltaic system with applications of electrical energy storage. *Renew. Sustain. Energy Rev.* **2017**, *78*, 439–451. [CrossRef]
2. Chatterjee, S.; Kumar, P.; Chatterjee, S. A techno-commercial review on grid connected photovoltaic system. *Renew. Sustain. Energy Rev.* **2018**, *81*, 2371–2397. [CrossRef]
3. Kabir, E.; Kumar, P.; Kumar, S.; Adelodun, A.A.; Kim, K.-H. Solar energy: Potential and future prospects. *Renew. Sustain. Energy Rev.* **2018**, *82*, 894–900. [CrossRef]
4. Adamas-Pérez, H.; Ponce-Silva, M.; Mina-Antonio, J.D.; Claudio-Sánchez, A.; Rodríguez-Benítez, O. Assessment of Energy Conversion in Passive Components of Single-Phase Photovoltaic Systems Interconnected to the Grid. *Electronics* **2023**, *12*, 3341. [CrossRef]
5. Ghosh, S.; Yadav, R. Future of photovoltaic technologies: A comprehensive review. *Sustain. Energy Technol. Assess.* **2021**, *47*, 101410. [CrossRef]
6. Marques Lameirinhas, R.A.; Torres, J.P.N.; de Melo Cunha, J.P. A photovoltaic technology review: History, fundamentals and applications. *Energies* **2022**, *15*, 1823. [CrossRef]
7. Rodríguez-Benítez, O.; Ponce-Silva, M.; Aqui-Tapia, J.A.; Rodríguez-Benítez, Ó.M.; Lozoya-Ponce, R.E.; Adamas-Pérez, H. Active Power-Decoupling Methods for Photovoltaic-Connected Applications: An Overview. *Processes* **2023**, *11*, 1808. [CrossRef]
8. Rasekh, N.; Hosseinpour, M. LCL filter design and robust converter side current feedback control for grid-connected Proton Exchange Membrane Fuel Cell system. *Int. J. Hydrogen Energy* **2020**, *45*, 13055–13067. [CrossRef]
9. Sedo, J.; Kascak, S. Design of output LCL filter and control of single-phase inverter for grid-connected system. *Electr. Eng.* **2017**, *99*, 1217–1232. [CrossRef]
10. Ruan, X.; Wang, X.; Pan, D.; Yang, D.; Li, W.; Bao, C. *Control Techniques for LCL-Type Grid-Connected Inverters*; Springer: Singapore, 2018.
11. Ibrahim, N.F.; Mahmoud, M.M.; Al Thaiban, A.M.; Barnawi, A.B.; Elbarbary, Z.S.; Omar, A.I.; Abdelfattah, H. Operation of grid-connected PV system with ANN-based MPPT and an optimized LCL filter using GRG algorithm for enhanced power quality. *IEEE Access* **2023**, *11*, 106859–106876. [CrossRef]
12. Cai, Y.; He, Y.; Zhou, H.; Liu, J. Design method of LCL filter for grid-connected inverter based on particle swarm optimization and screening method. *IEEE Trans. Power Electron.* **2021**, *36*, 10097–10113. [CrossRef]

13. Cittanti, D.; Mandrile, F.; Gregorio, M.; Bojoi, R. Design Space Optimization of a Three-Phase LCL Filter for Electric Vehicle Ultra-Fast Battery Charging. *Energies* **2021**, *14*, 1303. [CrossRef]
14. Osório, C.R.; Schuetz, D.A.; Koch, G.G.; Carnielutti, F.; Lima, D.M.; Maccari Jr, L.A.; Montagner, V.F.; Pinheiro, H. Modulated model predictive control applied to lcl-filtered grid-tied inverters: A convex optimization approach. *IEEE Open J. Ind. Appl.* **2021** *2*, 366–377. [CrossRef]
15. Park, K.-B.; Kieferndorf, F.D.; Drofenik, U.; Pettersson, S.; Canales, F. Optimization of LCL filter with integrated intercell transformer for two-interleaved high-power grid-tied converters. *IEEE Trans. Power Electron.* **2019**, *35*, 2317–2333. [CrossRef]
16. Hasan, F.A.; Rashad, L.J.; Humod, A.T. Integrating Particle Swarm Optimization and Routh-Hurwitz's Theory for Controlling Grid-Connected LCL-Filter Converter. *Int. J. Intell. Eng. Syst.* **2020**, *13*, 102–113. [CrossRef]
17. Gurrola-Corral, C.; Segundo, J.; Esparza, M.; Cruz, R. Optimal LCL-filter design method for grid-connected renewable energy sources. *Int. J. Electr. Power Energy Syst.* **2020**, *120*, 105998. [CrossRef]
18. Long, B.; Zhu, Z.; Yang, W.; Chong, K.T.; Rodríguez, J.; Guerrero, J.M. Gradient descent optimization based parameter identification for FCS-MPC control of LCL-type grid connected converter. *IEEE Trans. Ind. Electron.* **2021**, *69*, 2631–2643. [CrossRef]
19. Hernandez, O.; Mina, J.; Calleja, J.H.; Pérez, A.C.; de Leon, S.E. A multi-objective optimized design of LCL filters for grid connected voltage source inverters considering discrete components. *Int. Trans. Electr. Energy Syst.* **2021**, *31*, e12908. [CrossRef]
20. Azab, M. Multi-objective design approach of passive filters for single-phase distributed energy grid integration systems using particle swarm optimization. *Energy Rep.* **2020**, *6*, 157–172. [CrossRef]
21. Khan, D.; Qais, M.; Sami, I.; Hu, P.; Zhu, K.; Abdelaziz, A.Y. Optimal Lcl-Filter Design for a Single-Phase Grid-Connected Inverter Using Metaheuristic Algorithms. *Comput. Electr. Eng.* **2023**, *110*, 108857. [CrossRef]
22. Khan, M.A.; Haque, A.; Kurukuru, V.S.B.; Blaabjerg, F. Optimizing the Performance of Single-Phase Photovoltaic Inverter using Wavelet-Fuzzy Controller. *e-Prime-Adv. Electr. Eng. Electron. Energy* **2023**, *3*, 100093. [CrossRef]
23. Poongothai, C.; Vasudevan, K. Design of LCL filter for grid-interfaced PV system based on cost minimization. *IEEE Trans. Ind Appl.* **2018**, *55*, 584–592. [CrossRef]
24. Park, K.-B.; Kieferndorf, F.D.; Drofenik, U.; Pettersson, S.; Canales, F. Weight minimization of LCL filters for high-power converters: Impact of PWM method on power loss and power density. *IEEE Trans. Ind. Appl.* **2017**, *53*, 2282–2296. [CrossRef]
25. Aouichak, I.; Jacques, S.; Bissey, S.; Reymond, C.; Besson, T.; Le Bunetel, J.-C. A bidirectional grid-connected DC–AC converter for autonomous and intelligent electricity storage in the residential sector. *Energies* **2022**, *15*, 1194. [CrossRef]
26. Arab, N.; Vahedi, H.; Al-Haddad, K. LQR control of single-phase grid-tied PUC5 inverter with LCL filter. *IEEE Trans. Ind. Electron* **2019**, *67*, 297–307. [CrossRef]
27. Dursun, M.; Döşoğlu, M.K. LCL filter design for grid connected three-phase inverter. In Proceedings of the 2018 2nd International Symposium on Multidisciplinary Studies and Innovative Technologies (ISMSIT), Ankara, Turkey, 19–21 October 2018; pp. 1–4.
28. Jayalath, S.; Hanif, M. An LCL-filter design with optimum total inductance and capacitance. *IEEE Trans. Power Electron.* **2017** *33*, 6687–6698. [CrossRef]
29. Liu, Y.; See, K.-Y.; Yin, S.; Simanjorang, R.; Tong, C.F.; Nawawi, A.; Lai, J.-S.J. LCL filter design of a 50-kW 60-kHz SiC inverter with size and thermal considerations for aerospace applications. *IEEE Trans. Ind. Electron.* **2017**, *64*, 8321–8333. [CrossRef]
30. Jiao, Y.; Lee, F.C. LCL filter design and inductor current ripple analysis for a three-level NPC grid interface converter. *IEEE Trans Power Electron.* **2014**, *30*, 4659–4668. [CrossRef]
31. Liserre, M.; Blaabjerg, F.; Hansen, S. Design and control of an LCL-filter-based three-phase active rectifier. *IEEE Trans. Ind. Appl* **2005**, *41*, 1281–1291. [CrossRef]
32. Reznik, A.; Simões, M.G.; Al-Durra, A.; Muyeen, S. *LCL* filter design and performance analysis for grid-interconnected systems *IEEE Trans. Ind. Appl.* **2013**, *50*, 1225–1232. [CrossRef]
33. Wu, T.-F.; Misra, M.; Lin, L.-C.; Hsu, C.-W. An improved resonant frequency based systematic LCL filter design method for grid-connected inverter. *IEEE Trans. Ind. Electron.* **2017**, *64*, 6412–6421. [CrossRef]
34. Guan, Y.; Wang, Y.; Xie, Y.; Liang, Y.; Lin, A.; Wang, X. The dual-current control strategy of grid-connected inverter with LCL filter *IEEE Trans. Power Electron.* **2018**, *34*, 5940–5952. [CrossRef]
35. Prabaharan, N.; Palanisamy, K. Comparative analysis of symmetric and asymmetric reduced switch MLI topologies using unipolar pulse width modulation strategies. *IET Power Electron.* **2016**, *9*, 2808–2823. [CrossRef]
36. Sarker, R.; Datta, A.; Debnath, S. FPGA-based variable modulation-indexed-SPWM generator architecture for constant-output-voltage inverter applications. *Microprocess. Microsyst.* **2020**, *77*, 103123. [CrossRef]
37. Ahmed, M.; Orabi, M.; Ghoneim, S.S.; Al-Harthi, M.M.; Salem, F.A.; Alamri, B.; Mekhilef, S. General mathematical solution for selective harmonic elimination. *IEEE J. Emerg. Sel. Top. Power Electron.* **2019**, *8*, 4440–4456. [CrossRef]
38. Rai, N.; Chakravorty, S. Generalized formulations and solving techniques for selective harmonic elimination PWM strategy: A review. *J. Inst. Eng. (India) Ser. B* **2019**, *100*, 649–664. [CrossRef]
39. Ray, R.N.; Chatterjee, D.; Goswami, S.K. An application of PSO technique for harmonic elimination in a PWM inverter. *Appl. Soft Comput.* **2009**, *9*, 1315–1320. [CrossRef]
40. Wang, Q.; Hong, Z.; Deng, F.; Cheng, M.; Wu, Z.; Buja, G. A type of piecewise and modular energy storage topology achieved by dual carrier cross phase shift SPWM control. *IET Power Electron.* **2022**, *15*, 463–475. [CrossRef]
41. Liu, J.; Sun, Y.; Li, Y.; Fu, C. Theoretical harmonic analysis of cascaded H-bridge inverter under hybrid pulse width multilevel modulation. *IET Power Electron.* **2016**, *9*, 2714–2722. [CrossRef]

42. Memon, M.A.; Mekhilef, S.; Mubin, M.; Aamir, M. Selective harmonic elimination in inverters using bio-inspired intelligent algorithms for renewable energy conversion applications: A review. *Renew. Sustain. Energy Rev.* **2018**, *82*, 2235–2253. [CrossRef]
43. Etesami, M.H.; Vilathgamuwa, D.M.; Ghasemi, N.; Jovanovic, D.P. Enhanced metaheuristic methods for selective harmonic elimination technique. *IEEE Trans. Ind. Inform.* **2018**, *14*, 5210–5220. [CrossRef]
44. Mahlooji, M.H.; Mohammadi, H.R.; Rahimi, M. A review on modeling and control of grid-connected photovoltaic inverters with LCL filter. *Renew. Sustain. Energy Rev.* **2018**, *81*, 563–578. [CrossRef]
45. Said-Romdhane, M.B.; Naouar, M.W.; Belkhodja, I.S.; Monmasson, E. Simple and systematic LCL filter design for three-phase grid-connected power converters. *Math. Comput. Simul.* **2016**, *130*, 181–193. [CrossRef]
46. *IEEE Std 519-2022 (Revision of IEEE Std 519-2014)*; Standard for Harmonic Control in Electric Power Systems. IEEE Standard Association: Piscataway, NJ, USA, 2022.
47. Nagai, S.; Kusaka, K.; Itoh, J.-I. ZVRT capability of single-phase grid-connected inverter with high-speed gate-block and Minimized LCL filter design. *IEEE Trans. Ind. Appl.* **2018**, *54*, 5387–5399. [CrossRef]
48. Said-Romdhane, M.B.; Naouar, M.W.; Slama Belkhodja, I.; Monmasson, E. An improved LCL filter design in order to ensure stability without damping and despite large grid impedance variations. *Energies* **2017**, *10*, 336. [CrossRef]

technologies

MDPI

Review

A Review on the Nanofluids-PCMs Integrated Solutions for Solar Thermal Heat Transfer Enhancement Purposes

José Pereira *, Reinaldo Souza, António Moreira and Ana Moita

IN+ Center for Innovation, Technology and Policy Research, Instituto Superior Técnico, Universidade de Lisboa Avenida Rovisco Pais, 1049-001 Lisboa, Portugal; reinaldo.souza@tecnico.ulisboa.pt (R.S.); aluismoreira@tecnico.ulisboa.pt (A.M.); anamoita@tecnico.ulsiboa.pt (A.M.)

* Correspondence: sochapereira@tecnico.ulisboa.pt

Abstract: The current review offers a critical survey on published studies concerning the simultaneous use of PCMs and nanofluids for solar thermal energy storage and conversion processes. Also, the main thermophysical properties of PCMs and nanofluids are discussed in detail. On one hand the properties of these types of nanofluids are analyzed, as well as those of the general types of nanofluids, like the thermal conductivity and latent heat capacity. On the other hand, there are specific characteristics of PCMs like, for instance, the phase-change duration and the phase-change temperature. Moreover, the main improvement techniques in order for PCMs and nanofluids to be used in solar thermal applications are described in detail, including the inclusion of highly thermal conductive nanoparticles and other nanostructures in nano-enhanced PCMs and PCMs with extended surfaces, among others. Regarding those improvement techniques, it was found that, for instance nanofluids can enhance the thermal conductivity of the base fluids by up to 100%. In addition, it was also reported that the simultaneous use of PCMs and nanofluids enhances the overall, thermal and electrical efficiencies of solar thermal energy storage systems and photovoltaic-nano-enhanced PCM systems. Finally, the main limitations and guidelines are summarized for future research in the technological and research fields of nanofluids and PCMs.

Keywords: nanofluids; PCMs; heat transfer; thermal energy storage; photovoltaic/thermal systems

Citation: Pereira, J.; Souza, R.; Moreira, A.; Moita, A. A Review on the Nanofluids-PCMs Integrated Solutions for Solar Thermal Heat Transfer Enhancement Purposes. *Technologies* **2023**, *11*, 166. https://doi.org/10.3390/technologies11060166

Academic Editors: Jayanta Deb Mondol, Annamaria Buonomano and Biplab Das

Received: 1 November 2023
Revised: 18 November 2023
Accepted: 22 November 2023
Published: 24 November 2023

1. Introduction

The general usage of nanomaterials has been intensively studied in recent decades given that these materials have considerable potential for scientific and technological progress and offer some beneficial features that can be closely related to the exploration of nanomaterials in different applications. Nanotechnology can be implemented in distinct technological areas, including IT technologies, innovative materials, biomedicine, and heat transfer enhancement, among many others. In particular, the energy field is one of the most appealing application vectors of nanotechnology, as nanomaterials may lead to remarkable advancements in diverse engineering applications.

Nanoparticles and their applications have been one of the main research topics because of the numerous potential fields of implementation. This review is focused mainly on the energy-related field since the addition of nanoparticles to base fluids to compose nanofluids and to phase-change materials to compose nano-enhanced phase-change materials and the corresponding applications in thermal energy storage and conversion processes will be relevant.

The employment of binary nanofluids for solar absorption cooling was reported by the authors Nourafkan et al. [1]. Their work analyzed the photo-thermal conversion efficiency under laboratory conditions with the utilization of a solar simulator. Hybrid nanoparticles with 50% wt. of lithium bromide were dispersed in 50% wt. of water. The results confirmed that the addition of the nanoparticles led to light-trapping efficiency enhancements and consequently to an increase in the bulk temperature, ranging between 4.9% and 11.9%

Hence, the authors concluded that the nanoparticles used were very suitable for solar absorption cooling purposes.

In addition, the potential use of enhanced nanomaterials for thermoelectric generation was evaluated by the researchers Siddique et al. [2]. In their work, the Seebeck and Thomson effects of a nanostructured bulk alloy were determined with the addition of 0.1% vol. of silicon carbide nanoparticles, along with heat conduction capability and heat transfer loss into the surroundings. In view of the obtained results, it was confirmed that the inclusion of the silicon carbide nanoparticles increased the thermal efficiency to between 7.3% and 8.7%. Furthermore, the researchers Natividade et al. [3] studied the thermal efficiency of an evacuated tube solar collector system with a parabolic concentrator using a graphene nanofluid. The results demonstrated that the thermal efficiency of the solar collector increased from 31% to 76% with respect to that attained with the base fluid only.

Also, the usage of nanoparticles for photovoltaic/thermal purposes was investigated by the researchers Salem et al. [4]. In their work, the authors explored photovoltaic cell cooling with the enhancement of phase-change materials. In this direction, hybrid nanoparticles composed of an alumina phase-change material mixture were added to a water-based fluid in different concentration values. The obtained results revealed that it was possible to achieve a performance improvement in the delivered power output from the photovoltaic panel. The optimal concentration of the nanoparticles of 1% wt. provided the highest enhancement in the power output of the photovoltaic panel.

Also, the addition of silica nanoparticles to fatty acids was investigated by the researchers Martín et al. [5] to create a potential effective nano-enhanced phase-change material with increased thermal conductivity that is suitable for building applications. The experimental results confirmed that the addition of nanoparticles increased the thermal conductivity and heat transfer capability. The study also involved long-term performance experiments with cycling stability tests.

Additionally, the researchers Ren et al. [6] examined the phase-change material melting velocity for latent heat thermal energy storage, and the inclusion of nanoparticles was also analyzed. The results showed that the energy storage efficiency decreased with increasing volumetric concentration of nanoparticles, with the authors interpreting this result that way based on the enhanced viscosity and low energy storage capability of the phase-change material.

Balakin et al. [7] carried out an analysis of direct absorption collectors using the application of nanofluids. For this purpose, the authors conducted a computer fluid dynamics analysis in which the incorporation of nanoparticles induced an efficiency enhancement of the absorption collector by nearly 10%. In addition, the greatest collector performance enhancement was verified when magnetic nanoparticles were added, causing an increase in the efficiency of around 30%.

Bonab et al. [8] used aqueous copper oxide, aluminum oxide, and carbon nanotube nanofluids in a numeric simulation of a solar collector with absorbing pipes. The researchers reported that the thermal performance of the solar collector was enhanced with the incorporation of the nanoparticles in comparison to that of the one with water alone. The highest-efficiency enhancements were observed for concentrations inferior to 5% vol. in comparison to the ones obtained at higher concentrations.

Wang et al. [9] investigated the implementation of titanium nitride nanofluid as a plasmonic fluid in solar thermal applications. The obtained results showed that the titanium nitride nanofluid presented a superior photothermal conversion efficiency than those achieved with other commonly used nanofluids like, for example, carbon nanotubes or graphene, gold, or silver nanofluids.

A numerical analysis of the performance of gallium arsenide plasmonic solar cells with the addition of silver and gold nanoparticles was reported by the researchers Zhang et al. [10]. The research team verified that the incorporation of the nanoparticles enhanced the light absorption capability of the gallium arsenide solar cells. The highest increase in the optical–electrical conversion efficiency was achieved with the inclusion of gold nanoparticles. It was noticed that, with the proper optimization of the nanoparticle properties, it was possible to reduce the thermalization losses and increase the electrical performance.

An experimental work focusing on the analysis of different factors that affect the thermal performance of flat-plate solar collectors was presented by the researchers Zayed et al. [11], with an assumed application of nanofluids. The obtained results confirmed that the most adequate nanoparticles in terms of energy/exergy were carbon-based ones. The efficiency of the system increased from 6.3% to 37.3% with respect to that attained with conventional nanofluids with concentration values of up to 2% wt.

Abdelrazik et al. [12] carried out an analytical study on the performance amelioration of hybrid photovoltaic/thermal solar collectors. In this sense, the authors added nanoparticles above the photovoltaic module and on the back side of the photovoltaic panel, together with a nano-enhanced phase-change material. The numerical simulation analysis revealed that the overall performance of the photovoltaic/thermal system decreased by approximately 6.7% with the incorporation of the nanoparticles. Nonetheless, in the case of the optical filtration above the photovoltaic panel, the efficiency of the system was enhanced by between 6% and 12%.

Innovative emulsions of paraffin waxes and water were experimentally evaluated by the authors Agresti et al. [13] with concentrations ranging from 2% to 10%, and the stability over time of the emulsions was investigated. According to the results, the melting heat was reduced when compared to the base fluid; however, the thermal capacity improved by up to 40%.

The incorporation of organic nanoparticles in phase-change materials for thermal energy storage purposes was investigated by the authors Sheikholeslami et al. [14]. An external magnetic field was applied to the system, and in concentrations of up to 4% wt., a 14% decrease in the solidification time was verified. The application of the magnetic field over the nano-enhanced phase-change material was demonstrated to be a very effective option for considerably increasing the solidification in energy storage applications.

De Matteis et al. [15] examined nano-encapsulated phase-change materials as potential effective choices for thermal management purposes for residential buildings. The research team found that the use of a nano-encapsulated phase-change material could decrease the heating/cooling energy requirements in residential buildings to between 1% and nearly 4%. The toxicity of the proposed solution was also investigated, and no harmful effects were detected with respect to human exposure.

The previous briefly addressed research works proved the importance of nanotechnology in various engineering applications. The review presented herein mostly deals with the experimental research dealing with nano-enhanced fluid and nano-enhanced phase-change materials. Therefore, the current contribution investigates both types of materials. A detailed literature review of nanofluid phase-change materials integrating photovoltaic/thermal systems is carried out in the current review paper, which is anticipated to provide useful insights for the researcher community.

This review paper addresses advanced nanofluid phase-change materials integrated into photovoltaic/thermal systems and principles from recently reported works. In addition, the design, operation, and research findings of different types of photovoltaic/thermal systems are also addressed. Figure 1 schematically illustrates the overview methodology used in the present study.

The primary selection was obtained with respect to the targeted keywords. The subject area was limited to energy, science, and technology and to nanofluids and phase-change materials. Articles written in English with the year of publication between 2000 and 2023 were selected. After the primary selection, the scientific articles discussing nanofluids and phase-change materials were subjected to secondary selection, where the main objective was to group the papers considering the areas of renewable energy, heat transfer, and solar thermal energy storage and conversion.

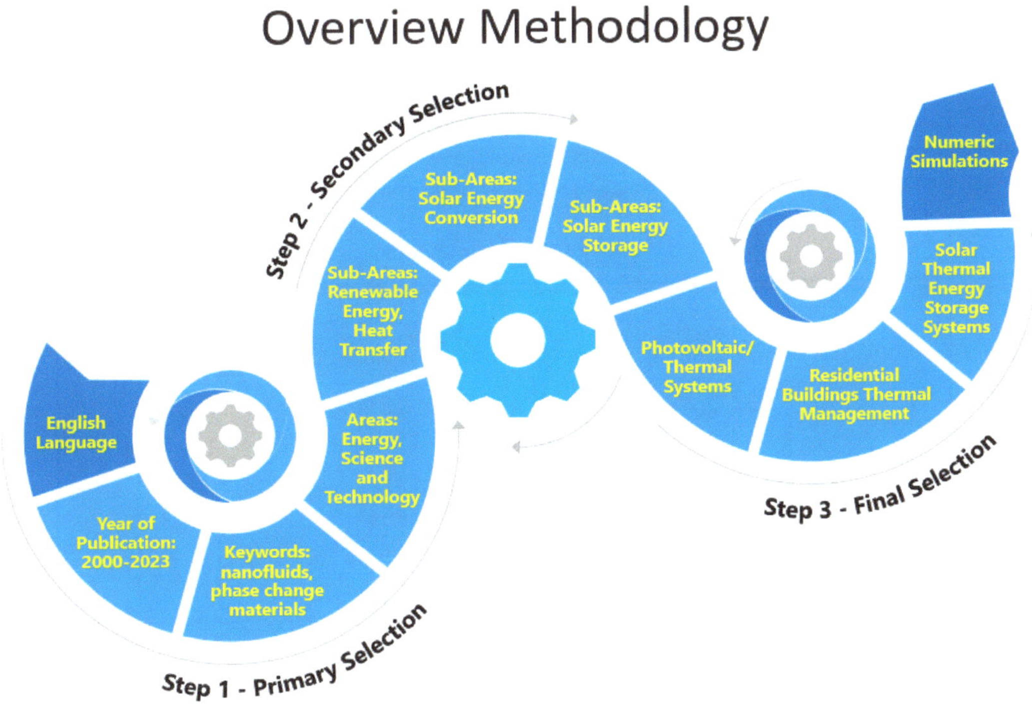

Figure 1. Schematic representation of the overview methodology followed in this work.

The final selection was carried out based on the combined exploration of the final solar thermal energy applications of nanofluids and phase-change materials. Therefore, it was not possible to describe each individual paper with respect to the specific applications, preparation techniques, and thermophysical property characterization. Instead, each article was classified with respect to the focus of its content considering the application, synthesis routes, and thermophysical characteristics. Depending on the specific selection criteria, some papers were addressed in multiple groups, based on the general content of the paper and the specific research findings.

The highest increase in interest regarding the released publications was in the period between 2015 and 2023 in the field of phase-change materials enhanced with nanoparticles in an amount superior to 100%.

The second highest increase in publications was in the case of energy storage applications in an amount of more than 90% from 2015 to 2023. In any case, the interest of the research community in nanotechnology applications is still progressing. The benefits related to the combined usage of nanofluids and phase-change materials and the addition of nanoparticles to these materials are remarkable and often closely related to the improvement of the performance of diverse systems, devices, and procedures. Nonetheless, the nature, size, and amount of the incorporated nanoparticles together with the preparation methods are the fundamental influencing factors. Since the potential implementation areas of nanoparticles are rather broad, this review focused on the simultaneous use of nanofluids and phase-change materials. Regarding the application area, the energy field was the focus of the review.

Also, the present work provides some guidelines for researchers dealing with phase-change materials, providing useful insights into the potential applications of combined nanofluids and phase-change materials. Figure 2 summarizes the main passive and active cooling techniques for the thermal management of photovoltaic/thermal systems.

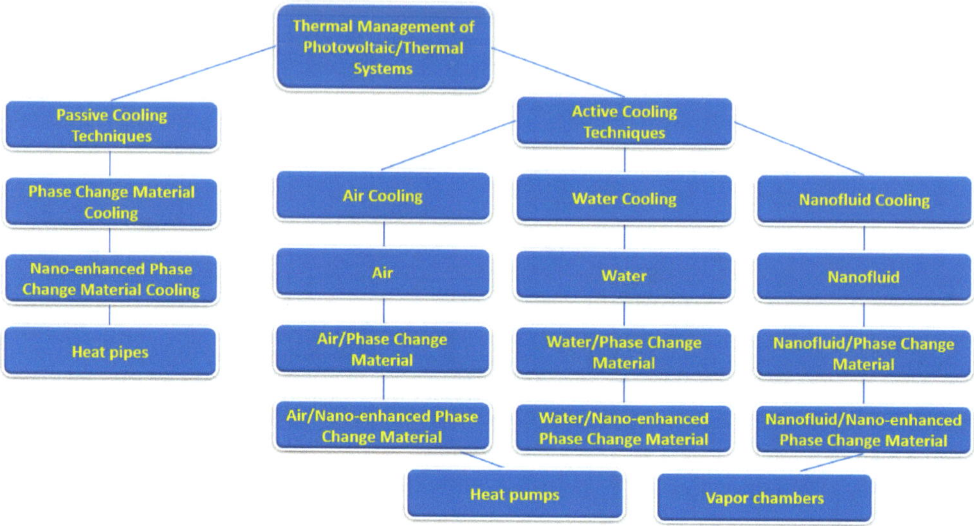

Figure 2. Main passive and active cooling techniques for photovoltaic/thermal systems.

2. Preparation Methodologies for Nanofluids

Nanofluids are normally employed as heat transfer fluids and thermal energy storage media. They also can be used as PCMs, and that is the case, for instance, for aqueous-based nanofluids. The nanofluids are synthesized through the dispersion of nanoparticles in a base fluid. First, a uniform and stable dispersion of nanoparticles is a critical issue concerning the stability over time of nanofluids. Consequently, there are several commonly followed strategies to improve the stability of nanofluids, such as the addition of surfactants, the application of strong ultrasonic force to break up the usual clustering of nanoparticles, and surface modification of nanoparticles. The one-step and two-step approaches are the main methods to produce nanofluids. Nonetheless, the emerging chemical processes are promising alternative methods of synthesis. In the one-step method, the procedures inherent to the drying, transportation, storage, and dispersion of nanoparticles are eliminated. For example, nanofluids can be synthesized via the physical vapor deposition technique in which evaporation and condensation are conducted directly in the base fluid. This methodology prepares a stable nanofluid with uniform nanoparticles with comparatively low levels of aggregation and sedimentation over time. Nonetheless, the fundamental problems associated with the one-step method are the existence of residual reactants left in the nanofluids and the relatively high investment cost. On the other hand, the cost-effective two-step method is very suitable for the large-scale production of nanofluids. In this method, nanoparticles are prepared according to distinct methods and, after that, are dispersed into the base fluids. The main associated limitation is the usual aggregation of the incorporated nanoparticles in the base fluid. To overcome such a drawback and to improve the stability over time of the nanofluid, the addition of a surfactant is often explored. Figure 3 schematically represents a typical two-step method to produce nanofluids.

The following paragraph briefly describes some examples of the possible nanofluid synthesis routes. For instance, the researchers Altohamy et al. [16] dispersed nanoparticles of gamma alumina in water and studied the cool storage behavior of this water-based nano-enhanced PCM. The nanoparticles were added in concentrations of 0.5% vol. to 2% vol. The mixing was conducted by means of a sonicator at 30 °C for 90 min. Moreover, the authors Shao et al. [17] investigated the solidification behavior of hybrid nanofluids with titanium oxide nanotubes and nanoplatelets. Hybrid nanofluids with different titanium nanoparticle percentages and titanium oxide concentrations of 0.1% wt. to 0.3% wt. were prepared. The nanofluids were based on deionized water that was mixed for 30 min with nanomaterials using firstly a magnetic stirrer and then an ultrasonic cleaner. It is not clear whether the 30 min period refers to a magnetic stirrer or an ultrasonic cleaner, nor is it indicated at which temperature, revolutions, and frequency the mixing was performed. The main research findings showed that different parameters affected the thermo-physical characteristics of the nanofluids and the mixing process. The size and shape of the nanoparticles were found to be critical, as was the sonication stage. It was also found that size and shape were mostly dependent on the preparation of the nanoparticles. The stability of the nanofluids was also affected by the surface charge, which was closely linked to the sonication step. It can be highlighted that the mixing process and the preparation of the nanoparticles were the most critical parameters and strongly affected the thermophysical properties of the nanofluids and their thermal stability and stability over time. Nanofluids present some disadvantages that have hindered their widespread use. Many of the discrepancies easily observed in the literature are related to differences in experimental results for the thermal conductivity of these mixtures [18]. An increase in thermal conductivity would be one of the main reasons for enhanced heat transfer in thermal systems. Although most researchers tend towards an increase in thermal conductivity with an increase in nanoparticle concentration, excess nanoparticles favor sedimentation and the formation of clusters in the base fluid. In thermal systems involving fluid flow, higher pumping energy may be required, especially in compact systems [19–21]. In these cases, in addition to stability-related issues, effects on viscosity and flow in channels with more complex geometries may suffer from clogging and pressure drops [22].

Figure 3. Typical two-step preparation method for nanofluids.

To overcome the problems related to long-term sedimentation, a series of techniques are employed, such as the use of stabilizing agents (surfactants), ultrasound, pH control, modification of the nanoparticle surface (such as size and shape), functionalization of the nanoparticles, and the use of different base fluids and concentrations of nanofluids in varying amounts. Surfactants act on the particle surface, preventing them from aggregating, which generally improves the wettability of the nanoparticles and reduces their surface tension [23].

The effect of nanoparticle size on nanofluids is still a highly controversial topic in the literature [21]. Whereas some authors argue that reducing the size of the nanoparticles increases the effective surface area for thermal exchange and intensifies the Brownian motion, resulting in an increase in thermal conductivity [24], other authors have reported a linear increase in conductivity with increasing particle size [18,25]. Ultrasound is part of the so-called physical or mechanical methods, during which nanofluids are subjected to vibrations for a predetermined period to homogenize the mixtures and break up agglomerated nanoparticles or clusters, as they are commonly known as [26].

With proper pH control, nanoparticle agglomeration can be avoided, as particle electric charges are related to the pH of the medium, allowing for the alteration of repulsive and attractive forces between nanoparticles [27,28]. The functionalization of nanoparticles in nanofluids is a process that involves modifying nanoparticles with specific functional groups [21,29]. These functional groups aim to improve the stability, dispersion, and properties of the nanofluid according to its application. Furthermore, such modifications can increase the affinity of nanoparticles with the base fluid, promoting a more uniform dispersion and preventing long-term sedimentation.

Although the use of nanofluids may pose challenges, especially regarding long-term sedimentation due to the gain in thermal conductivity, interest in techniques to enhance stability and pursue better thermal properties of these colloidal mixtures has been intensifying.

Regarding the direct comparison between the two main preparation methods, it is noteworthy to mention the experimental work conducted by the authors Mohammadpoor et al. [30], who synthesized a copper–ethylene glycol nanofluid using different methods. The researchers compared the stability and heat transfer properties of nanofluids prepared by the one-step method and the two-step method. They confirmed that the nanofluid prepared with the one-step prepared was more stable without the incorporation of any stabilizer. They also verified that the single step nanofluid increased thermal conductivity by around 21%, whereas the two-step nanofluid increased it by 39.4% at a concentration of 0.01% wt. In general, the two-step method is preferred for the preparation of oxide nanoparticle nanofluids, whereas the one-step method is preferred for metal nanoparticles [31].

3. General Types and Characteristics of PCMs

The capability of storing substantial quantities of latent energy of between 100 MJ/m^3 and 250 MJ/m^3 and energy densities of from 150 MJ/m^3 to 430 MJ/m^3 for organic and inorganic PCMs [32], respectively, makes them very promising alternatives for thermal energy storage purposes. The increased latent heat capacity is the primary factor for a PCM to be applied in thermal energy storage applications. PCMs operate in narrow phase-change temperature ranges, which makes the evaluation of their suitability based on the operating temperatures and corresponding energy storage capacities. Figure 4 summarizes the main types of PCMs.

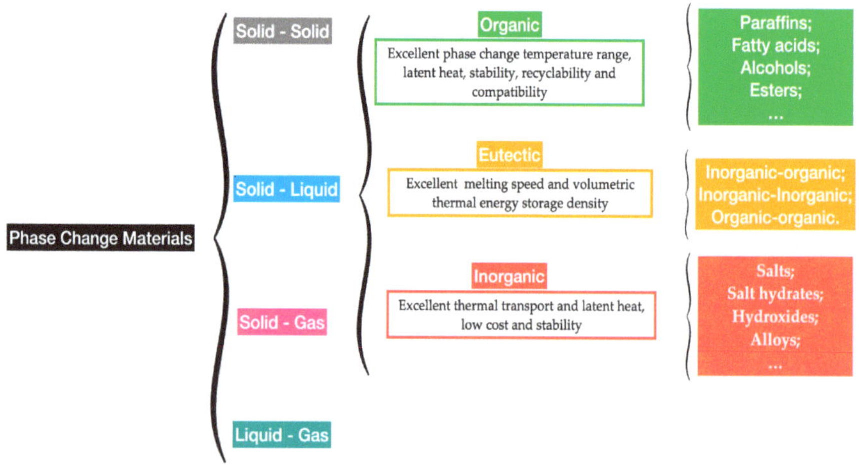

Figure 4. Main types of PCMs.

Nonetheless, PCMs possess relatively poor thermal conductivity, which is a limitation that negatively affects the charging rate of these materials. This thermal conductivity shortcoming is more noticeable in organic PCMs in comparison to inorganic PCMs [33]. Nevertheless, recent research efforts have been made to overcome such limitations by enhancing the thermal conductivity of PCMs. The transient hotwire method, transient plane-source method, laser flash method, and 3ω method are the most used thermal conductivity measurement techniques for PCMs. In the transient hotwire method, a hotwire is buried into solid, powder, or fluid samples, and the heat transfer process is initiated by a heating pulse from the hotwire to the sample. By monitoring the temperature response of the hotwire, the thermophysical parameters of the sample PCM can be rapidly acquired. Examples include multi-walled carbon nanotubes and graphene nanoparticles in stearic acid nano-enhanced PCMs [34]. Additionally, this method is also widely used for the measurement of PCM-based colloidal suspensions, including double-walled carbon nanotubes in palmitic acid [35] and carbon nanotubes in paraffin wax [36]. Nonetheless, the method poses some practical drawbacks, such as the damage inflicted on the internal structure of the samples, which is provoked by the insertion of the hotwire into them. Moreover, the transient plane-source method is also widely explored for the determination of the thermal conductivity of PCMs. The thermophysical features of the samples are detected through the heat flow from a plane-shape sensor. The sensor heats the samples and detects the heat flow rate, which is affected by the thermal transport ability of the sample PCMs. A relevant issue for the test structure is that both sides of the sensor must be in direct contact with the samples. In addition, the uniformity of the samples is also vital, given that the heating is one dimensional. For instance, the transient plane-source method measures the thermal conductivity of graphite nanoparticle nano-enhanced PCMs [37] and titanium oxide foam skeleton-reinforced PCMs [38]. As a practical restriction, this method makes it hard to determine with accuracy the thermal conductivity of uneven samples. Furthermore, the laser flash method is a non-contact method that heats PCM samples via a short energy pulse on one side of them. The thermophysical properties can be determined by measuring the temperature increase on the other side of the samples. On one hand, this method has the advantage of saving the sample preparation stages needed for contact measurement techniques, which may involve the risk of breaking up the internal structure of the samples. On the other hand, the method requires the use of uniformly opaque samples, small-width energy pulses, total absorption of the energy pulse by the illuminated side of the sample, and that there are only minor heat losses on the surface of the sample being analyzed. The laser flash method is mainly used to measure the thermal conductivity of bulk PCMs, such as, for example, aluminum–carbon nanoparticle paraffin wax PCMs [39] and expanded graphite–paraffin PCMs [40]. Additionally, the 3ω method uses a metal strip to heat the bulk material. The main beneficial feature is that a small sinusoidal current at angular frequency ω is fed into the metal strip, and the voltage response of the strip contains the first and third harmonics of the voltage, which improves the quality of the measurement of the thermophysical characteristics of the PCM. The 3ω method has the benefits of involving a minimized radiation effect and allowing for quick determination of the temperature dependence of the thermal conductivity over the steady-state method, although it usually takes more time than other measuring techniques. The 3ω method measures the thermal conductivity of PCMs in the form of nanopowders, bulk, coatings, fluids, and nanoparticles at an extended temperature range of between 10 K and 800 K. This method has also been employed, for instance, for the measurement of the thermal conductivity of adipic acid/sebacic acid–graphene nanoparticle PCMs [41] and multilayer film PCMs [42]. Moreover, the researchers Qiu et al. [43] proposed a freestanding sensor-based 3ω method with enhanced applicability to a broad range of dense and porous materials via non-destructive measurement and extraction of the thermal contact resistance. Urea formaldehyde paraffin PCM microcapsules were characterized with the freestanding sensor-based 3ω method. This measurement provided a reliable basis for the study of the thermal conduction mechanisms of PCM microcapsules. Figure 5 illustrates the main measuring methods for the thermal conductivity of PCMs. Figure 6 summarizes the fundamental strategies for the thermal conductivity enhancement of PCMs. Most PCMs perform poorly in latent thermal energy systems due to their low thermal conductivity. PCMs' low thermal conductivity leads to a lower heat transfer rate from heat transfer fluid to PCMs. In consequence, it reduces the energy storage and release capacity of the latent thermal energy storage system and increases the time required for complete melting and solidification processes as well. To overcome this challenge, researchers have introduced new techniques, namely, extended surface, composites, multiple PCMs, and encapsulation which are helpful for increasing the heat transfer rate between the heat transfer fluid and the PCM. As is known, during the melting process, convective heat transfer is dominant and conductive heat transfer vanishes as the process progresses, which is contrary to the solidification process. If fin configuration is used along with a PCM, it accelerates the melting and solidification processes by means of encouraging conductive heat transfer in both processes throughout. Numerical investigation on heat transfer enhancement of PCMs with a fin array system was studied for carrying out the effect of the fin pitch, the thickness of the module, the vertical/horizontal orientation of the modules, and the driving temperature gradient [44]. The obtained results indicated that a decrease in fin spacing caused a significant decrease in the amount of time required for complete melting for both horizontal and vertical modules. In addition, the authors Castell et al. [45] evaluated the heat transfer coefficient using external vertical fins and found that the vertical fins in PCM modules decreased the heat transfer coefficient more than the one attained with the PCM modules without any fins. But, in contrast, the exploration of vertical fins increased the heat transfer rate between the PCM and the heat transfer fluid because of the increased heat transfer area of the vertical fins. Also, the authors Nayak et al. [46] numerically evaluated the effectiveness of fins in improving the thermal conductivity of PCMs for cooling electronics. In view of the results, it was found that an increase in the number of fins used in the PCM increased the heat transfer rate. Nonetheless, beyond increasing a certain number of fins, there was no noticeable increase in the heat transfer rate. It was also found that rod-shaped fins performed better than plate-shaped fins, as they were able to maintain a better temperature distribution within the PCM. Moreover the researchers Zhao et al. [47] inferred the thermal performance of PCMs embedded with metal foams experimentally and numerically. Compared to the results of the pure PCM the effect of metal foams on the heat transfer rate was increased by up to 20 times and 10 times during the solidification and melting processes, respectively. Additionally, the heat transfer enhancement by metal beads was investigated for energy storage and release in single spheres during the solidification and melting of paraffin wax-enclosed spherical capsules [48]. The experiments were conducted as a function of the number and diameter of the metal beads. The combined thermal conductivity of the PCM and the metal beads increased significantly, thereby accelerating the heat transfer rate, which, in turn, provoked a decrease in the solidification and melting times. High-thermal-conductivity materials like graphite can be impregnated with PCMs with low thermal conductivity to achieve improved heat transfer features and phase-change performance of those PCMs. Furthermore the authors Sari and Karaipekli [49] prepared palmitic acid and an expandable graphite composite to form a stable PCM via vacuum impregnation. The thermal conductivity of the form-stable composite of the palmitic acid-expandable graphite was 2.5 times higher than that of the pure palmitic acid of 0.17 $Wm^{-1}K^{-1}$. Also, the researchers Rao and Zhang [50] prepared graphite–paraffin nano-enhanced PCMs and found that, with higher weight concentrations of graphite, there was a decrease in the latent heat storage capability. For this reason, it was necessary to find the ideal concentration of graphite that would induce the maximum enhancement in the thermal conductivity of the PCM. Moreover, the researchers Wang et al. [51] synthesized a stearic acid–polymethylmethacrylate PCM for latent heat thermal storage purposes through ultraviolet curing dispersion polymerization. The melting and freezing temperatures and the latent heats of the stable form of the stearic acid–polymethylmethacrylate PCM were determined to be 60.4 °C and 50.6 °C and

92.1 J/g, and 95.9 J/g, respectively. Based on the obtained results, the authors concluded that nano-enhanced PCMs can be successfully explored as latent heat storage materials for passive solar building applications. In the multiple-PCM approach, the PCMs are usually arranged in order of decreasing melting points, maintaining an almost constant temperature difference during the melting process, even though the heat transfer fluid temperature decreases in the flow direction [52]. This leads to an almost constant heat flux to the PCM. During discharge, if the heat transfer fluid flow direction is just opposite to charging, then the PCMs remain in increasing order of their melting points. The overall energy efficiency can be enhanced by using multiple PCMs rather than only one PCM in a photovoltaic/thermal system. This research work can also help with determining the required number of PCMs and their respective melting point temperatures to improve system efficiency in the practical design of thermal storage systems. In another research work, similar work was carried out for multiple PCMs with fins in a latent heat storage unit [53]. The authors summarized that the multiple-PCM model resulted in a considerable amount of stored energy in the form of latent heat compared to the single-PCM model. A nearly uniform outlet temperature of the heat transfer fluid was obtained with the multiple-PCM latent heat storage modules compared to the single-PCM module. Moreover, the authors Shaikh and Lafdi [54] performed a numerical analysis on a combined diffusion and convection heat transfer in varied configurations of composite slabs employing multiple PCMs. It was observed that the total energy charge rate could be considerably enhanced by using composite PCMs compared to a single PCM. The authors Cui et al. [55] performed a numeric simulation carried out on a solar receiver thermal storage module operating with three different PCMs and a single PCM separately, in which the maximum heat transfer temperature, fluid outlet temperature, and liquid PCM fraction of the total heat transfer tube were obtained and compared with those attained with the single PCM. Compared with the single PCM, the multiple PCMs enhanced the energy rate, improving the performance of the receiver and profoundly decreasing the fluctuation of the fluid outlet temperature. Furthermore, in the PCM encapsulation method, micro-sized PCM particles are enclosed in a sphere or cylinder. The PCM inside the capsule is designated as the core and the solid capsule as the shell. The shell can be made of a wide range of materials, including natural and synthetic polymers. The encapsulation of PCMs can be performed chemically through, for instance, co-acervation, and physically through, for instance, spray drying. The authors Hawlader et al. [56] prepared an encapsulated paraffin wax and inferred its phase-change behavior during energy storage and release processes. Both complex coacervation and spray-drying methods were employed to prepare the encapsulated paraffin particles. The microcapsules had high energy storage and release capacities in the range of 145 to 240 J/g. Also, polymethylmethacrylate microcapsules containing n-octacosane as a phase-change material for thermal energy storage were investigated [57]. The melting and freezing points and the latent heats of the microencapsulated octacosane as a PCM were found to be 50.6 °C and 53.2 °C and 86.4 J/g and −88.5 J/g, respectively. In addition, thermal gravimetric analysis indicated that the microencapsulated octacosane had good chemical stability. It was understood that the prepared microencapsulated octacosane has good energy storage potential. Though the microencapsulated PCMs exhibited good energy storage and release capacities, it was not able to perform well under repeated cycling, even though the large particles of the microencapsulated PCM not only increased the viscosity of the fluid but were also often crushed during pumping. For this reason, it is necessary to prepare nano-encapsulated PCMs with smaller nanoparticle sizes of up to 150 nm.

Also, organic PCMs exhibit high flammability, tend to leak in the course of the melting process, and have low volumetric heat storage capabilities [58]. On the other hand, inorganic PCMs exhibit enhanced volumetric heat storage capacities [59], with metals and salt hydrates demonstrating improved thermal conductivity and latent heat capacities, making inorganic PCMs a very suitable choice, especially for applications in solar thermal energy systems and devices [60]. Nevertheless, these materials often present supercooling and incongruent phase transitions. Figure 7 summarizes the fundamental beneficial features of PCMs that determine their selection for specific applications.

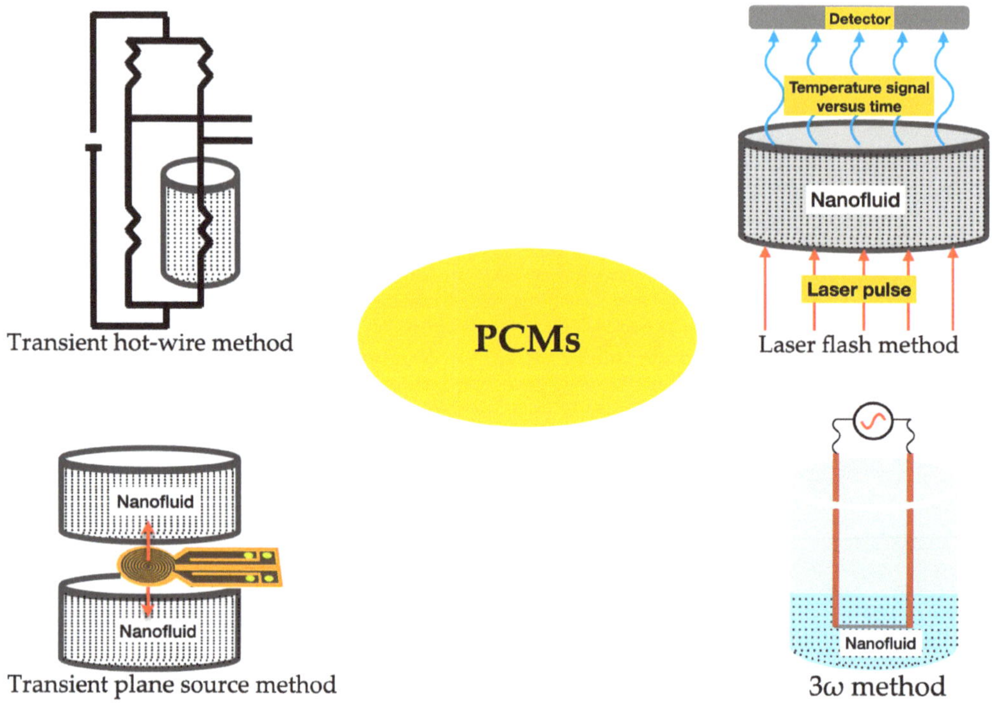

Figure 5. Main methods of measuring the thermal conductivity of PCMs.

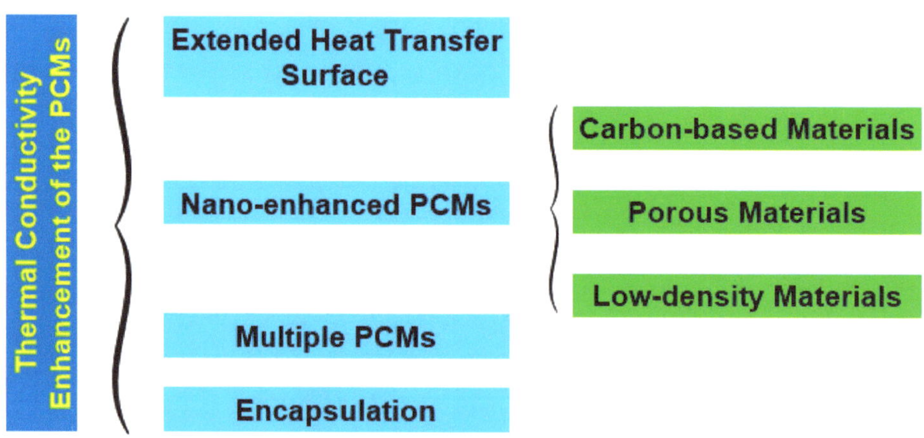

Figure 6. Main strategies for the enhancement of the thermal conductivity of PCMs.

Still concerning the beneficial features of PCMs, recently published research studies attempted to produce environmentally friendly PCMs [58]. These green PCMs are derived from bio-degradable and eco-friendly sources, including soya-oil, palm oil, and coconut oil [61]. The published research efforts in this field include the studies by the authors Boussaba et al. [62], who revealed that the exploration of palm kernel vegetable oil for thermal energy management of residential buildings is feasible. The use of coconut oil was also suggested in [63] for air temperature regulation. Bio-based PCMs are highly promising yet are represented less in the literature compared to paraffin-based PCMs, and consequently, further studies on the exploration of eco-friendly PCMs for thermal energy storage applications are welcome. The solid–liquid type of PCMs is greatly employed in thermal energy storage systems because of their enhanced latent heat capacity and facile reversibility to either state during the charge and discharge processes [64]. Organic PCMs exhibit desirable latent heat of fusion, congruent fusion, thermal stability, minimal supercooling, recyclability, low toxicity, and thermal insulation due to their low thermal conductivity [65]. Organic PCMs are also compatible with the building materials and polymeric materials of latent heat storage tanks. Paraffins are organic PCMs with the general chemical formula C_xH_{2x+2}. They have a volumetric latent heat capacity of 179 MJ/m^3 and a thermal conductivity of 0.2 W/m °C [60]. Non-paraffin PCMs include esters, glycols, fatty acids, alcohols, and other organic PCMs. Inorganic PCMs include water, salts, hydrated salts, and metallic alloys in association with additives. The general form of salts is A_xB_y and the general form of hydrated salts is $A_uB_v \cdot n(H_2O)$, and they are widely studied for thermal energy storage purposes. In addition, inorganic PCMs often present significant heat of fusion per unit volume; better thermal conductivity than that of their organic counterparts, being twice that of paraffin-based PCMs; minimal volume change during phase change; non-flammability; compatibility with polymeric materials; and cost-effectiveness, making them very attractive options for thermal energy storage applications. The main problems associated with the use of inorganic PCMs are their supercooling effect, poor thermal stability, comparatively poor nucleation, and segregated phase change [66]. Also, it should be stated that the metallic alloy PCMs require further research work, since their elevated weight may hinder their applicability in thermal energy storage purposes [64]. On the other hand, eutectic mixtures are obtained from mixing two or more compounds, either organic or inorganic or both, such that they take on a suitable operating temperature [64]. The main benefits of eutectic mixtures comes from their high phase temperature for the intended application [59]. Eutectic mixtures also demonstrate increased enthalpy, do not undergo supercooling, and have congruent phase transition, unlike inorganic PCMs. Nonetheless, these mixtures are costly, at two or three times the price of their individual constituent PCMs. Hence, further studies should be conducted to overcome this price limitation, and the challenges of cost-effective feasibility and scalability of the exploration of eutectic mixture PCMs in energy storage applications should be better understood. It should be noted that the published literature in the field also covers the solid–gas and solid–solid classes. That is the case, for example, of the study performed by the authors Das et al. [67], who studied the application of liquid–gas PCMs in photovoltaic/thermal solar energy systems. Moreover, the researchers Kong et al. [68] studied the use of solid–solid PCMs for thermal energy storage purposes. The investigated PCMs exhibited a latent heat of 111.7 J/g and improved thermal stability. Also, the authors Du et al. [69] applied flame-retardant solid–solid phase-change composite nanosheets in solar thermal energy storage technology, resulting in enhancements of 88.5% and 69.4% in the energy efficiency and thermal conductivity, respectively. The current review pays special attention to the solid–liquid PCMs that absorb considerable thermal energy amounts during the melting process (charging stage) and release this energy during the solidification process when a cold inlet airstream is applied over it (discharging stage) [70]. Such practical characteristics are very suitable for thermal applications using the temperature changes between day and night. Many studies are completely focused on the general exploration of solid–liquid PCMs because of their improved properties in practical applications. Nonetheless, the gap still remains in overcoming their fundamental problems [71], hindering their best thermal performance potential in thermal energy storage systems.

Figure 7. Fundamental benefits of PCMs.

4. Thermophysical Properties of Nanofluids and PCMs

4.1. Thermal Conductivity

The thermal conductivity of nanofluids depends on various factors like the type, morphology, and concentration of the added nanoparticles; the type of base fluid; the addition of surfactants; and the operating temperature. Nonetheless, the recent findings about their contribution to the effective thermal conductivity of nanofluids are inconsistent and often persist with some discrepancies and somewhat unexpected results. For example, the researchers Philip et al. [72] investigated thermal conductivity changes with a volumetric concentration of oleic acid-coated magnetite nanoparticles dispersed in kerosene-based fluid. No thermal conductivity increase was reported up to a concentration value of 1.7% vol. Beyond this value, the thermal conductivity of the developed nanofluid changed linearly with the concentration of nanoparticles. The authors stated that the thermal conductivity remained unchanged at low concentrations because of the uniformly dispersed nanoparticles, whereas at higher concentrations the existing nanoparticle clusters in the kerosene could have been the responsible for the verified thermal conductivity increase.

Chon et al. [73] evaluated the impact of nanoparticles added in the sizes of 11 nm, 47 nm, and 150 nm on the thermal conductivity of alumina aqueous nanofluids. The authors observed a greater thermal conductivity enhancement when they added the smaller nanoparticles of 11 nm. This was interpreted based on the enhanced Brownian motion of the smaller nanoparticles.

Beck et al. [74] found that the thermal conductivity of the nanofluids increased with increasing nanoparticle size. The research team argued that the verified thermal conductivity decrease for the smaller particles was caused by the enhanced phonon scattering at the interface between the nanoparticles and the base fluid. Indeed, when dealing with nanoparticle based fluid suspensions, the actual size of the nanoparticles is their hydrodynamic size because of the contribution from the ordered liquid layer around the nanoparticles and the surfactant molecules around the surface of the dispersed nanoparticles.

Jeong et al. [75] studied the thermal conductivity of aqueous nanofluids containing zinc oxide nanoparticles with different morphologies and in different concentrations. The obtained results revealed a 12% increase in the thermal conductivity of the nanofluids with spherical nanoparticles, whereas the increase was of 18% when the authors added rectangular zinc oxide nanoparticles to water.

Murshed et al. [76] observed an increase in thermal conductivity of 17.5% with respect to that of ethylene glycol alone in the nanofluid composed of 5% vol. titanium oxide spherical nanoparticles dispersed in ethylene glycol-based fluid. The increase was of 20% when the same concentration of titanium oxide cylindrical nanoparticles was added. Also, the authors Zhu et al. [77] verified a higher thermal conductivity enhancement of 38% for iron oxide nanofluids than for titanium oxide, copper oxide, and alumina nanofluids (30% increase), though the thermal conductivity of the iron oxide of 7 W/mK was lower than that of the titanium oxide (11.7 W/mK), copper oxide (20 W/mK), and alumina (36 W/mK). The observed enhancement for Fe_3O_4-based nanofluid was attributed to the alignment of the nanoparticles as clusters, where the thermal conductivity increased with increasing concentration because of the increased length of the aligned particles.

Hong et al. [78] reported a higher thermal conductivity increase for iron nanofluids than for copper nanofluids, even though the bulk iron had a lower thermal conductivity of 80 W/mK than the bulk copper, of 384 W/mK. In fact, the iron nanofluids demonstrated an 18% increase in the thermal conductivity at a 0.55% vol. of nanoparticles, whereas the copper nanofluid at the same concentration value exhibited a 14% increase.

The last two studies demonstrated that, contrary to what was expected, the incorporation of highly thermal conductive nanoparticles is not always the best choice for enhancing the thermal conductivity of nanofluids. Instead, the wetting of the surface of the nanoparticles and the heat transfer capability at the solid–liquid interface are usually determining factors for the thermal conductivity enhancement verified in the nanofluids.

On the other hand, the researchers Dadwal et al. [79] examined the influence of the size of the nanoparticles on the thermal conductivity of magnetite nanoparticles suspended in kerosene and toluene. The researchers reported distinct tendencies of the thermal conductivity evolution with the size of the nanoparticles in the two base fluids, given that a thermal conductivity increase with increasing nanoparticle size was observed for the kerosene, whereas, with toluene as base fluid, the smaller nanoparticles promoted a thermal conductivity increase. The authors argued that the different evolutions of the size-dependent thermal conductivity could have been caused by the different nanoparticle-based-fluid interactions. The interaction between the nanoparticles and the molecules of the base fluid determines the thickness of the interfacial layer. This interaction is different for different base fluids, and it fundamentally decides the density of the nanolayer at the solid–liquid interface, which in turn affects the heat transfer performance at this interface.

Accordingly, the authors Kamalvand et al. [80] analyzed the enhancement in thermal conductivity of nanofluids with the adsorption of the molecules of the solvent onto the nanoparticles at distinct temperature values. The investigation team found a linear increase in thermal conductivity with the adsorption of the base fluid molecules onto the surface of the dispersed nanoparticles. Also, the authors confirmed that the adsorption level of the molecules of the base liquid increased with increasing bulk temperature.

Altan et al. [81] investigated the impact of the base fluid on the thermal conductivity of iron oxide nanoparticles coated with capric acid and oleic acid surfactants and suspended in hexane, heptane, and mineral oil. The research team verified a greater thermal conductivity increase when using hexane as the base fluid, followed by heptane and mineral oil. The researchers stated that the greater increase was not directly linked to the low thermal conductivity of the base fluid, and that, instead, the interactions between the base fluid and the added surfactant at the solid–liquid interface could be the main reason behind the thermal conductivity enhancement.

Shao et al. [17] evaluated an aqueous hybrid nanofluid containing nanoparticles of titanium oxide under the form of nanotubes of 9 nm to 10 nm and nanoplatelets of 50 nm to 80 nm. It was confirmed that the nanofluid at 0.1% wt. of titanium oxide with 25% of titanium nanoparticles showed the greatest enhancement in thermal conductivity, of 22.3%, in comparison to that of the water-based fluid. The impact of the super-cooling degree of the hybrid nanofluids decreased by up to 4.97 ± 0.2 °C and 5.27 ± 0.2 °C respectively, compared to mono nanofluids. The freezing time for the hybrid nanofluid was also diminished by up to 54.9% in comparison to the titanium nanoplatelet nanofluid and by 56.4% compared with the nanofluid with titanium nanoparticles. Concerning the thermal conductivity of the phase-change materials with the incorporation of nanoparticles, it can be stated that most of the available papers on the subject focus on thermal conductivity enhancement as a direct effect of the addition of nanomaterials to phase-change materials.

The only exception was the study conducted by Colla et al. [82], where the addition of alumina nanoparticles to paraffin caused a 7–8% lower thermal conductivity compared with that of pure paraffin. There is no clear explanation for this phenomenon in the mentioned study. Generally, due to the improved thermal conductivity of nanocomposites and nanofluids, a reduction in heating, melting, and solidification times can be expected. Usually, the expected improvement in thermal conductivity is in the range of 20–100%, but there are examples of even higher improvements being reported—for instance, in the case of the lauric acid phase-change nanocomposite with graphene nanoplatelet [83], carbon-based phase-change material [84], or magnesium chloride hexahydrate phase-change material composites [85]. Generally, it can be concluded that an increase in the concentration of nanomaterials dispersed in the phase-change materials increases the thermal conductivity. This may be misleading, since there is a limit to the amount of nanomaterial in the phase-change material—i.e., nanomaterials in larger fractions tend to agglomerate. Agglomeration may lead to a decrease or increase in thermal conductivity. It is hard to predict when agglomeration will occur since it depends on many parameters, such as the mass or volume fraction, size, and morphology of the nanomaterials, among others. Furthermore, the type, phase, and temperature of the phase-change material may also have a significant influence on this phenomenon.

Sharma et al. [86] prepared a nano-enhanced phase-change material composed of palmitic acid and nanoparticles of titanium oxide. The authors stated that the increase in the mass fraction of the nanoparticles causes the nano-enhanced phase-change material thermal conductivity to increase. A maximum increase in thermal conductivity of about 80% was recorded for 5% wt. loading of nanoparticles. For the same concentration of nanoparticles, there was a 15.5% decrease in latent heat of fusion with respect to the pure phase-change material.

Mayilvelnathan and Arasu [87] reported on the thermal conductivity measured by the laser flash method of pure erythritol phase-change material, i.e., erythritol with 0.1%, 0.5%, and 1% of graphene nanoparticles before as well as after thermal-cycling periods. The thermal conductivity of pure erythritol was 0.733 W/mK, whereas, when considering the erythritol with 0.1% wt., 0.5% wt., and 1% wt. of graphene nanoparticles, the thermal conductivity increased to 1.074, 1.095, and 1.122 W/mK. The thermal conductivity was 0.692, 0.899, 0.921, and 1.020 W/mK for pure erythritol without and with 0.1% wt., 0.5% wt., and 1% wt. of graphene nanoparticles, respectively. Additionally, the authors Putra et al. [88] studied the thermophysical characteristics of nano-enhanced phase-change materials based on Rubitherm paraffin wax, RT22 HC, and graphene nanoplatelets. The thermal conductivity of the phase-change material was increased by around 90% with the addition of 0.3% wt. of graphene.

Nourani et al. [89] modified paraffin with different mass fractions of alumina nanoparticles. The authors in the article claim that the character of the relationship between the effective thermal conductivity and increasing the alumina mass fraction was nonlinear for both the solid and the liquid phase. The enhancement ratio of the effective thermal conductivity was 31% in the solid phase and 13% in the liquid phase for the nano-enhanced phase-change material at 10% wt. of alumina. A reduction of 27% of the heating and melting times was also detected for the same concentration of nanoparticles in the phase-change material.

Wang et al. [90] produced a stable OP10E and water emulsion enhanced by the inclusion of graphite nanoparticles. The authors argued that the supercooling effect of the emulsion could be prevented by the addition of graphite nanoparticles with a concentration superior to 2% wt. Furthermore, it was stated that the addition of graphite nanoparticles to the OP10E/water emulsion did not affect the latent heat. In the same study, it was found that 2% wt. of graphite nanoparticles resulted in a thermal conductivity increase of 88.9%.

4.2. Specific Heat

The specific heat of nanofluids depends primarily on the type, morphology, and concentration of the incorporated nanoparticles in the base fluid. Considering water as a base fluid, the published studies confirmed a considerable reduction in the specific heat of the water-based nanofluids, with the fundamental influencing factor for that to happen being the concentration of the nanoparticles added to the water. In this scope, the authors Zhou and Ni [91] prepared an aqueous alumina nanofluid at different concentrations ranging from 1.4% vol. to 21.7% vol. The differential scanning calorimetry technique was used to measure the specific heat of the nanofluids in a temperature range of between 20 °C and 45 °C. The research team observed decreases in the specific heat capacity of between 6% and 45% at concentrations of 1.4% vol. and 21.7% vol., respectively. Several other studies have investigated aqueous nanofluids with silica particles with dimensions of 32 nm [92] and 20 nm [93] and copper oxide particles with dimensions of 30 nm [94] and 23–37 nm [95] at various particle concentrations. These studies reported large decreases in the specific heat value with increasing nanoparticle concentrations. These lower specific heat values could have been a result of agglomerated nanoparticles in these samples, but a comprehensive characterization was reported. Like water-based nanofluids, ethylene glycol and ethylene glycol–water nanofluids showed decreases in specific heat from the values of pristine fluid. Although ethylene glycol had a lower specific heat value than water, it was still much higher than that of the particle materials. Thus, the proper trend of decreasing specific heat at higher particle concentrations was upheld in these studies. Additionally, the researchers Teng et al. [96] evaluated the specific heat of multi-walled carbon nanotubes dispersed in a mixture of ethylene glycol and water nanofluid at concentrations of 0.1% wt., 0.2% wt., and 0.4 wt.%, employing chitosan as a surfactant. The authors found that the inclusion of the multi-walled carbon nanotubes reduced the specific heat capacity of the base fluid mixture by between 2% and 8%, with the highest reduction shown by the most concentrated nanofluid. Furthermore, the researchers De Robertis et al. [97] analyzed the specific heat of a copper–ethylene glycol nanofluid at a 0.5% wt. of nanoparticles and at two different pH values. The pH of the primary nanofluid was 2.2, and a solution of sodium hydroxide was used to increase the pH to 10. The specific heat measurement revealed a decrease in specific heat in the nanofluids compared to the ethylene glycol-based fluid. The experiments showed a decrease of around 4% for the copper–ethylene glycol with the pH adjusted to 2.2 and a 12% decrease when the pH was adjusted to 10. It is likely that an increase in the pH moved the solution around the isoelectric point, perhaps resulting in an agglomeration of the nanoparticles in this case. From these available studies, it can be concluded that water and ethylene glycol nanofluids exhibit a decrease in specific heat with an increase in particle concentration and an increase in specific heat with an increase in temperature. In a more comprehensive pursuit of enhanced nanofluid-specific heat, a study by Starace et al. [98] explored various nanoparticle and base fluid combinations at different concentrations. The study used 36 different nanofluids with negligible changes in the heat capacity with respect to the base fluid. The authors used nanoparticles of silica, fumed silica, alumina needles, aluminum nitride, exfoliated graphite, iron core/iron oxide shell, bismuth, and mesoporous silica. The nanoparticles were dispersed in mineral oil, poly-α olefin, ethylene glycol, 60/40 by mass ethylene glycol/water, and calcium nitrate tetrahydrate. As some nanoparticles were suspended easily in some fluids and others required the use of surfactants, the researchers defined specific mixing procedures for each combination. The results showed only small changes in the specific heat considering the overall scatter of measured specific heats over the temperature range investigated. It was suggested that a change in overall specific heat could be possible only if a rearrangement of the base fluid molecules to a higher heat capacity configuration was caused by the nanoparticles. Molecular dynamic simulations of copper nanoparticles in ethylene glycol [99] have shown an ordering layer of 0.75 nm thick around the nanoparticle; however, the specific heat of this predicted ordered layer has not been studied. Despite the high number of available experimental methods for evaluating the thermal transport of PCMs during the phase change process, numerical simulations are also powerful tools for the optimized design of PCMs. Numeric simulations can accurately describe the phase-change heat transfer process by exploring diverse models like the energy balance method, finite element simulation, or commercially available computational fluid dynamics (CFD) software such as Ansys Fluent or COMSOL Multiphysics with Heat Transfer Module. The structural and operational parameters after optimization can provide guidance for the thermal enhancement of PCMs and thermal energy storage systems containing PCMs. As the most common analytical methods for solving engineering problems, the energy balance method and the enthalpy method are widely used to simulate the thermal transport behavior of PCMs during the solidification/melting stages. To simulate the thermal transport of PCMs in thermal energy storage systems more efficiently, the finite element method and the CFD modeling method are introduced, which can easily accomplish the simulation. The energy balance method is usually used to study the flow of energy based on energy conservation and to analyze the thermal enhancement effect in PCMs. The energy equation for a system without energy generation can be given by Equation (1):

$$E_{st} = E_{in} - E_{out} \tag{1}$$

where, in the cases of solar energy modules operating with PCMs, E_{st} is the energy transfer and energy retention between the different modules of the system, E_{in} is the received solar energy, and E_{out} is the energy loss of the system. When E_{st} of the PCMs needs to be considered, the energy equation between the PCMs and other parts should be used. Based on the flow and change in energy, the thermal enhancement effect can be reliably addressed and regulated, which further improves the thermal energy storage ability of PCMs. The enthalpy method is based on the enthalpy modifications of the different modules of a solar thermal energy system. The fundamental benefit of the enthalpy method is closely related to the capability of avoiding direct front tracking when dealing with materials that melt or solidify over a range of temperatures, which enable numeric simulation using complex geometries. The enthalpy equation for the solidification and melting processes can be expressed by Equation (2):

$$\frac{\partial(\rho H)}{\partial \tau} + \nabla\cdot(\rho v H) = \nabla\cdot(k\nabla T) + S \tag{2}$$

where H is the total enthalpy, ρ is the density of the operating fluid, v is the velocity of the operating fluid, and S is the source term defined by the condition of the sample. The total enthalpy can be determined by Equation (3):

$$H = h + \Delta H \tag{3}$$

where h is the sensible heat enthalpy and ΔH is the latent heat enthalpy. The sensible heat enthalpy is expressed by Equation (4):

$$h = h_{ref} + \int_{T_{ref}}^{T} C_p dT \tag{4}$$

where C_p is the specific heat capacity of the operating fluid and H_{ref} and T_{ref} are reference values. The initial condition $T_i = T(x,y,z,\tau_0)$ and boundary conditions like $k \cdot \partial T/\partial n = q_0$ and $-k \cdot \partial T/\partial n = h_s(T - T_o)$ should be used in the numeric simulation, where q_0 is the heat flux, h_s is the air surface heat transfer coefficient, and T_o is the ambient temperature. The governing equation of the enthalpy method is similar to the single-phase equation, providing an easier solution to the phase-change problems. Also, in the enthalpy method, there are no conditions to be satisfied at the solid–liquid interface, and this method yields a solution involving both solid and liquid phases [100]. For example, the researchers Han et al. studied the thermal energy storage and release processes in multi-cavity-microstructured PCM microcapsules and found that the charge and discharge velocity of PCMs were accelerated by increasing the number of cavities and cavity interlayers in the microcapsules [101]. The feature is that the enthalpy method possesses great suitability to be applied in the field of engineering thermal energy storage systems, providing a reliable evaluation for engineering applications of PCMs. The finite element simulation method subdivides a large phase-change energy storage system into smaller and simpler elements and simulates them separately, which is called the discretization strategy. There are many types of discretization strategies, such as the h-version and the p-version. Both of these versions are numerical methods for solving partial differential equations. Their main difference is that the polynomial degrees of the elements are fixed and the mesh is refined in the h-version, whereas the finite element mesh is fixed and the polynomial degrees of elements are enhanced in the p-version [102]. After the simulation, the subdivided elements are then assembled into a larger system, which makes it easier to model the thermal transport problem during the phase-change process. The finite element simulation method offers many options for regulating the complexity of both modeling and analysis in a heat transfer system. At the same time, it can also balance the necessary accuracy and computational processing time that can assess many of the concerns associated with engineering applications. The thermal performance of PCMs, as one of the most studied research topics in the technological field, can be reliably described via finite element simulations [103]. For instance, the researchers Wang et al. [104] studied the impact of the porosity and thermal conductivity of a metal foam on the thermal conductivity and melting of PCMs. The optimum fin structure was attained, which considerably increased the rate of solidification of the PCMs, opposite to the velocity increase via the action of the dispersed copper nanoparticles studied by the authors Lohrasbi et al. [105]. The finite element simulation method provides accurate prediction of the thermal enhancement effect for micro-/nano-PCMs. To achieve a highly efficient finite element simulation, CFD modeling was developed to analyze engineering situations with inherent complex problems, such as the solidification and melting processes in PCMs for solar thermal purposes. This method also involves fluid mechanics such as turbulence models and two-phase flow, which focus on the flow of gas and liquid and can resolve the intricate restriction of the fluid flows in solar thermal research and technology, like the flow of heat transfer fluids at various flow rates. Therefore, to analyze the thermal enhancement during the solidification process of PCMs or the heat exchange process of colloidal suspensions, CFD is undoubtedly an ideal numerical simulation method. Moreover, the researchers Mahdi et al. [106] combined nanoparticles with metal foam or metal fins in a thermal energy storage system containing triplex-tube PCMs and confirmed the thermal enhancement provoked by this solution for the improvement of the thermal energy storage system. Further work on PCM thermal energy storage analysis will undoubtedly require the development of CFD modeling with enhanced computing processing capacity. Also, to seize the internal and external thermal energy exchange of PCMs or PCMs operating in thermal energy storage systems, the combined exploration of experimental findings and numeric simulations is a vital tendency that should be maintained. For thermal conductivity and the drastic augmentation of the latent heat of PCMs or the use of the thermal energy in a thermal energy storage system, this combination provides accurate data and reliable evidence for the optimized design of thermal energy storage systems using PCMs. Figure 8 summarizes the fundamental formulations for the numerical simulations involving PCMs and nano-enhanced PCMs.

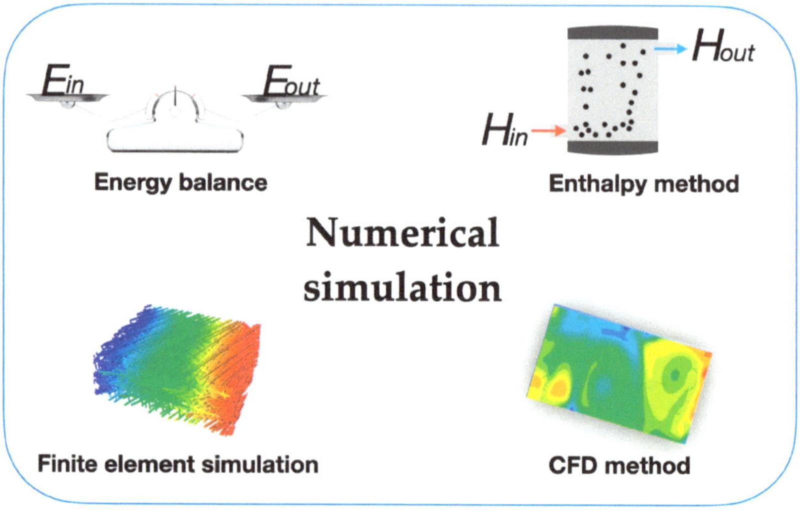

Figure 8. Main formulations for numerical simulations involving PCMs.

Moreover, the authors Starace et al. [98] also referred to many experimental works investigating the ordered layer of the water around proteins and hydrophobic molecules where the water molecules are almost immobile in that layer, providing an increase in the specific heat of the system proportional to the volume of the ordered layer interface Hence, the authors concluded that, to have an increase in the specific heat of nanofluids the layered structure should have a specific heat many times higher than that of the base fluid. This ordered interfacial structure will be discussed in more detail in the next section. Furthermore, the authors Shin and Banerjee [107] and Tiznobaik and Shin [108] investigated the binary eutectic salt Li_2CO_3-K_2CO_3 (62:38) with the addition of 1% wt. of silica nanoparticles in different sizes. Their experimental works reported enhancements of around 25% in the specific heat regardless of the size of the nanoparticles. In addition the obtained TEM images of the nano-enhanced salts after a melting and freezing process showed the formation of needle-like nanostructures in the salt, which could be the main reason behind the improved specific heat. Nonetheless, these special nanostructures within the molten salt were not observed in other published studies. Moreover, the researchers Dudda and Shin [109] investigated the binary eutectic salt $NaNO_3$-KNO_3 (60:40), commonly designated as solar salt given its suitability for commercial solar thermal power plants as a thermal energy storage medium, with the incorporation of silica nanoparticles. In their first study, the authors added 1% wt. of silica nanoparticles in two different sizes and confirmed a 19% enhancement in the specific heat for the 5 nm nanoparticles and a 25% increase when adding 30 nm nanoparticles. In their subsequent experimental work [110], the researchers added the same fraction of silica nanoparticles and found enhancements in the specific heat of 8%, 12%, 19%, and 27% with nanoparticles in sizes of 5 nm, 10 nm, 30 nm, and 60 nm respectively. Furthermore, the authors Lu and Huang [111] evaluated the specific heat of molten $NaNO_3$-KNO_3 (60:40) salt containing alumina nanoparticles 9 nm and 13 nm in size. The concentrations used were superior to 1% wt., and a decrease in the specific heat capacity of the nanofluid was found. Also, it was verified by the authors that the specific heat decreased with increasing concentration and size of the nanoparticles. Additionally the researchers Chieruzzi et al. [112] studied $NaNO_3$-KNO_3 (60:40) salt with the addition of alumina nanoparticles at concentrations of between 0.5% wt. and 1.5% wt. and reported an increase in the specific heat of 5.9% at 1% wt. and decreases in the specific heat at 0.5% wt and 1.5% wt. Also, the researchers Ho and Pan [113] confirmed a nearly 20% enhancement in the specific heat at a very reduced concentration of 0.063% wt. of alumina nanoparticles in the ternary eutectic salt KNO_3-$NaNO_2$-$NaNO_3$. Interestingly, increases in temperature were shown to decrease the specific heat of the salt, which is not in agreement with any other study, all of which showed constant or increasing specific heat with temperature. The specific heat enhancement of the molten salts with the addition of nanoparticles was attributed by the authors Shin and Banerjee [114,115] to three different possible mechanisms. The first one is related to the higher specific heat capacities of nanoparticles themselves, which is possible when the size of the particles is decreased. At present, nanoparticle-specific heat values have only been investigated up to 350 K [116], indicating that their properties at the high operating temperatures of molten salts is still largely unknown—a topic that should be investigated further. Also, the authors suggested that a high surface area per unit mass of nanoparticles increases the interfacial thermal resistance between nanoparticles and the surrounding liquid molecules. This high interfacial thermal resistance acts as additional thermal storage due to the interfacial interaction of the vibration energies between nanoparticle atoms and the interfacial molecules. This phenomenon could cause an increase in the specific heat of the salts [115]. Finally, the semi-solid behavior of layered liquid molecules at the surface of solid particles must be considered. The semi-solid layer can be visualized as a liquid–solid phase change at the surface and has been shown to have higher thermal properties than the bulk liquid and be equivalent to a latent phase change heat. Also, the researchers Jung and Banerjee [117] carried out molecular dynamics simulations to investigate this mechanism and introduced an analytical model to determine the specific heat of the salts. The empirical model for this mechanism simply assumes values for the semi-solid layer thickness, which could vary for different salts and particle materials and sizes. Moreover, the authors Shin and Banerjee [115] estimated that the semi-solid layer of liquid molecules on a crystalline surface can reach 2–5 nm in thickness—with this range depending on the surface energy of the crystalline interface. Thus, the mass fraction of the semi-solid layer should increase proportionally with the reduction in the size of the nanoparticles (for the same concentration of nanoparticles). Consequently, smaller nanoparticles should cause a greater enhancement in the specific heat of nanofluids. However, constant enhancement with different sizes of nanoparticles [115] and more enhancements with larger particles [112] have been reported, which seems to contradict this part of the theory. It is possible, though, that these unexpected trends could be explained by clustering at high temperatures, resulting in different particle size distributions during testing. The ionic liquid-based solvation force stability, as previously described, is also based on the existence of layered structures of ions in the vicinity of the solid–liquid boundary. Therefore, the common concept of the ionic liquid interfacial structure and semi-solid layers indicate a plausible mechanism to produce a stable colloidal system with enhanced specific heat. However, this is unproven for molten salts and still requires further investigation. On the other hand, the nanocomposites had lower specific heat compared to the base PCM, such as, for example, in the case of the improvement in photo-thermal performance [90], for PCM-filled cylinders [88], and in thermoelectric applications [118]. On the contrary, the authors Chieruzzi et al. [119] found that nanocomposites had a higher specific heat than that of the corresponding base PCM. The researchers prepared a nanofluid based on a nitrate salt mixture of 60% wt. $NaNO_3$ and 40% wt. KNO_3, T_m 220 °C. To create the nano-enhanced PCM, silica (7 nm), alumina (13 nm), and a mixture of silica and alumina (2–200 nm) nanoparticles at 1% wt. were added to the nitrate salt mixtures. The best results were achieved for the nanofluid with 1% wt. of silica/alumina nanoparticles. The specific heat in the solid phase was improved by around 52 and in the liquid phase by around 19%. The stored heat was increased by 13.5% compared with that of the PCM itself. On the other hand, the researchers Liu and Yang [120] enhanced the thermal properties of inorganic hydrate salt with titanium oxide–P25 nanoparticles with a dimension of 21 nm. The specific heat was improved by 83.5% in the solid phase and by 15.1% in the liquid phase by adding 0.3% wt. of nanoparticles. Furthermore, the latent heat was increased by 6.4%.

4.3. Latent Heat

In most studies, the addition of nanomaterials caused a slight decrease in the latent heat, except in [119] in relation to the passive cooling application and in [120] for the case of hydrate salts, where increases of 10.8% and 6.4% were reported, respectively, whereas in [121], the addition of graphite nanoparticles to an OP10E/water emulsion did not affect the latent heat. Also, the authors Warzoha et al. [122] found that the latent heat of fusion of organic paraffin decreased with the percentage of herringbone graphite nanofibers. The latent heat of fusion of the pure paraffin PCM was 271.6 J/g, whereas the latent heat of fusion of the nano-enhanced PCM with 11.4 vol% of graphite nanofibers was 242.7 J/g. The thermal conductivity and diffusivity increased with the increase in volume fraction of the HGNF. Furthermore, the authors Colla et al. [82] added alumina and carbon black nanoparticles to the paraffin waxes RUBITHERM®RT20 and RUBITHERM®RT25. The authors claimed that the addition of 1% wt. of alumina nanoparticles caused a degradation in thermal conductivity in both PCMs by 7–8%, whereas 1% wt. of carbon black nanoparticles increased the thermal conductivity by more than 25%. When alumina nanoparticles were added to RT20, the improvement in the latent heat was 10.8%, whereas in the case of carbon black nanoparticles the improvement was only 3.4%. The addition of carbon black nanoparticles to RT25 caused a reduction in latent heat of 11.6%. Moreover, the authors Muthoka et al. [123] prepared barium chloride dehydrate solutions in which nanoparticles of magnesium oxide and multi-walled carbon nanotubes were added. The authors stated that a 7% reduction in latent heat was detected for 1% wt. of multi-walled carbon nanotube0-enhanced fluid and a 5.2% reduction for 1% wt. magnesium oxide-enhanced fluid. Furthermore, the researchers Ebadi et al. [118] enhanced coconut oil with nanoparticles of copper oxide to create biobased nano-enhanced PCMs. A nanocomposite with 1% wt. of nanoparticles had 7.5% higher thermal conductivity than the base PCM, whereas the specific heat and latent heat of fusion decreased by 0.75% and 8.2%, respectively.

In Section 3, the literature results related to the most relevant thermophysical properties of nanofluids were presented, and several discrepancies were discussed. Regarding thermal conductivity, such divergences may have arisen from the techniques used to measure this property, as exemplified in Figure 5 of this paper. Many of these techniques are adaptations of methods traditionally used to measure solids, powders, and gases, as mentioned in [124]. Another factor to be considered concerns the methodology used by researchers in the literature; the possibilities for preparing nanofluids can vary considerably between different studies, as seen in Figure 3. Even in two independent studies that use the same nanofluids, with identical concentrations and particle diameters, the lack of a common protocol that considers the pH, the dispersion technique adopted, whether surfactants were used, and the type of base fluid can influence the measurements of the thermal properties.

5. Main Solar Thermal Conversion and Harvesting Applications

5.1. Solar Thermal Energy Storage Systems

Most nano-enhanced PCMs are very suitable for the thermal energy storage technological field. Thermal energy storage systems based on a nano-enhanced PCM operate according to the passive energy storage principle of latent heat. Such systems allow an active thermal management of the release and storage of latent heat. The main limiting parameter is the phase-change temperature value or temperature value range. Also, the choice of the most suitable nano-enhanced PCM and nanofluid for a given system should be based on the ease of the preparation method and the cost, as well as on their improved thermophysical properties. The fundamental interest of researchers was directed towards the amelioration of the thermal conductivity of nano-enhanced PCMs. The authors Lin and Al-Kayiem [125] conducted an experimental study on the thermophysical properties of nano-enhanced PCMs used in a solar collector. The experimental results demonstrated that the addition of 1% wt. of copper nanoparticles to the base paraffin wax increased the solar collector efficiency by 1.7% compared to that attained with a pure PCM. Also, the authors Natividade et al. [3] studied the thermal efficiency of a solar collector system with a parabolic concentrator. In their experiments, a nanofluid composed of water and multilayer graphene at extremely low volume fractions of 0.00045% vol. and 0.00068% vol. to prevent agglomeration and sedimentation was used. Despite having only a shallow description of the preparation method, with no references about the synthesis parameters or the equipment used, the nanofluid at 0.00045% vol. increased the thermal efficiency by 31% and the nanofluid at 0.00068% increased the same parameter by 76% compared to the results obtained with the water-based fluid alone. Additionally, the researchers Altohamy et al. [16] studied the inclusion of an aqueous alumina nanofluid in a latent heat thermal energy storage system. The authors confirmed that the use of the nanofluid had an impact on the solidified mass fraction and charging time of the system, which decreased considerably. Moreover, the authors found appreciable improvements in the charging rate and amount of energy stored. The charging time was reduced by up to 30% at a volume fraction of 2% vol. compared to that achieved with pure water. Additionally, the authors Kannan and Nadaraj [126] improved the efficiency of a solar flat-plate collector simultaneously employing nanofluids and PCMs. The researchers used the solar water heater during the off-shine hours using a PCM, which maintained the heat even after the sunshine was gone. Therefore, during off-sunshine hours, it yielded a better thermal performance. The performance of the solar collector could be improved by increasing the absorber plate area and the size of the water tubes. The performance of the solar water heater with a PCM was improved by selecting an appropriate PCM with a high thermal storage capacity. It was also improved by increasing the quantity of the PCM and heat transfer area by fixing the absorber with better performance. By using an alumina nanofluid as a heat transfer fluid at a concentration of 0.01% wt., the instantaneous efficiency of the collector was enhanced from 0.9% to 3.4% and from 0.1 to 3.5 under mass flow rates of 20 L per h and 40 L per h. The thermal efficiency of the collector was increased from 1% to 2.9% and from 0.05 to 1.9 under the same mass flow rates. Figure 9 shows the schematic representation of the adopted photovoltaic/thermal system configuration.

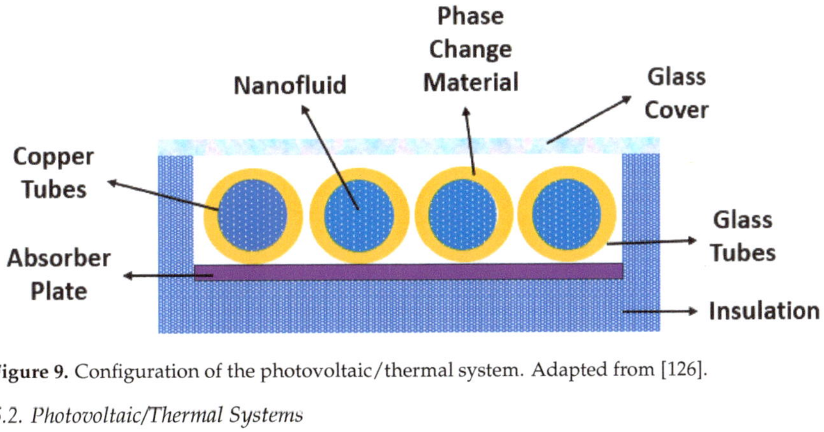

Figure 9. Configuration of the photovoltaic/thermal system. Adapted from [126].

5.2. *Photovoltaic/Thermal Systems*

The incoming solar radiation on the surface of the photovoltaic/thermal equipment may produce a large amount of electric power. Nonetheless, the absorbed solar radiation increases the temperature of the photovoltaic module and, consequently, decreases its overall efficiency. It is often stated in the published literature in the field that an increment in the temperature of photovoltaic panels causes an efficiency reduction of nearly 0.5% per degree [127], depending on the explored photovoltaic technology. Figure 10 schematically represents a typical photovoltaic/thermal system.

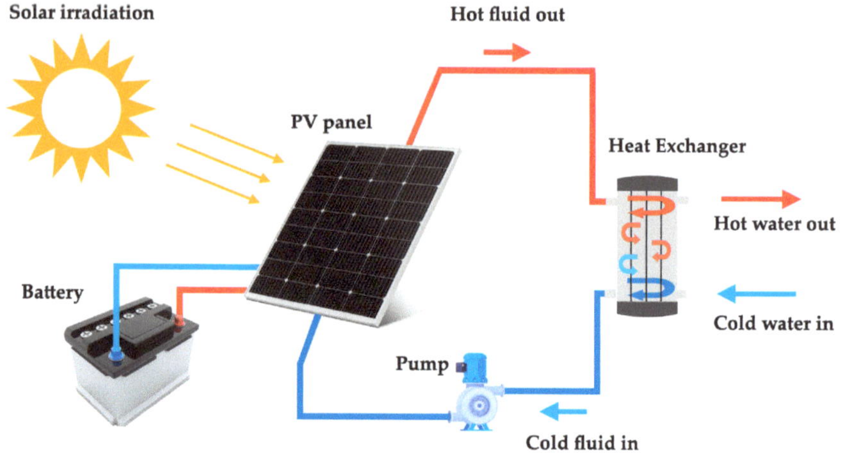

Figure 10. Schematic representation of a typical photovoltaic/thermal system.

Figure 11 schematically represents the energy and exergy associated with the control volume of a photovoltaic/thermal system.

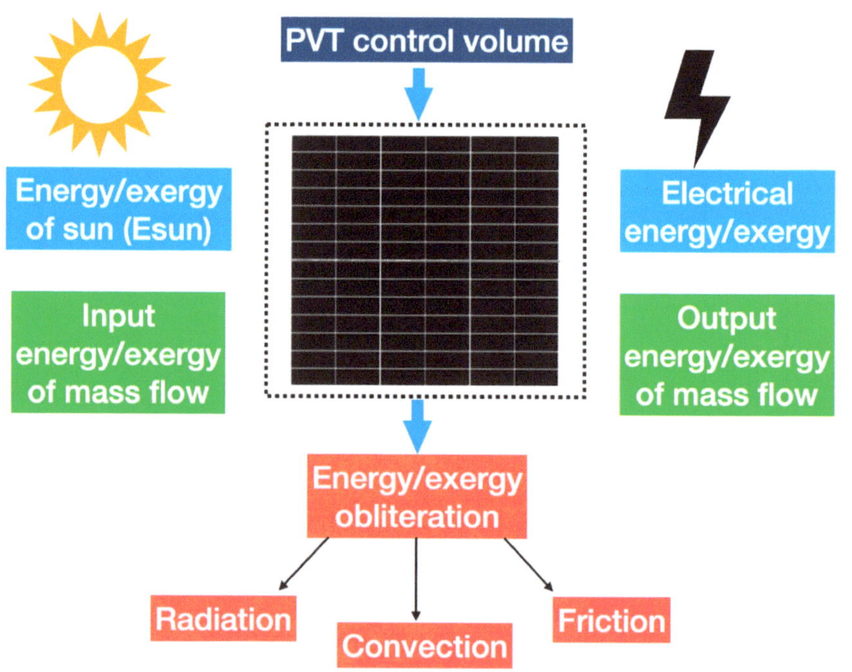

Figure 11. Energy and exergy of the control volume of a photovoltaic/thermal system.

The overall performance of a photovoltaic/thermal system can be evaluated based on a thermodynamic approach from the viewpoint of the first and second laws. The first and second laws may be referred to as energy and exergy analyses, respectively. Whereas an energetic analysis determines the quantity of the energy, an exergetic analysis determines its quality. Hence, considering the photovoltaic module and the thermal collector as a single control volume, the input energy is the quantity of solar irradiation and the output energies are the electrical and thermal ones. Assuming a steady-state condition, the control volume energy balance can be given by Equation (5):

$$E_{in} = E_{el} + E_{th} - E_{loss} \tag{5}$$

where E_{in} is the incident solar irradiation to the photovoltaic/thermal system, E_{el} is the output electrical power, E_{th} is the thermal energy gained from the collector, and E_{loss} is the energy loss for the control volume. Moreover, the overall efficiency of a photovoltaic thermal system η_{PV} is the ratio of the output energies to the input energy during a defined period and is given by Equation (6) [128]:

$$\eta_{PV} = \frac{E_{th} + E_{el}}{E_{in}} = \frac{\int_{t_1}^{t_2} \left(A_c E'_{th} + A_{pv} E'_{el}\right) dt}{A_c \int_{t_1}^{t_2} G'_{eff} dt} = \eta_{th} + r\eta_{el} \tag{6}$$

where A_c is the area of the collector, A_{pv} is the area of the photovoltaic cells, E'_{th} is the rate of output thermal energy per unit area of the collector, E'_{el} is the rate of output electrical energy per unit area of photovoltaic cells, G_{eff} is the rate of the effective incident radiation per unit area of the collector, η_{th} is the thermal efficiency of the system, η_{el} is the electrical efficiency of the system, and r is the packing factor given by the ratio A_{pv}/A_c. Furthermore, the term E_{th} can be determined by Equation (7):

$$E_{th} = m \cdot C_p \cdot (T_{fo} + T_{fi}) \tag{7}$$

where m is the mass flow rate through the collector of the operating fluid, C_p is the specific heat capacity of the operating fluid, T_{f0} is the outlet temperature of the fluid, and T_{fi} is the inlet temperature of the fluid. The electrical efficiency η_{th} can be expressed by Equation (8):

$$\eta_{el} = \frac{E_{el}}{E_{in}} = \frac{VOC \times ISC \times FF}{G_{eff}} \tag{8}$$

where VOC is the open-circuit voltage, ISC is the short-circuit current, and FF is the filled factor that defines the maximum power conversion efficiency of the photovoltaic module, which can be determined by Equation (9) [129]:

$$FF = \frac{P_m}{VOC \times ISC} \tag{9}$$

where P_m is the ideal output power. In Equation (8), G'_{eff} can be given by Equation (10):

$$G_{eff} = \tau_g \cdot \alpha_{cell} \cdot G \tag{10}$$

where τ_g is the glass surface transmissivity, α_{cell} is the cell absorptivity, and G is the rate of the incident irradiation. To infer the thermal energy of the photovoltaic/thermal systems, the output electrical energy should be converted into thermal energy using a conversion factor c_f. For most of the developed photovoltaic/thermal fluid systems, C_f values of 0.35 to 0.40 have been proposed [130]. Therefore, the overall equivalent photovoltaic/thermal system can be given by Equation (11):

$$\eta_{PV,TH} = \eta_{th} + r\frac{\eta_{el}}{C_f} \tag{11}$$

Similar to the energy analysis, by considering the photovoltaic module and the thermal collector as a unified control volume, the exergy balance can be expressed by Equation (12):

$$Ex_{in} = Ex_{el} + Ex_{th} - Ex_{loss} \tag{12}$$

The terms in Equation (12) are defined similarly to the terms of the energy balance but in an exergy expression. In addition, the overall exergetic efficiency of the photovoltaic/thermal system can be given by Equation (13) [128]:

$$\varepsilon_{pv} = \frac{Ex_{th} + Ex_{el}}{Ex_{in}} = \frac{\int_{t_1}^{t_2} (A_c Ex'_{th} + A_{PV} Ex'_{el})dt}{A_c \int_{t_1}^{t_2} Ex'_{sun} dt} = \varepsilon_{th} + r\varepsilon_{el} \quad (13)$$

where ε_{pv} is the overall exergetic efficiency, ε_{th} is the thermal exergetic efficiency, ε_{el} is the electrical exergetic efficiency, Ex'_{th} is the rate of output thermal exergy per unit area of the collector, Ex'_{el} is the rate of output electrical exergy per unit area of the cells, and Ex'_{sun} is the rate of the incident irradiation exergy per unit area of the collector. In the equation above, the thermal and electrical output exergies can be linked with the output energies by Equations (14) and (15), respectively:

$$Ex_{el} = E_{el} \quad (14)$$

$$Ex_{th} = E_{th} \cdot (1 - \frac{T_a}{T_{fo}}) \quad (15)$$

where the temperatures are expressed in Kelvin and T_a is the ambient temperature. The received input exergy can be determined by Equation (16) [131]:

$$Ex_{sun} = G\left[1 - \frac{4}{3}\frac{T_a}{T_{sun}}(1 - \cos\beta)^{\frac{1}{4}} + \frac{1}{3}\left(\frac{T_a}{T_{sun}}\right)^4\right] \quad (16)$$

where T_{sun} is the equivalent temperature of the sun as a blackbody of nearly 5800 K and β is the half-angle of the cone subtended by the sun disc, where β is around 0.0047 radian on a clear day for a beam sunlight and around $\pi/2$ radian for diffuse sunlight [132]. Figure 12 summarizes the fundamental thermal mechanisms associated with a photovoltaic/thermal system. The temperature at which a photovoltaic module operates is at equilibrium between the heat generated by the photovoltaic module and the heat loss to the surrounding environment. The various thermal processes of heat loss are reflection, radiation, convection, and conduction. The conductive heat loss is derived from the different temperatures between the photovoltaic module and the other materials with which the photovoltaic module is in contact. The capability of the photovoltaic module to transport heat to its surroundings can be symbolized by thermal resistance. The convective heat transfer arises from the transport of heat away from a surface as the result of one material moving across the surface of another. In a photovoltaic module, the forced convective heat transfer is generated by the wind blowing across the surface of the cell. The free or natural convection exchange of the panel with the area surrounding it has a smaller impact. This natural convection is caused by the air temperature and can be defined by Newton's law of cooling. The free convection becomes considerable in cases where only a little wind blows under cold climates, in which the temperature gradient between the surface of the photovoltaic module and the ambient air is relatively large. Also, the photovoltaic module may transfer heat to the surrounding environment through reflection and radiation. In the exchange of heat via radiation, long-wave radiation may be involved. To determine its contribution to the heat balance, it is necessary to consider the value of the radiative conductance or emissivity between the surfaces of both sides of the photovoltaic panel with both the ground and the sky.

Hence, considering the efficiency and economic feasibility of photovoltaic/thermal systems, it is reasonable to consider that their innovative thermal regulation and the beneficial thermophysical characteristics of nano-enhanced PCMs may be very suitable for this end. In recent years, many different new photovoltaic/thermal technologies have emerged other than those of air, water, and employing two thermal fluids with some desirable output. The heat transfer capability depends on the thermal conductivity of the cooling media. Dispersing highly conductive nanoparticles into the cooling fluid improves the thermal conductivity of the cooling medium, and the heat transfer rate is also increased. Figure 13 summarizes the fundamental design of and operating parameters for a photovoltaic/thermal system to operate with nanofluids and PCMs.

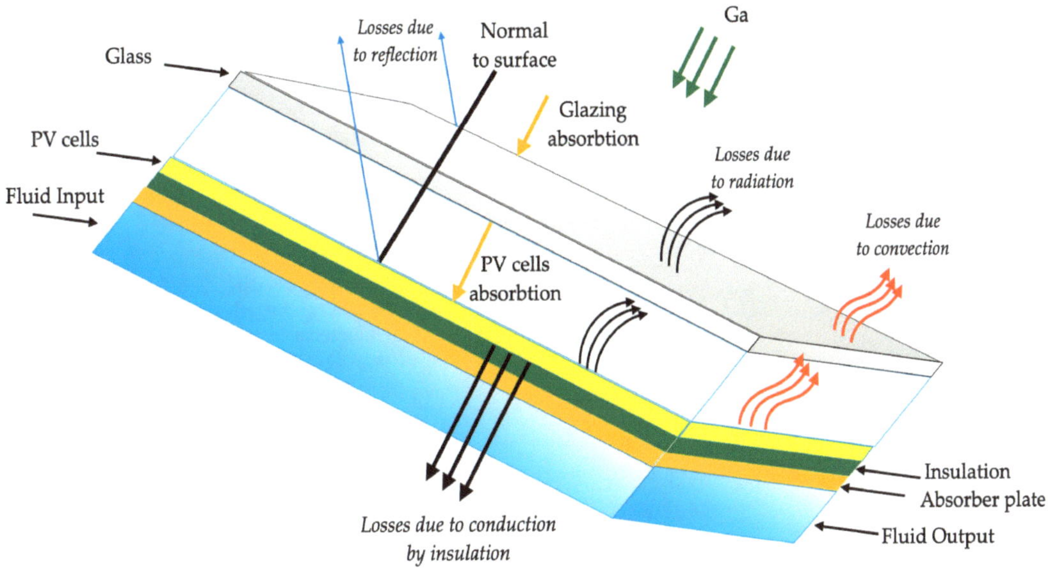

Figure 12. Main heat transfer mechanisms associated with a photovoltaic/thermal system.

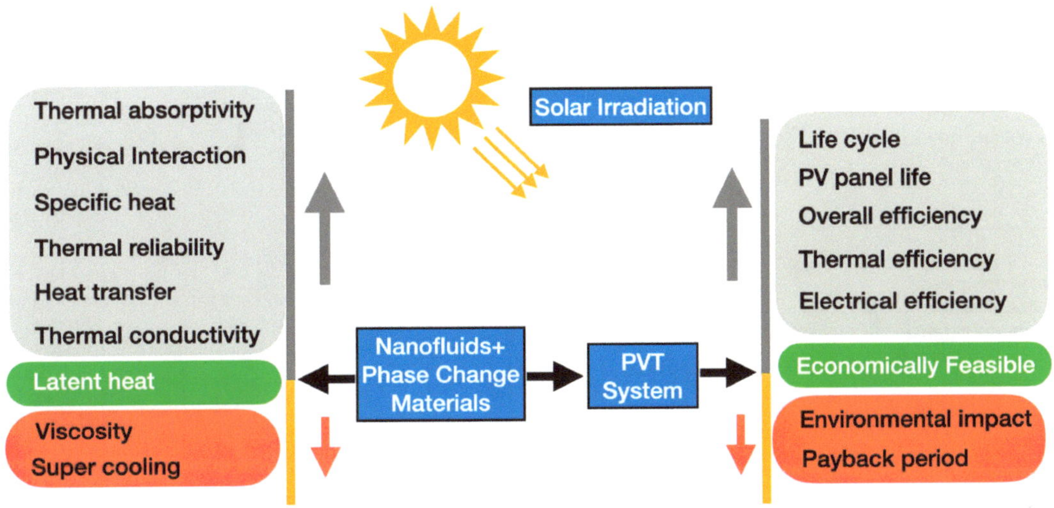

Figure 13. Main design parameters for a photovoltaic/thermal system operating with nanofluids and PCMs.

Table 1 summarizes the benefits and disadvantages of the different cooling media in photovoltaic/thermal systems.

Table 1. Main benefits and disadvantages of the different cooling media in photovoltaic/thermal systems.

Photovoltaic/Thermal System	Cooling Medium	Benefits	Problems	Applications
PCM-integrated photovoltaic/thermal system	Air	Low investment cost Easy to install and maintain Less bulky system Air availability	Poor thermal conductivity Low heat transfer capability Low photovoltaic panel temperature reduction Low thermal performance Fan is required for circulating the air Some of the PCMs may leak during melting	HVAC Residential buildings Thermal space management
PCM-integrated photovoltaic/thermal system	Water	Higher heat transfer capability than the one with air as a coolant Suitable for different fluidic coolants	High investment cost Low thermal conductivity High maintenance cost Risk of equipment corrosion Freezing Leakage High power consumption	HVAC Residential buildings Thermal space management
PCM-integrated photovoltaic/thermal system	Nanofluid	Advanced technology Enhanced heat transfer capability Broad range of commercially available nanoparticles	High pumping power is required Stability of the dispersion of nanoparticles in the base fluid High cost of nanoparticles Leakage and fouling risks	HVAC Residential buildings Thermal space management
Nano-enhanced PCM-integrated photovoltaic/thermal system	Air	Advanced technology	Fan or blower is required Only average thermal performance	HVAC Water heating Snow and ice removal All-season thermal comfort Clothing washing and drying
Nano-enhanced phase-change material-integrated photovoltaic/thermal system	Water	Well -developed technologySuitable for different fluidic coolants Enhanced heat transfer capability	Settling of nanoparticles after several operating cycles Leakage	HVAC Water heating Snow and ice removal All-season thermal comfort Clothing washing and drying
Nano-enhanced phase-change material-integrated photovoltaic/thermal system	Nanofluid	Well-developed technology Very high thermal performance Broad range of commercially available nanoparticles	High pumping power is required Stability over time of the dispersion of the nanoparticles in the PCM and base fluid Settling of nanoparticles	HVAC Water heating Snow and ice removal All-season thermal comfort Clothing washing and drying

In this direction, the authors Nada and El-Nagar [133] evaluated the efficiency of PCMs and nano-enhanced PCMs in the thermal management of integrated photovoltaic modules. In view of the obtained results, it was found that, for the integrated modules, the inclusion of a PCM composed of paraffin wax and Rubitherm® RT55 and a nano-enhanced PCM with the addition of 2% vol. of alumina nanoparticles resulted in an increase in the photovoltaic efficiency of the system. The researchers reported that the daily efficiency was increased from nearly 5.7% using the PCM and from around 13.1% employing the nano-enhanced PCMs. Moreover, the researchers Al-Waeli et al. [134] added silicon carbide nanoparticles with concentrations of up to 4% wt. to a paraffin PCM to determine the thermal and electrical performance of a hybrid photovoltaic/thermal panel. The authors found that the heat dissipation from the photovoltaic/thermal panel was more uniform when the PCM was incorporated. Also, the PCM enhanced the efficiency of the system from 7.1% to 13.7% in comparison to common photovoltaic systems. Furthermore, the combined effect of aqueous zinc oxide nanofluids and a paraffin wax PCM was experimentally investigated by the authors Sardarabadi et al. [135] in a photovoltaic/thermal system. The concentration of the nanofluids varied from 0.1% wt. and 0.4% wt. and the researchers observed that the electrical efficiency of the system was increased by 13%. The efficiency of the photovoltaic/thermal system incorporating the nanofluids was 5% higher than the one with water. Also, the paraffin wax further improved the efficiency of the photovoltaic panel by 9%. Figure 14 schematically illustrates a typical configuration of the combined use of nanofluids and PCMs in a photovoltaic/thermal system.

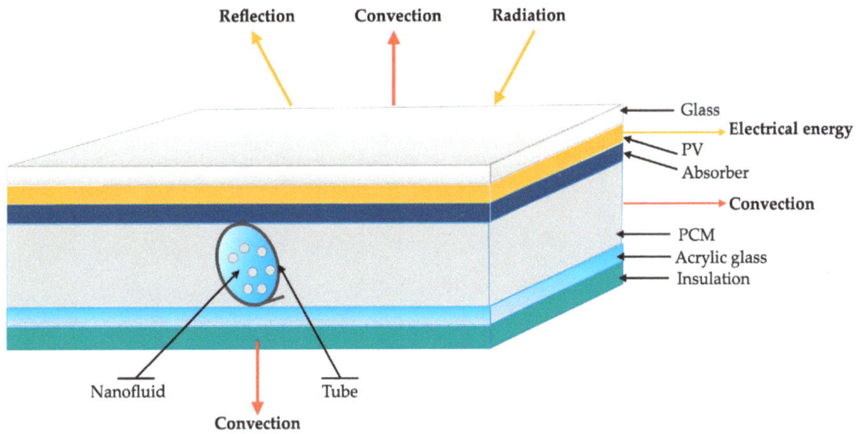

Figure 14. Typical configuration of a photovoltaic thermal system operating simultaneously with nanofluids and PCMs.

Additionally, the authors Hosseinzadeh et al. [136] evaluated the energy and exergy efficiency of a photovoltaic module, a photovoltaic/thermal system using a nanofluid, and a photovoltaic/thermal system operating with a nanofluid and a PCM. A zinc oxide–water nanofluid and a paraffin PCM were used. The obtained results indicated that the electrical efficiencies of the photovoltaic module, the nanofluid photovoltaic/thermal system, and the nanofluid and PCM photovoltaic/thermal system were around 12.6%, 13.4%, and 14.1%, respectively. The overall efficiencies were approximately 12.6%, 53.3%, and 65.7%, respectively. The thermal efficiencies of the nanofluid photovoltaic/thermal system and the nanofluid and PCM photovoltaic/thermal system were around 39.9% and 51.7%, respectively. Also, the exergy efficiencies of the possible alternatives were about 10.7%, 13.6%, and 12.4%, respectively. In addition, the nanofluid and PCM photovoltaic/thermal equipment provided an increase of 26% in the thermal efficiency in comparison to that attained with the conventional system. Moreover, the researchers Sardarabadi et al. [135] evaluated the performance of a PCM photovoltaic/thermal system operating according to distinct configurations and with diverse heat transfer fluids. A conventional water operating system, a zinc oxide aqueous nanofluid operating system, a water- and PCM integrated system, and, finally, a zinc oxide nanofluid- and PCM-integrated system were studied. The obtained results indicated that this last approach exhibited the greatest overall thermal and electrical efficiencies and the most reduced photovoltaic panel surface temperature. The research team concluded that the nanofluid–PCM-integrated system provided increased electrical and overall efficiencies of 13% and 23%, respectively, with respect to those achieved with a conventional photovoltaic system without requiring any extra consumption of energy. This system also yielded a 48% increase in thermal energy in comparison with the one provided by the nanofluid operating system. Also the temperature reduction of the photovoltaic panel was of only 10 °C when the system operated with water cooling, whereas the use of the nanofluid–PCM-integrated system reduced the homologous temperature by 16 °C. Figure 15 schematically presents the general view of the experimental setup developed by the authors.

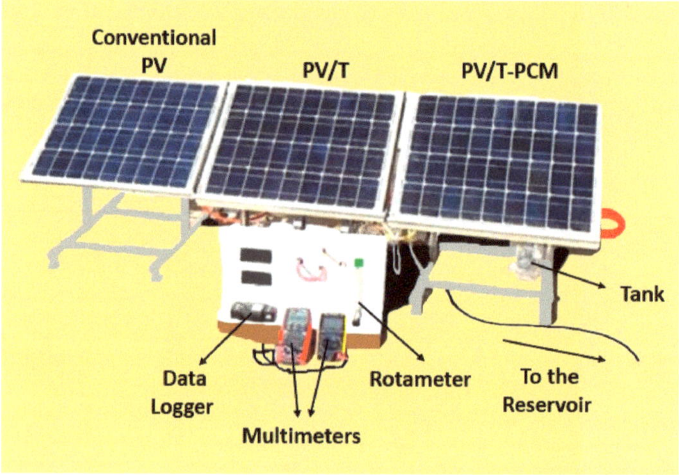

Figure 15. General view of the experimental setup developed by the authors Sardarabadi et al. Adapted from [135].

Moreover, the authors Hassan et al. [137] investigated the performance of a photovoltaic/thermal system using a PCM and nanofluids simultaneously. Flow rates of between 20 and 40 L per min and volumetric concentrations of nanofluids of 0.05% vol. to 0.15% vol. were imposed. A conventional photovoltaic system, a PCM operating system, a water- and PCM-integrated operating system, and a nanofluid- and PCM-integrated operating system were examined. The investigation team concluded that the nanofluid–PCM-integrated system achieved the best performance at 0.1% vol. and with a flow rate of 40 L per minute. This system attained electrical energy, thermal energy, and overall performances of 14%, 45.8%, and 60.3%, respectively.

The researchers Al-Waeli et al. [134] developed nanofluid nano-enhanced PCM photovoltaic/thermal equipment with an imposed flow rate of 0.17 kg/s. The authors verified that there was a considerable decrease in the temperature of the surface of the photovoltaic panel. The temperature of the photovoltaic panel was reduced by up to 30 °C during the peak solar irradiation time. Also, the obtained results showed that the open-circuit voltage, power, and electrical efficiency increased from 13 V to 21 V, from 61.1 W to 120.7 W and from 7.1% to 13.7%, respectively. The researchers also reported a 72% increase in the thermal efficiency of the device. Furthermore, the researchers Al-Waeli et al. [138] evaluated various cooling methodologies for a photovoltaic/thermal system using experimental data and artificial neural network software. The results confirmed that the integrated nano-enhanced PCM–nanofluid approach exhibited superior performance with respect to the other photovoltaic/thermal systems. Additionally, the integrated nanofluid–PCM system had superior performance, and the electrical current and electrical efficiency increased from 3.69 A to 4.04 A and from 8.07% to 13.2%, respectively, compared to the conventional photovoltaic systems. Also, the authors compared the experimental results with artificial neural network software and discovered that the neural model was in good agreement with the experimental results. On the other hand, the researchers Sarafraz et al. [139] analyzed the experimental and theoretical work of a photovoltaic/thermal PCM system. In their study, the authors analyzed the optimum pH value of the solution and flow rate of the cooling fluid to obtain a uniform dispersion and enhanced performance. The highest electrical and thermal performance was attained at 0.2% vol. of multi-walled carbon nanotubes for the nano-enhanced PCM and nanofluid. The results showed that the thermal and electrical performances were improved by 130% and 20%, respectively using the multi-walled carbon nanotubes dispersed in paraffin as a nano-enhanced PCM and multiwalled carbon nanotubes dispersed in WEG50 as nanofluids. The authors also found a considerable reduction in the temperature of the photovoltaic panel. Also, the equivalent electrical and thermal powers at the maximum time were 276.3 W/m^2 and 307.9 W/m^2, respectively.

The researchers Al-Waeli et al. [140] analyzed the enhancement of the power production performance of a photovoltaic/thermal device using an artificial neural network and a mathematical model with three different cooling methods. The authors proposed three linear models. The first model is to find the best parameters to predict the production of power. The second one is to obtain identities that are closer to experimental results, and it is represented by its speed, low cost, and ease of use. The third model is to speed up the reflection of the best conditions for any innovative system to be designed or developed. The developed model aids to minimize the errors of further studies and find the best conditions for the photovoltaic/thermal device to operate. Moreover, the researchers Al-Waeli et al. [141] studied an artificial neural network model and experimentally evaluated the photovoltaic/thermal device under the same environmental conditions. They examined a conventional photovoltaic system, a water photovoltaic/thermal system, a nanofluid operating photovoltaic/thermal system, and a nanofluid–PCM-integrated system. The investigation team noted that the nanofluid–PCM system yielded the best performance, with thermal and electrical efficiencies of 72% and 13.3%, respectively. Additionally, the artificial neural network model results were found to be consistent with the obtained experimental results and with those published earlier. Al-Waeli et al. [142] developed a photovoltaic thermal system to ameliorate the performance and reduce the number of required materials and parts and compared them with the proposed experimentally recorded values and modeled results. It was concluded that the electrical and thermal performance of the simulation and experimental results were 13.7% and 13.2%, and 72% and 71.3%, respectively. The overall efficiency was measured at 85.7%, and the peak photovoltaic panel temperature was observed to be 41.2 °C. The modeled outcomes were in good agreement with the experimentally obtained ones.

A novel approach of comparing the performance of a conventional photovoltaic system and a nanofluid photovoltaic/thermal system with a nanofluid–PCM-integrated system through experimental investigation was proposed by the authors Sardarabadi et al. [143]. For this purpose, monocrystalline silicon photovoltaic panels of 40 W for all alternatives were used outdoors in Mashhad, Iran, for several days in August and September. A zinc oxide aqueous nanofluid at 0.2% wt. flowing at 30 kg/h in a copper sheet and a tube collector attached to the rear of the photovoltaic panel were used. The employed PCM was paraffin wax with a melting point in the temperature range of between 42 °C and 72 °C, and it was positioned in the surroundings of the third system's tubes. The authors reported a maximum temperature decrease of approximately 7 °C and 16 °C for the nanofluid photovoltaic/thermal system and the nanofluid–PCM-integrated system, respectively, with respect to that of the conventional photovoltaic system. The PCM absorbed an additional amount of heat from the photovoltaic panel in its latent heat, lowering and stabilizing the average surface temperature of the photovoltaic panel. The zinc oxide nanofluid also absorbed a considerable amount of heat because of its high thermal conductivity derived from the incorporation of nanoparticles, and this absorption of heat also resulted in enhanced thermal power. The increases in the electrical efficiency enhancements were of 8% and 23% for the system working with nanofluid and for the nanofluid–PCM-integrated system, respectively, in comparison to the traditional photovoltaic system. Hence, the research team concluded that the nanofluid–PCM operating system was the best performing one. Moreover, the researchers Al-Waeli et al. [144] determined the overall efficiency of a photovoltaic/thermal system using a nanofluid and a nano-enhanced PCM for the efficient cooling of the photovoltaics with a focus on improving the electrical efficiency Silicon carbide nanoparticles were added to water and to paraffin wax. The nano-enhanced PCM was synthesized in an ultrasonic mixer with the addition of from 0.1% wt. to 4% wt of silicon carbide. At 3% wt., the maximum thermal conductivity increase was reached, which, consequently, was the concentration used for the nano-enhanced PCM. Similarly, the prepared nanofluid also had a 3% mass fraction of SiC in water. The photovoltaic average surface temperature decreased with increasing flowrate in the copper tubes, with an imposed flow rate of 0.170 kg/s. The incorporation of the nanoparticles enhanced the heat transfer capability of both the water and the paraffin. The PCM first absorbed the extra photovoltaic panel heat and then transferred it to nanofluid flowing in tubes, leading to more heat extraction from the photovoltaic module. By simultaneously exploring a nanofluid and a nano-enhanced PCM in the photovoltaic/thermal system, a reduction in the temperature of the photovoltaic cell of around 17 °C was obtained. Also, this system provided an approximately 42.5% heat gain increase in comparison with a photovoltaic/thermal system operating with water and paraffin.

The combined use of nanofluids with PCMs has been an influential thermal management methodology for photovoltaic/thermal systems in the recent years. Many experimental works and numerical simulations have been performed in this area. PCMs together with nanofluids have been a very attractive alternative to researchers in comparison with other materials, mainly due to the synergistic effect derived from the enhanced thermal conductivity and diffusivity offered by nanofluids and the enhanced heat storage capability of PCMs. The latter are integrated with nanofluids to overcome the poor heat transfer coefficient and insufficient contact area problem associated with PCMs. Furthermore, the authors Salem et al. [4] studied the cooling effect provided by an alumina/PCM mixture on a photovoltaic thermal system. The researchers circulated the alumina/PCM mixture and/or water at nanoparticle concentrations of up to 1% wt. and mass fluxes of up to 5.31 kg/s.m^2 in aluminum channels underneath the photovoltaic cells. The authors stated that the cooling process using this compound solution performed better than cooling with ordinary water. The highest photovoltaic performance of 110% achieved with the mixture was translated into the highest photovoltaic electrical output. Additionally, the researchers Al-Waeli et al. [140] evaluated the thermal performance of a photovoltaic/thermal system deploying diverse heat transfer enhancement routes: a PCM with a silicon carbide nanofluid, a PCM and water, and only water. The authors found that the nano-enhanced PCM and the nanofluid compound solution enhanced the electrical efficiency from 8% to around 13.3% and the electrical current output from nearly 3.7% to 4% compared with those attained with a common photovoltaic/thermal system. Also, it should be emphasized that the authors arrived at a consistent comparison between the experimentally obtained results and the ones estimated through the artificial neural network, multilayer perceptron, and support vector machine methodologies. The proposed model results were compared with the experimental ones with coefficient of determination R2, MSE, and RMSE values of 0.99, 0.006, and 0.009, respectively. Similarly, the authors Al-Waeli et al. [144] proposed a hybrid photovoltaic/thermal system using a PCM and a nanofluid to enhance the overall efficiency of the system. The developed nano-enhanced PCM was one composed of paraffin with a dispersion of silicon carbide nanoparticles. The experiment was conducted outdoors, employing a fluid flow rate of around 0.2 Kg/s. The electrical performance with this technological solution was enhanced from nearly 7% to 14%, the circuit voltage was enhanced from 11 V to 13 V to 20–21 V, and the power output increased from nearly 61 W to around 121 W. Also, there was a considerable cell temperature decrease, and the thermal efficiency reached 72%.

The authors Hassan et al. [137] reported on an experimental study on the simultaneous usage of a nanofluid and a PCM for thermal regulation in photovoltaic modules. The authors used an aqueous graphene nanofluid and a TR-35HC PCM with varying graphene nanostructure concentrations of 0.05% vol. to 0.15% vol., and different flow rates of between 20 L/min and 40 L/min. The authors found that the best thermal performance was attained using a graphene concentration of 0.1% vol. and a flow rate of 40 L/min. The average temperature decrease of the photovoltaic panels was of nearly 24 °C, 16 °C, and 12 °C with the system working with water and a PCM, a nano-enhanced PCM, and a PCM, respectively. The performance of the system employing a nano-enhanced PCM was improved by 12% with respect to that attained with water and a PCM, and the electrical efficiency was approximately 24% superior to the one for the common photovoltaic/thermal system. Furthermore, the authors Naghdbishi et al. [145] studied the combined effect of multi-walled carbon nanotubes dispersed in water nanofluid and paraffin wax on one hybrid photovoltaic/thermal PCM system. The maximum electrical energetic and exergetic efficiencies were attained in the cases where the multi-walled carbon nanotubes/water nanofluid was employed. The enhanced thermal fluid increased the electrical and thermal efficiencies of the system to nearly 4.2% and 23.6%, respectively, with respect to those obtained with a traditional photovoltaic module. Similarly, the researchers Salari et al. [146] elaborated a numerical simulation on a nanofluid photovoltaic/thermal system integrated with PCMs. In this direction, magnesium oxide nanoparticles, multi-walled carbon nanotubes, and a combination of both were added to water and used as working fluids. The obtained results showed that the surface temperature of the system using the multi-walled carbon nanotube nanofluid decreased by 0.3 °C when the concentration of the carbon nanotubes was increased from 3% wt. to 6% wt. Also, energy efficiencies of nearly 60.7%, 61.1%, 60.1%, and 55.2% were achieved using the hybrid form of the nanofluid, only multi-walled carbon nanotube fillers, only magnesium oxide fillers, and pure water, respectively. The authors also found that the outflow as well as the surface temperatures were diminished with the increase in the PCM layer thickness. Table 2 summarizes the fundamental recent published works on the simultaneous use of nanofluids and PCMs in photovoltaic/thermal systems.

Table 2. Main published works on the combined usage of nanofluids and PCMs in photovoltaic/thermal systems.

Authors	Nanofluid	Concentration	Phase-Change Material	Main Findings	Reference
Salem et al.	Alumina–water	Up to 1% wt.	Calcium chloride hexahydrate	The best performance was achieved with the compound cooling technique with an alumina–PCM mixture.	[4]
Abdelrahman et al.	Alumina–water	0.1 to 0.8% vol.	RT35HC	The mixture of the alumina nanoparticles into the RT35HC PCM with an increase in nanoparticle concentration from 0.11 to 0.77% vol. reduced the surface temperature of the photovoltaic cell by 53.2%.	[147]
Abdollahi and Rahimi	Boehmite–oil	0.9% wt.	Coconut oil/sunflower oil mixture	The highest efficiency was achieved using the nano-enhanced PCM with an increase in power output of up to 48.2%.	[148]
Hassan et al.	Graphene–water	0.05 to 0.15% vol.	RT35HC	The effectiveness of the nano-photovoltaic/thermal PCM device was 17.5% higher and the overall efficiency was 12% higher than those of the water photovoltaic/thermal PCM system.	[137]

Table 2. *Cont.*

Authors	Nanofluid	Concentration	Phase-Change Material	Main Findings	Reference
Al-Waeli et al.	Silicon carbide–water	0.1 to 4% wt.	Paraffin wax	Combined silicon carbide nanofluid and a nano-PCM was the best cooling option, with a 13.3% electrical efficiency compared with only 8.1% in the conventional photovoltaic system.	[141]
Al-Waeli et al.	Silicon carbide–water	0.1 to 4% wt.	Paraffin wax	The performance of the proposed combined nano-PCM and nanofluid photovoltaic/thermal system enhanced the electrical efficiency from 7.1% to 13.7%, the power from 61.1 W to 129.7 W, and the open-circuit voltage from 11 to 13 V to 20–21 V, and the thermal efficiency reached 72%.	[134]
Al-Waeli et al.	Silicon carbide–water	0.1 to 4% wt.	Paraffin wax	The technoeconomic evaluation of the nano-enhanced PCM and nanofluid-based PVT system revealed that the specific yield, the capacity factor, the efficiency of the inverter, the cost of electricity, and the payback period were 190.4 kWh/kWp, 25%, 97.3%, 0.125 USD/kWh, and 5–6 years, respectively. The ANN model findings were found to be consistent with those of recent published experimental works, demonstrating the reliability of this model.	[138]
Al-Waeli et al.	Silicon carbide–water	0.1 to 4% wt.	Paraffin wax	Proposition of linear projection models to reduce the error of the potential outcomes and determine the optimal conditions for every solar thermal energy system	[140]
Hosseinzadeh et al.	Zinc oxide–water		Paraffin wax	The overall exergy efficiencies of the photovoltaic, nanofluid photovoltaic, and nanofluid–PCM-integrated system were 10.7%, 13.6%, and 12.4%, respectively. There was a decrease in entropy generation of 1.6% and 3.2% in the nanofluid and nanofluid–PCM systems, respectively. An electrical power efficiency of 13.6% and a thermal power efficiency of 29.6% were attained.	[136]
Sardarabadi et al.	Zinc oxide–water		Paraffin wax	The photovoltaic/thermal system operating with a PCM and zinc oxide nanofluid had higher thermal, electrical, and overall efficiencies and more photovoltaic panel surface temperature reduction. It had a thermal power efficiency 46% greater than that of the photovoltaic system with water as a coolant.	[135]
Sarafraz et al.	MWCNTs–water	0.2% wt.	Paraffin wax	The best thermal and electrical performance was attained at 0.2% of the multi-walled carbon nanotubes for the nano-enhanced PCM and nanofluid. An electrical power efficiency of around 276 W/m^2 and a thermal power efficiency of around 308 W/m^2 were attained.	[139]

5.3. Thermal Management of Residential Buildings

To meet the ever-increasing thermal energy demands in the residential building universe, the researchers Lari and Sahin [149] developed an upgrade to a nanofluid photovoltaic/thermal retrofitting PCM thermal battery. The authors conducted daily and yearly technoeconomic analyses of the system using the Engineering Equation Solver (EES) software considering the annual performance of the system without heat storage and with uncooled photovoltaics. Compared to the uncooled photovoltaic systems, the researchers found an almost 12% increase in the electric output of the developed thermal battery and better thermal regulation in the photovoltaic panel. It was concluded that the proposed system could fulfil the power demand with 77% electrical power and 27.3% thermal power

6. Limitations and Prospects for Further Research

The main limitations and prospects for further investigation into the combined use of PCMs and nanofluids for solar thermal heat transfer enhancement can be summarized in the following points:

- The technological approaches described herein are very promising but exhibit some disadvantages, including the absence of standard preparation methods for nanofluids and nano-enhanced PCMs and some inconsistencies in the measurement methodologies for the different factors to determine the behavior of PCMs. Moreover, nanofluids and nano-enhanced PCMs may have agglomeration and settling problems after only a limited number of cycles of operation. Additionally, some potential PCMs and nano-enhanced PCMs possess leakage problems inherent to the solid–liquid phase transition.
- There is a lack of knowledge and internationally recognized standards for monitoring and testing photovoltaic/thermal systems operating with nanofluids and PCMs simultaneously, which limits the growth of this technological approach.
- There is a very limited number of commercially available photovoltaic/thermal systems due to some key challenges that need to be overcome, such as their thermal and electrical efficiencies, long-term performance details, and the compatibility of the thermal systems with the different types of photovoltaic panels. This may hinder improved knowledge of the practical technical design limitations and long-term real-life reliability studies of photovoltaic/thermal systems in general and of those assisted with the combined employment of nanofluids and PCMs.
- Further experimental works should be conducted and more accurate preparation routes should be developed for thermal management systems using PCMs and nanofluids. In the available published studies about this hybrid heat transfer route, some relevant details are still missing for the sake of the repeatability of the results and representative sampling, including the different types of base fluids and optimal concentration of nanofluids, synthesis methodologies, and safety procedures.
- A limited number of researchers investigating PCMs have carried out specific heat determination. Additionally, the scarce experimental works regarding the specific heat are not consistent with each other, normally presenting considerable variations. In some of the published studies, the specific heat of PCMs increased with increasing concentration of nanoparticles, and in others, the opposite evolution was observed. Considering these facts, further studies on the specific heat capacity of PCMs and the influencing factors are most welcome.
- The major concerns about the synthesis, characterization, and employment of PCMs together with nanofluids are the initial investment cost and the inherent economic analysis. Nonetheless, the economic viability and overall cost are issues that have still not yet been sufficiently addressed, and, consequently, further in-depth studies on the subject are needed.
- It is highly recommended to conduct further analysis on the environmental impact of nanofluid and phase-change material combinations to acquire enriched knowledge on the topic. Most of the scientific articles lack an explanation of the environmental impact in the synthesis, utilization, and final disposal stages of these materials. Also, the available literature does not present extensive guidelines for the safety procedures associated with the handling, use, and characterization of nanofluids and PCMs. Hence, it is strongly suggested to publish an environmental impact evaluation through life cycle assessment analysis and a description of the safety procedures to ensure a safe working environment for the researchers and potential users of nanofluids and PCMs.
- Additional issues that should be addressed include the need for a decrease in the cost of PCM synthesis methods and active equipment and the thermal stability and stability over time of nanofluids explored in solar thermal energy systems.

- Most of the PCMs employed in the technological area of solar energy storage and conversion are single materials like, for instance, paraffin waxes, and, consequently, further experimental works involving mixtures of different PCMs along with the use of nanofluids should be carried out. These works would provide useful insights into the synergistic benefits coming from the high thermal energy storage density and stability of such mixtures.
- The high-grade exergy and environmental effects of the developed photovoltaics/thermal systems with the combined usage of nanofluids and PCMs should be evaluated. For this purpose, it is suggested to conduct further life cycle investigation and comparison analysis between the conventional and enhanced systems. Such studies would enable the determination of the accumulated exergies and carbon dioxide emissions during the diverse working stages of the system's lifespan.
- The search for more environmentally friendly PCMs with less toxicity should be continued. Also, one of the future research topics should address attempts to improve the thermal properties of PCMs, namely, their relatively poor thermal conductivity.

7. Conclusions

The main concluding remarks of this review on the combined usage of nanofluids and PCMs in energy storage and conversion systems are highlighted in the following points:

- PCMs and nanofluids are very suitable to being applied in solar energy recovery systems because of their intrinsic beneficial features like improved thermal stability, recyclability, and lack of supercooling. Hence, it is predictable that the improved thermal energy storage equipment and systems using PCMs and nanofluids simultaneously will have an important role in future thermal solar energy conversion and harvesting processes.
- The combination of PCMs and nanofluids is the most effective photovoltaic thermal management choice with respect to the separated contribution of PCMs and nanofluids because of the additional heat dissipation of photovoltaic panels. Improved solar thermal energy management is attained in cases where nanofluid is used simultaneously with a nano-enhanced PCM and used as working fluid in photovoltaic/thermal systems.
- The combined usage of nanofluids and PCMs appreciably increases the thermal energy storage capacity and extends the working time of solar thermal energy storage systems considerably. Also, photovoltaic/thermal systems using nanofluids and PCMs simultaneously can considerably reduce the overreliance on fossil fuels to produce electricity and decrease carbon dioxide and greenhouse gas emissions.
- The simultaneous exploration of nanofluids and PCMs significantly increases the thermal conductivity and diffusivity and the convective heat transfer coefficient of photovoltaic/thermal systems. These benefits entail much lower convective and radiant losses derived from the overall improvement of the heat transfer performance. Nonetheless, it should be noted that increasing the concentration of the nanoparticles in the nanofluids increases the viscosity and pressure drops of the solar thermal system, requiring additional pumping power, which results in a higher overall investment cost of the produced electricity.
- The combined usage of nanofluids and PCMs has proven to be more effective for photovoltaic/thermal systems cooling than the individual exploration of either PCMs or nanofluids. Such a synergetic route normally gives rise to extra heat removal capability for photovoltaic panels because the heat is extracted in sequence by the PCM and nanofluid. The combination of the PCM and nanofluid lowers the surface temperature and, at the same time, improves the temperature uniformity of the photovoltaic panels. Such effects mainly derive from the uniform contact of the PCM with the panels.
- The not-converted incident thermal solar energy in photovoltaic/thermal systems can be stored by PCMs in the form of latent heat, which may reduce the average surface temperature of the panels by more than 30 °C. Additionally, the adoption of a particular PCM should be based on many factors, including the environment typical temperature

values and latitude, solar irradiation intensity, and wind velocity, among others, given that the effectiveness of a PCM is more intense during the summer than the winter because the PCM absorbs more heat in summer, leading to an increased efficiency.
- The thermal performance of photovoltaic/thermal systems using PCMs and nanofluids simultaneously can be improved with different approaches, using, for instance, different PCMs, nanoparticles and base fluids, flow rates, channel configurations, operating fluid inlet temperatures, and heat dissipation methodologies of photovoltaic panels. An extra measure could be the jet impingement of nanofluids for the thermal management of photovoltaic panels and to extract a great quantity of energy in the form of heat.
- The increase in the overall electrical efficiency and thermal performance of photovoltaic/thermal systems may lead to a reduction in the corresponding payback periods. Innovative methods like the tolerance capital cost method accurately predict the economic feasibility of photovoltaic/thermal systems. Nonetheless, negative effects like the deposition over time of dust on photovoltaic panels can diminish the output of the system's life cycle cost.
- PCMs should be carefully selected to operate in photovoltaic/thermal systems installed in areas with hot climates. In these cases, among the diverse possibilities, salt hydrates have already been demonstrated to perform better and make the systems technically and financially viable when implemented in hot climates.
- The artificial neural network, multilayer perceptron, and support vector machine predicting methodologies can be explored for an accurate estimation of the efficiencies of photovoltaic/thermal systems employing PCMs and nanofluids. These methodologies are also very useful for optimizing the operating parameters of the systems to attain superior thermal performance.

Author Contributions: Conceptualization, J.P. and R.S.; methodology, J.P. and A.M. (Ana Moita); software, A.M. (Ana Moita); validation, A.M. (Ana Moita) and A.M. (António Moreira); formal analysis, A.M. (Ana Moita); investigation, J.P. and R.S.; resources, A.M. (António Moreira); data curation, J.P. and R.S.; writing—original draft preparation, J.P. and R.S.; writing—review and editing, J.P. and R.S.; supervision, A.M. (Ana Moita) and A.M. (António Moreira); project administration, A.M. (António Moreira); funding acquisition, A.M. (António Moreira). All authors have read and agreed to the published version of the manuscript.

Funding: The authors are grateful to the Fundação para a Ciência e a Tecnologia (FCT), Avenida D. Carlos I, 126, 1249-074 Lisboa, Portugal, for partially financing the project "Estratégias interfaciais de arrefecimento para tecnologias de conversão com elevadas potências de dissipação", Ref. PTDC/EMETED/7801/2020, António Luís Nobre Moreira, Associação do Instituto Superior Técnico para a Investigação e o Desenvolvimento (IST-ID). José Pereira also acknowledges FCT for his PhD Fellowship (Ref. 2021.05830.BD). The authors are also grateful for FCT funding through 2022.03151.PTD and LA/P/0083/2020 IN + -IST-ID.

Institutional Review Board Statement: Not applicable.

Informed Consent Statement: Not applicable.

Data Availability Statement: Data sharing is not applicable.

Conflicts of Interest: The authors declare no conflict of interest.

Sample Availability: Not applicable.

References

1. Nourafkan, E.; Asachi, M.; Jin, H.; Wen, D.; Ahmed, W. Stability and Photo-Thermal Conversion Performance of Binary Nanofluids for Solar Absorption Refrigeration Systems. *Renew. Energy* **2019**, *140*, 264–273. [CrossRef]
2. Siddique, R.M.A.; Kratz, F.; Mahmud, S.; Van Heyst, B. Energy Conversion by Nanomaterial-Based Trapezoidal-Shaped Leg of Thermoelectric Generator Considering Convection Heat Transfer Effect. *J. Energy Resour. Technol.* **2019**, *141*, 082001. [CrossRef]

3. Natividade, P.S.G.; de Moraes Moura, G.; Avallone, E.; Bandarra Filho, E.P.; Gelamo, R.V.; Gonçalves, J.C.d.S.I. Experimenta Analysis Applied to an Evacuated Tube Solar Collector Equipped with Parabolic Concentrator Using Multilayer Graphene-Basec Nanofluids. *Renew. Energy* **2019**, *138*, 152–160. [CrossRef]
4. Salem, M.; Elsayed, M.; Abd-Elaziz, A.; Elshazly, K. Performance enhancement of the photovoltaic cells using Al_2O_3/PCM mixture and/or water cooling-techniques. *Renew. Energy* **2019**, *138*, 876–890. [CrossRef]
5. Martín, M.; Villalba, A.; Inés Fernández, A.; Barreneche, C. Development of New Nano-Enhanced PCMs (NEPCM) to Improve Energy Efficiency in Buildings: Lab-Scale Characterization. *Energy Build.* **2019**, *192*, 75–83. [CrossRef]
6. Ren, Q.; Xu, H.; Luo, Z. PCM Charging Process Accelerated with Combination of Optimized Triangle Fins and Nanoparticles. *Int J. Therm. Sci.* **2019**, *140*, 466–479. [CrossRef]
7. Balakin, B.V.; Zhdaneev, O.V.; Kosinska, A.; Kutsenko, K.V. Direct Absorption Solar Collector with Magnetic Nanofluid: CFD Model and Parametric Analysis. *Renew. Energy* **2019**, *136*, 23–32. [CrossRef]
8. Bonab, H.B.; Javani, N. Investigation and Optimization of Solar Volumetric Absorption Systems Using Nanoparticles. *Sol. Energy Mater. Sol. Cells* **2019**, *194*, 229–234. [CrossRef]
9. Wang, L.; Zhu, G.; Wang, M.; Yu, W.; Zeng, J.; Yu, X.; Xie, H.; Li, Q. Dual Plasmonic Au/TiN Nanofluids for Efficient Solar Photothermal Conversion. *Sol. Energy* **2019**, *184*, 240–248. [CrossRef]
10. Zhang, J.J.; Qu, Z.G.; Maharjan, A. Numerical Investigation of Coupled Optical-Electrical-Thermal Processes for Plasmonic Solar Cells at Various Angles of Incident Irradiance. *Energy* **2019**, *174*, 110–121. [CrossRef]
11. Zayed, M.E.; Zhao, J.; Du, Y.; Kabeel, A.E.; Shalaby, S.M. Factors Affecting the Thermal Performance of the Flat Plate Solar Collector Using Nanofluids: A Review. *Sol. Energy* **2019**, *182*, 382–396. [CrossRef]
12. Abdelrazik, A.S.; Al-Sulaiman, F.A.; Saidur, R.; Ben-Mansour, R. Evaluation of the Effects of Optical Filtration and NanoPCM on the Performance of a Hybrid Photovoltaic-Thermal Solar Collector. *Energy Convers. Manag.* **2019**, *195*, 139–156. [CrossRef]
13. Agresti, F.; Fedele, L.; Rossi, S.; Cabaleiro, D.; Bobbo, S.; Ischia, G.; Barison, S. Nano-Encapsulated PCM Emulsions Prepared by a Solvent-Assisted Method for Solar Applications. *Sol. Energy Mater. Sol. Cells* **2019**, *194*, 268–275. [CrossRef]
14. Sheikholeslami, M.; Mahian, O. Enhancement of PCM Solidification Using Inorganic Nanoparticles and an External Magnetic Field with Application in Energy Storage Systems. *J. Clean. Prod.* **2019**, *215*, 963–977. [CrossRef]
15. De Matteis, V.; Cannavale, A.; Martellotta, F.; Rinaldi, R.; Calcagnile, P.; Ferrari, F.; Ayr, U.; Fiorito, F. Nano-Encapsulation of PCMs: From Design to Thermal Performance, Simulations and Toxicological Assessment. *Energy Build.* **2019**, *188–189*, 1–11 [CrossRef]
16. Altohamy, A.A.; Abd Rabbo, M.F.; Sakr, R.Y.; Attia, A.A. Effect of water based Al_2O_3 nanoparticle PCM on cool storage performance. *Appl. Therm. Eng.* **2015**, *84*, 331–338. [CrossRef]
17. Shao, X.-F.; Mo, S.-P.; Chen, Y.; Yin, T.; Yang, Z.; Jia, L.-S.; Cheng, Z.-D. Solidification behavior of hybrid TiO_2 nanofluids containing nanotubes and nanoplatelets for cold thermal energy storage. *Appl. Therm. Eng.* **2017**, *117*, 427–436. [CrossRef]
18. Philip, J.; Shima, P.D. Thermal properties of nanofluids. *Adv. Colloid Interface Sci.* **2012**, *183*, 30–45. [CrossRef]
19. Ouabouch, O.; Kriraa, M.; Lamsaadi, M. Stability, Thermophysical Properties of Nanofluids, and Applications in Solar Collectors: A Review. *AIMS Mater. Sci.* **2021**, *8*, 659–684. [CrossRef]
20. Pinto, R.V.; Fiorelli, F.A.S. Review of the Mechanisms Responsible for Heat Transfer Enhancement Using Nanofluids. *Appl. Therm Eng.* **2016**, *108*, 720–739. [CrossRef]
21. Gonçalves, I.; Souza, R.; Coutinho, G.; Miranda, J.; Moita, A.; Pereira, J.E.; Moreira, A.; Lima, R. Thermal Conductivity of Nanofluids: A Review on Prediction Models, Controversies and Challenges. *Appl. Sci.* **2021**, *11*, 2525. [CrossRef]
22. Lomascolo, M.; Colangelo, G.; Milanese, M.; de Risi, A. Review of Heat Transfer in Nanofluids: Conductive, Convective and Radiative Experimental Results. *Renew. Sustain. Energy Rev.* **2015**, *43*, 1182–1198. [CrossRef]
23. Souza, R.R.; Gonçalves, I.M.; Rodrigues, R.O.; Minas, G.; Miranda, J.M.; Moreira, A.L.N.; Lima, R.; Coutinho, G.; Pereira, J.E. Moita, A.S. Recent Advances on the Thermal Properties and Applications of Nanofluids: From Nanomedicine to Renewable Energies. *Appl. Therm. Eng.* **2022**, *201*, 117725. [CrossRef]
24. Ambreen, T.; Kim, M.-H. Influence of Particle Size on the Effective Thermal Conductivity of Nanofluids: A Critical Review. *Appl Energy* **2020**, *264*, 114684. [CrossRef]
25. Qiu, L.; Zhu, N.; Feng, Y.; Michaelides, E.E.; Żyła, G.; Jing, D.; Zhang, X.; Norris, P.M.; Markides, C.N.; Mahian, O. A Review of Recent Advances in Thermophysical Properties at the Nanoscale: From Solid State to Colloids. *Phys. Rep.* **2020**, *843*, 1–81 [CrossRef]
26. Ghadimi, A.; Saidur, R.; Metselaar, H.S.C. A Review of Nanofluid Stability Properties and Characterization in Stationary Conditions. *Int. J. Heat Mass Transf.* **2011**, *54*, 4051–4068. [CrossRef]
27. Jia-Fei, Z.; Zhong-Yang, L.; Ming-Jiang, N.; Ke-Fa, C. Dependence of Nanofluid Viscosity on Particle Size and PH Value. *Chin Phys. Lett.* **2009**, *26*, 66202. [CrossRef]
28. Hwang, Y.; Lee, J.-K.; Lee, J.-K.; Jeong, Y.-M.; Cheong, S.; Ahn, Y.-C.; Kim, S.H. Production and Dispersion Stability of Nanoparticles in Nanofluids. *Powder Technol.* **2008**, *186*, 145–153. [CrossRef]
29. Hwang, Y.; Lee, J.K.; Lee, C.H.; Jung, Y.M.; Cheong, S.I.; Lee, C.G.; Ku, B.C.; Jang, S.P. Stability and Thermal Conductivity Characteristics of Nanofluids. *Thermochim. Acta* **2007**, *455*, 70–74. [CrossRef]
30. Mohammadpoor, M.; Sabbaghi, S.; Zerafat, M.M.; Manafi, Z. Investigating heat transfer properties of copper nanofluid in ethylene glycol synthesized through single and two-step routes. *Int. J. Refrig.* **2019**, *99*, 243–250. [CrossRef]

31. Kiani, M.R.; Meshksar, M.; Makarem, M.A.; Rahimpour, M.R. Chapter 2—Preparation, stability, and characterization of nanofluids. In *Nanofluids and Mass Transfer*; Rahimpour, M.R., Makarem, M.A., Kiani, M.R., Sedghamiz, M.A., Eds.; Elsevier: Amsterdam, The Netherlands, 2022; pp. 21–38. [CrossRef]
32. Lizana, J.; Chacartegui, R.; Barrios-Padura, A.; Valverde, J.M. Advances in Thermal Energy Storage Materials and Their Applications towards Zero Energy Buildings: A Critical Review. *Appl. Energy* **2017**, *203*, 219–239. [CrossRef]
33. Yang, X.H.; Huang, C.H.; Ke, H.B.; Chen, L.; Song, P. Evaluation of Thermal Control Performance of PCMs for Thermal Shock Protection of Electronics. *J. Phys. Conf. Ser.* **2021**, *2045*, 12032. [CrossRef]
34. Choi, D.H.; Lee, J.; Hong, H.; Kang, Y.T. Thermal conductivity and heat transfer performance enhancement of phase change materials (PCM) containing carbon additives for heat storage application. *Int. J. Refrig.* **2014**, *42*, 112–120. [CrossRef]
35. Wang, J.; Xie, H.; Xin, Z.; Li, Y.; Chen, L. Enhancing thermal conductivity of palmitic acid based phase change materials with carbon nanotubes as fillers. *Sol. Energy* **2010**, *84*, 339–344. [CrossRef]
36. Yu, Z.; Fang, X.; Fan, L.; Wang, X.; Xiao, Y.; Zeng, Y.; Xu, X.; Hu, Y.; Cen, K. Increased thermal conductivity of liquid paraffinbased suspensions in the presence of carbon nano-additives of various sizes and shapes. *Carbon* **2013**, *53*, 277–285. [CrossRef]
37. Li, M. A nano-graphite/paraffin phase change material with high thermal conductivity. *Appl. Energy* **2013**, *106*, 25–30. [CrossRef]
38. Li, Y.; Li, J.; Deng, Y.; Guan, W.; Wang, X.; Qian, T. Preparation of paraffin/porous TiO_2 foams with enhanced thermal conductivity as PCM, by covering the TiO_2 surface with a carbon layer. *Appl. Energy* **2016**, *171*, 37–45. [CrossRef]
39. Chen, Y.; Luo, W.; Wang, J.; Huang, J. Enhanced thermal conductivity and durability of a paraffin wax nanocomposite based on carbon-coated aluminum nanoparticles. *J. Phys. Chem. C* **2017**, *121*, 12603–12609. [CrossRef]
40. Cheng, F.; Wen, R.; Huang, Z.; Fang, M.; Liu, Y.; Wu, X.; Min, X. Preparation and analysis of lightweight wall material with expanded graphite (EG)/paraffin composites for solar energy storage. *Appl. Therm. Eng.* **2017**, *120*, 107–114. [CrossRef]
41. Seki, Y.; Ince, S.; Ezan, M.A.; Turgut, A.; Erek, A. Graphite nanoplates loading into eutectic mixture of Adipic acid and Sebacic acid as phase change material. *Sol. Energy Mater. Sol. Cells* **2015**, *140*, 457–463. [CrossRef]
42. Huang, Y.; Hsieh, T.E. Effective thermal parameters of chalcogenide thin films and simulation of phase-change memory. *Int. J. Therm. Sci.* **2015**, *87*, 207–214. [CrossRef]
43. Qiu, L.; Zheng, X.; Su, G.; Tang, D. Design and application of a freestanding sensor based on 3ω technique for thermal-conductivity measurement of solids, liquids, and nanopowders. *Int. J. Thermophys.* **2013**, *34*, 2261–2275. [CrossRef]
44. Gharebaghi, M.; Sezai, I. Enhancement of heat transfer in latent heat storage modules with internal fins. *Numer. Heat Transf. Part A* **2008**, *53*, 749–765. [CrossRef]
45. Castell, A.; Sole, C.; Medrano, M.; Roca, J.; Cabeza, L.F.; Garcia, D. Natural convection heat transfer coefficients in phase change material (PCM) modules with external vertical fins. *Appl. Therm. Eng.* **2008**, *28*, 1676–1686. [CrossRef]
46. Nayak, K.C.; Saha, S.K.; Srinivasan, K.; Dutta, P. A numerical model for heat sinks with phase change materials and thermal conductivity enhancers. *Int. J. Heat Mass Transf.* **2006**, *49*, 1833–1844. [CrossRef]
47. Zhao, C.Y.; Lu, W.; Tian, Y. Heat transfer enhancement for thermal energy storage using metal foams embedded within phase change materials (PCMs). *Solar Energy* **2010**, *84*, 1402–1412. [CrossRef]
48. Ettouney, H.M.; Alatiqi, I.; Al-Sahali, M.; Al-Hajirie, K. Heat transfer enhancement in energy storage in spherical capsules filled with paraffin wax and metal beads. *Energy Convers. Manag.* **2006**, *47*, 211–228. [CrossRef]
49. Sari, A.; Karaipekli, A. Preparation, thermal properties and thermal reliability of palmitic acid/expanded graphite composite as form-stable PCM for thermal energy storage. *Sol. Energy Mater. Sol. Cells* **2009**, *93*, 571–576. [CrossRef]
50. Rao, Z.H.; Zhang, G.Q. Thermal Properties of Paraffin Wax-based Composites Containing Graphite. *Energy Sources Part A* **2011**, *33*, 587–593. [CrossRef]
51. Wang, Y.; Xia, T.D.; Feng, H.X.; Zhang, H. Stearic acid/polymethylmethacrylate composite as form-stable phase change materials for latent heat thermal energy storage. *Renew. Energy* **2011**, *36*, 1814–1820. [CrossRef]
52. Jegadheeswaran, S.; Pohekar, S.D. Performance enhancement in latent heat thermal storage system: A review. *Renew. Sustain. Energy Rev.* **2009**, *13*, 2225–2244. [CrossRef]
53. Seeniraj, R.V.; Narasimhan, N.L. Performance enhancement of a solar dynamic LHTS module having both fins and multiple PCMs. *Sol. Energy* **2008**, *82*, 535–542. [CrossRef]
54. Shaikh, S.; Lafdi, K. Effect of multiple phase change materials (PCMs) slab configurations on thermal energy storage. *Energy Convers. Manag.* **2006**, *47*, 2103–2117. [CrossRef]
55. Cui, H.; Yuan, X.; Hou, X. Thermal performance analysis for a heat receiver using multiple phase change materials. *Appl. Therm. Eng.* **2003**, *23*, 2353–2361. [CrossRef]
56. Hawlader, M.N.A.; Uddin, M.S.; Khin, M.M. Microencapsulated PCM thermal-energy storage system. *Appl. Energy* **2003**, *74*, 195–202. [CrossRef]
57. Sari, A.; Alkan, C.; Karaipekli, A. Preparation, characterization and thermal properties of PMMA/n-heptadecane microcapsules as novel solid–liquid microPCM for thermal energy storage. *Appl. Energy* **2010**, *87*, 1529–1534. [CrossRef]
58. Muzhanje, A.T.; Hassan, M.A.; Ookawara, S.; Hassan, H. An Overview of the Preparation and Characteristics of PCMs with Nanomaterials. *J. Energy Storage* **2022**, *51*, 104353. [CrossRef]
59. Yadav, A.; Barman, B.; Kumar, V.; Kardam, A.; Shankara Narayanan, S.; Verma, A.; Madhwal, D.; Shukla, P.; Jain, V.K. *A Review on Thermophysical Properties of Nanoparticle-Enhanced PCMs for Thermal Energy Storage BT—Recent Trends in Materials and Devices*; Jain, V.K., Rattan, S., Verma, A., Eds.; Springer International Publishing: Cham, Switzerland, 2017; pp. 37–47.

60. Fullenkamp, K.; Montané, M.; Cáceres, G.; Araya-Letelier, G. Review and Selection of EPCM as TES Materials for Building Applications. *Int. J. Sustain. Energy* **2019**, *38*, 561–582. [CrossRef]
61. Kahwaji, S.; White, M.A. Edible Oils as Practical PCMs for Thermal Energy Storage. *Appl. Sci.* **2019**, *9*, 1627. [CrossRef]
62. Boussaba, L.; Makhlouf, S.; Foufa, A.; Lefebvre, G.; Royon, L. Vegetable Fat: A Low-Cost Bio-Based PCM for Thermal Energy Storage in Buildings. *J. Build. Eng.* **2019**, *21*, 222–229. [CrossRef]
63. Wonorahardjo, S.; Sutjahja, I.M.; Kurnia, D.; Fahmi, Z.; Putri, W.A. Potential of Thermal Energy Storage Using Coconut Oil for Air Temperature Control. *Buildings* **2018**, *8*, 95. [CrossRef]
64. Da Cunha, S.R.L.; de Aguiar, J.L.B. PCMs and Energy Efficiency of Buildings: A Review of Knowledge. *J. Energy Storage* **2020**, *27*, 101083. [CrossRef]
65. Al-Yasiri, Q.; Szabó, M. Incorporation of PCMs into Building Envelope for Thermal Comfort and Energy Saving: A Comprehensive Analysis. *J. Build. Eng.* **2021**, *36*, 102122. [CrossRef]
66. Akeiber, H.; Nejat, P.; Majid, M.Z.A.; Wahid, M.A.; Jomehzadeh, F.; Zeynali Famileh, I.; Calautit, J.K.; Hughes, B.R.; Zaki, S.A. A Review on PCM (PCM) for Sustainable Passive Cooling in Building Envelopes. *Renew. Sustain. Energy Rev.* **2016**, *60*, 1470–1497. [CrossRef]
67. Das, S.S.; Kumar, P.; Sandhu, S.S. Hybrid Photovoltaic-Thermal Systems Utilizing Liquid-Gas PCM. Energy Sources, Part A Recover. *Util. Environ. Eff.* **2021**, *43*, 2896–2914. [CrossRef]
68. Kong, W.; Fu, X.; Liu, Z.; Zhou, C.; Lei, J. A Facile Synthesis of Solid-Solid PCM for Thermal Energy Storage. *Appl. Therm. Eng.* **2017**, *117*, 622–628. [CrossRef]
69. Du, X.; Qiu, J.; Deng, S.; Du, Z.; Cheng, X.; Wang, H. Flame-Retardant and Solid-Solid Phase Change Composites Based on Dopamine-Decorated BP Nanosheets/Polyurethane for Efficient Solar-to-Thermal Energy Storage. *Renew. Energy* **2021**, *164*, 1–10. [CrossRef]
70. Sikiru, S.; Oladosu, T.L.; Amosa, T.I.; Kolawole, S.Y.; Soleimani, H. Recent advances and impact of PCMs on solar energy: A comprehensive review. *J. Energy Storage* **2022**, *53*, 105200. [CrossRef]
71. Lebedev, V.A.; Amer, A.E. Limitations of using PCMs for thermal energy storage. *IOP Conf. Ser. Earth Environ. Sci.* **2019**, *378*, 012044. [CrossRef]
72. Philip, J.; Shima, P.D.; Raj, B. Evidence for Enhanced Thermal Conduction through Percolating Structures in Nanofluids. *Nanotechnology* **2008**, *19*, 305706. [CrossRef]
73. Chon, C.H.; Kihm, K.D.; Lee, S.P.; Choi, S.U.; Chon, C.H.; Kihm, K.D. Empirical correlation finding the role of temperature and particle size for nanofluid (Al_2O_3) thermal conductivity enhancement. *Appl. Phys. Lett.* **2005**, *87*, 153107. [CrossRef]
74. Beck, M.P.; Yuan, Y.; Warrier, P.; Teja, A.S. The Effect of Particle Size on the Thermal Conductivity of Alumina Nanofluids. *J. Nanoparticle Res.* **2009**, *11*, 1129–1136. [CrossRef]
75. Jeong, J.; Li, C.; Kwon, Y.; Lee, J.; Kim, S.H.; Yun, R. Particle Shape Effect on the Viscosity and Thermal Conductivity of ZnO Nanofluids. *Int. J. Refrig.* **2013**, *36*, 2233–2241. [CrossRef]
76. Murshed, S.M.S. Simultaneous Measurement of Thermal Conductivity, Thermal Diffusivity, and Specific Heat of Nanofluids. *Heat Transf. Eng.* **2012**, *33*, 722–731. [CrossRef]
77. Zhu, H.; Zhang, C.; Liu, S.; Tang, Y.; Yin, Y. Effects of Nanoparticle Clustering and Alignment on Thermal Conductivities of Fe_3O_4 Aqueous Nanofluids. *Appl. Phys. Lett.* **2006**, *89*, 23123. [CrossRef]
78. Hong, T.-K.; Yang, H.-S.; Choi, C.J. Study of the Enhanced Thermal Conductivity of Fe Nanofluids. *J. Appl. Phys.* **2005**, *97*, 064311. [CrossRef]
79. Dadwal, A.; Joy, P.A. Particle Size Effect in Different Base Fluids on the Thermal Conductivity of Fatty Acid Coated Magnetite Nanofluids. *J. Mol. Liq.* **2020**, *303*, 112650. [CrossRef]
80. Kamalvand, M.; Karami, M. A Linear Regularity between Thermal Conductivity Enhancement and Fluid Adsorption in Nanofluids. *Int. J. Therm. Sci.* **2013**, *65*, 189–195. [CrossRef]
81. Altan, C.L.; Gurten, B.; Sommerdijk, N.A.J.M.; Bucak, S. Deterioration in Effective Thermal Conductivity of Aqueous Magnetic Nanofluids. *J. Appl. Phys.* **2014**, *116*, 224904. [CrossRef]
82. Colla, L.; Fedele, L.; Mancin, S.; Danza, L.; Manca, O. Nano-PCMs for Enhanced Energy Storage and Passive Cooling Applications. *Appl. Therm. Eng.* **2017**, *110*, 584–589. [CrossRef]
83. Harish, S.; Orejon, D.; Takata, Y.; Kohno, M. Thermal Conductivity Enhancement of Lauric Acid Phase Change Nanocomposite with Graphene Nanoplatelets. *Appl. Therm. Eng.* **2015**, *80*, 205–211. [CrossRef]
84. Sarı, A.; Biçer, A.; Hekimoğlu, G. Effects of Carbon Nanotubes Additive on Thermal Conductivity and Thermal Energy Storage Properties of a Novel Composite PCM. *J. Compos. Mater.* **2018**, *53*, 2967–2980. [CrossRef]
85. Yadav, A.; Barman, B.; Kardam, A.; Narayanan, S.S.; Verma, A.; Jain, V.K. Thermal Properties of Nano-Graphite-Embedded Magnesium Chloride Hexahydrate Phase Change Composites. *Energy Environ.* **2017**, *28*, 651–660. [CrossRef]
86. Sharma, R.K.; Ganesan, P.; Tyagi, V.V.; Metselaar, H.S.C.; Sandaran, S.C. Thermal Properties and Heat Storage Analysis of Palmitic Acid-TiO_2 Composite as Nano-Enhanced Organic PCM (NEOPCM). *Appl. Therm. Eng.* **2016**, *99*, 1254–1262. [CrossRef]
87. Vivekananthan, M.; Amirtham, V.A. Characterization and Thermophysical Properties of Graphene Nanoparticles Dispersed Erythritol PCM for Medium Temperature Thermal Energy Storage Applications. *Thermochim. Acta* **2019**, *676*, 94–103. [CrossRef]
88. Putra, N.; Amin, M.; Kosasih, E.A.; Luanto, R.A.; Abdullah, N.A. Characterization of the Thermal Stability of RT 22 HC/Graphene Using a Thermal Cycle Method Based on Thermoelectric Methods. *Appl. Therm. Eng.* **2017**, *124*, 62–70. [CrossRef]

89. Nourani, M.; Hamdami, N.; Keramat, J.; Moheb, A.; Shahedi, M. Thermal Behavior of Paraffin-Nano-Al2O3 Stabilized by Sodium Stearoyl Lactylate as a Stable PCM with High Thermal Conductivity. *Renew. Energy* **2016**, *88*, 474–482. [CrossRef]
90. Wang, F.; Liu, J.; Fang, X.; Zhang, Z. Graphite Nanoparticles-Dispersed Paraffin/Water Emulsion with Enhanced Thermal-Physical Property and Photo-Thermal Performance. *Sol. Energy Mater. Sol. Cells* **2016**, *147*, 101–107. [CrossRef]
91. Zhou, S.Q.; Ni, R. Measurement of the specific heat capacity of water-based Al_2O_3 nanofluid. *Appl. Phys. Lett.* **2008**, *92*, 093123. [CrossRef]
92. O'Hanley, H.; Buongiorno, J.; McKrell, T.; Hu, L.W. Measurement and model correlation of specific heat capacity of water-based nanofluids with silica, alumina and copper oxide nanoparticles. *ASME Int. Mech. Eng. Congr. Expo.* **2011**, *54969*, 1209–1214.
93. Vajjha, R.S.; Das, D.K. Specific heat measurement of three nanofluids and development of new correlations. *J. Heat Transfer.* **2009**, *131*, 071601. [CrossRef]
94. O'Hanley, H.; Buongiorno, J.; McKrell, T.; Hu, L.W. Measurement and model validation of nanofluid specific heat capacity with differential scanning calorimetry. *Adv. Mech. Eng.* **2012**, *4*, 181079. [CrossRef]
95. Barbés, B.; Páramo, R.; Blanco, E.; Casanova, C. Thermal conductivity and specific heat capacity measurements of CuO nanofluids. *J. Therm. Anal. Calorim.* **2014**, *115*, 1883–1891. [CrossRef]
96. Teng, T.P.; Lin, L.; Yu, C.C. Preparation and Characterization of Carbon Nanofluids by Using a Revised Water-Assisted Synthesis Method. *J. Nanomater.* **2013**, *2013*, 133. [CrossRef]
97. De Robertis, E.; Cosme, E.H.H.; Neves, R.S.; Kuznetsov, A.Y.; Campos, A.P.C.; Landi, S.M.; Achete, C.A. Application of the modulated temperature differential scanning calorimetry technique for the determination of the specific heat of copper nanofluids. *Appl Therm Eng* **2012**, *41*, 10–17. [CrossRef]
98. Starace, A.K.; Gomez, J.C.; Wang, J.; Pradhan, S.; Glatzmaier, G.C. Nanofluid heat capacities. *J. Appl. Phys.* **2011**, *110*, 124323. [CrossRef]
99. Lin, Y.S.; Hsiao, P.Y.; Chieng, C.C. Roles of nanolayer and particle size on thermophysical characteristics of ethylene glycol-based copper nanofluids. *Appl. Phys. Lett.* **2011**, *98*, 153105. [CrossRef]
100. Wang, C.; Lin, T.; Li, N.; Zheng, H. Heat transfer enhancement of phase change composite material: Copper foam/paraffin. *Renew. Energy* **2016**, *96*, 960–965. [CrossRef]
101. Han, P.; Zheng, X.; Hou, W.; Qiu, L.; Tang, D. Study on heat-storage and release characteristics of multi-cavity-structured phase change microcapsules. *Phase Transit.* **2015**, *88*, 704–715. [CrossRef]
102. Babuska, I.; Szabo, B.A.; Katz, I.N. The p-version of the finite element method. *SIAM J. Numer. Anal.* **1981**, *18*, 515–545. [CrossRef]
103. Reddy, J. *An Introduction to the Finite Element Method*; McGraweHill: New York, NY, USA, 2013.
104. Wang, G.; Wei, G.; Xu, C.; Ju, X.; Yang, Y.; Du, X. Numerical simulation of effective thermal conductivity and pore-scale melting process of PCMs in foam metals. *Appl. Therm. Eng.* **2019**, *147*, 464–472. [CrossRef]
105. Lohrasbi, S.; Sheikholeslami, M.; Ganji, D.D. Multi-objective RSM optimization of fin assisted latent heat thermal energy storage system based on solidification process of phase change Material in presence of copper nanoparticles. *Appl. Therm. Eng.* **2017**, *118*, 430–447. [CrossRef]
106. Mahdi, J.M.; Lohrasbi, S.; Ganji, D.D.; Nsofor, E.C. Simultaneous energy storage and recovery in the triplex-tube heat exchanger with PCM, copper fins and Al_2O_3 nanoparticles. *Energy Convers. Manag.* **2019**, *180*, 949–961. [CrossRef]
107. Shin, D.; Banerjee, D. Enhanced Specific Heat of Silica Nanofluid. *ASME. J. Heat Transfer.* **2011**, *133*, 024501. [CrossRef]
108. Tiznobaik, H.; Shin, D. Enhanced specific heat capacity of high-temperature molten salt-based nanofluids. *Int. J. Heat Mass Transf.* **2013**, *57*, 542–548. [CrossRef]
109. Dudda, B.; Shin, D. Investigation of molten salt nanomaterial as thermal energy storage in concentrated solar power. In Proceedings of the ASME 2012 International Mechanical Engineering Congress & Exposition, IMECE2012, Houston, TX, USA, 9–15 November 2012.
110. Dudda, B.; Shin, D. Effect of nanoparticle dispersion on specific heat capacity of a binary nitrate salt eutectic for concentrated solar power applications. *Int. J. Therm. Sci.* **2013**, *69*, 37–42. [CrossRef]
111. Lu, M.C.; Huang, C.H. Specific heat capacity of molten salt-based alumina nanofluid. *Nanoscale Res. Lett.* **2013**, *8*, 292. [CrossRef]
112. Chieruzzi, M.; Cerritelli, G.F.; Miliozzi, A.; Kenny, J.M. Effect of nanoparticles on heat capacity of nanofluids based on molten salts as PCM for thermal energy storage. *Nanoscale Res. Lett.* **2013**, *8*, 448. [CrossRef]
113. Ho, M.X.; Pan, C. Optimal concentration of alumina nanoparticles in molten Hitec salt to maximize its specific heat capacity. *Int. J. Heat Mass Transf.* **2014**, *70*, 174–184. [CrossRef]
114. Shin, D.; Banerjee, D. Enhancement of specific heat capacity of high-temperature silica-nanofluids synthesized in alkali chloride salt eutectics for solar thermal-energy storage applications. *Int. J. Heat Mass Transf.* **2011**, *54*, 1064–1070. [CrossRef]
115. Shin, D.; Banerjee, D. Experimental investigation of molten salt nanofluid for solar thermal energy application. *ASME/JSME Therm. Eng. Jt. Conf.* **2011**, *38921*, T30024.
116. Likhachev, V.N.; Vinogradov, G.A.; Alymov, M.I. Anomalous heat capacity of nanoparticles. *Phys. Lett. A* **2006**, *357*, 236–239. [CrossRef]
117. Jung, S.; Banerjee, D. A simple analytical model for specific heat of nanofluid with tube shaped and disc shaped nanoparticles. *ASME/JSME Therm. Eng. Jt. Conf.* **2011**, *38921*, T30023.
118. Ebadi, S.; Tasnim, S.H.; Aliabadi, A.A.; Mahmud, S. Geometry and nanoparticle loading effects on the bio-based nano-PCM filled cylindrical thermal energy storage system. *Appl. Therm. Eng.* **2018**, *141*, 724–740. [CrossRef]

119. Chieruzzi, M.; Cerritelli, G.F.; Miliozzi, A.; Kenny, J.M.; Torre, L. Heat capacity of nanofluids for solar energy storage produced by dispersing oxide nanoparticles in nitrate salt mixture directly at high temperature. *Sol. Energy Mater. Sol. Cells* **2017**, *167*, 60–69. [CrossRef]
120. Liu, Y.; Yang, Y. Investigation of specific heat and latent heat enhancement in hydrate salt based TiO_2 nanofluid PCM. *Appl. Therm. Eng.* **2017**, *124*, 533–538. [CrossRef]
121. Wang, F.; Zhang, C.; Liu, J.; Fang, X.; Zhang, Z. Highly stable graphite nanoparticle-dispersed phase change emulsions with little supercooling and high thermal conductivity for cold energy storage. *Appl. Energy* **2017**, *188*, 97–106. [CrossRef]
122. Warzoha, R.J.; Weigand, R.M.; Fleischer, A.S. Temperature-dependent thermal properties of a paraffin PCM embedded with herringbone style graphite nanofibers. *Appl. Energy* **2015**, *137*, 716–725. [CrossRef]
123. Muthoka, M.J.; Xuelai, Z.; Yuyang, Y.; Yue, C.; Xiaofeng, X. Latent heat of fusion prediction for nanofluid based PCM. *Appl. Therm. Eng.* **2018**, *130*, 1590–1597. [CrossRef]
124. Souza, R.R.; Faustino, V.; Gonçalves, I.M.; Moita, A.S.; Bañobre-López, M.; Lima, R. A Review of the Advances and Challenges in Measuring the Thermal Conductivity of Nanofluids. *Nanomaterials* **2022**, *12*, 2526. [CrossRef]
125. Lin, S.C.; Al-Kayiem, H.H. Evaluation of copper nanoparticles—Paraffin wax compositions for solar thermal energy storage. *Sol. Energy* **2016**, *132*, 267–278. [CrossRef]
126. Kannan, P.C.D.; Nadaraj, P. Improving the Efficiency of Solar Flat Plate Solar Collector Using PCM and Nanofluids. *Int. Res. J. Adv. Sci. Hub* **2020**, *11*, 14.
127. Hasan, A.; McCormack, S.J.; Huang, M.J.; Norton, B. Evaluation of PCMs for thermal regulation enhancement of building integrated photovoltaics. *Sol. Energy* **2010**, *84*, 1601–1612. [CrossRef]
128. Chow, T.T.; Pei, G.; Fong, K.F.; Lin, Z.; Chan, A.L.S.; Ji, J. Energy and exergy analysis of photovoltaic-thermal collector with and without glass cover. *Appl. Energy* **2009**, *86*, 310–316. [CrossRef]
129. Hu, C.; White, R.M. *Solar Cells from Basic to Advanced Systems*; McGraw-Hill: New York, NY, USA, 1983.
130. Kumar, R.; Rosen, M.A. Performance evaluation of a double pass PV/T solar air heater with and without fins. *Appl. Therm. Eng.* **2011**, *31*, 1402–1410. [CrossRef]
131. Bejan, A. *Entropy Generation Minimization*; Wiley: New York, NY, USA, 1982.
132. Petela, R. Energy of heat radiation. *Heat Transf.* **1964**, *86*, 187–192. [CrossRef]
133. Nada, S.A.; El-Nagar, D.H. Possibility of using PCMs in temperature control and performance enhancements of free stand and building integrated PV modules. *Renew. Energy* **2018**, *127*, 630–641. [CrossRef]
134. Al-Waeli, A.H.A.; Sopian, K.; Chaichan, M.T.; Kazem, H.A.; Ibrahim, A.; Mat, S.; Ruslan, M.H. Evaluation of the Nanofluid and Nano-PCM Based Photovoltaic Thermal (PVT) System: An Experimental Study. *Energy Convers. Manag.* **2017**, *151*, 693–708. [CrossRef]
135. Sardarabadi, M.; Passandideh-Fard, M.; Maghrebi, M.-J.; Ghazikhani, M. Experimental Study of Using Both ZnO/Water Nanofluid and PCM (PCM) in Photovoltaic Thermal Systems. *Sol. Energy Mater. Sol. Cells* **2017**, *161*, 62–69. [CrossRef]
136. Hosseinzadeh, M.; Sardarabadi, M.; Passandideh-Fard, M. Energy and Exergy Analysis of Nanofluid Based Photovoltaic Thermal System Integrated with PCM. *Energy* **2018**, *147*, 636–647. [CrossRef]
137. Hassan, A.; Wahab, A.; Qasim, M.A.; Janjua, M.M.; Ali, M.A.; Ali, H.M.; Jadoon, T.R.; Ali, E.; Raza, A.; Javaid, N. Thermal Management and Uniform Temperature Regulation of Photovoltaic Modules Using Hybrid PCMs-Nanofluids System. *Renew. Energy* **2020**, *145*, 282–293. [CrossRef]
138. Al-Waeli, A.H.A.; Sopian, K.; Kazem, H.A.; Yousif, J.H.; Chaichan, M.T.; Ibrahim, A.; Mat, S.; Ruslan, M.H. Comparison of prediction methods of PV/T nanofluid and nano-PCM system using a measured dataset and artificial neural network. *Sol. Energy* **2018**, *162*, 378–396. [CrossRef]
139. Sarafraz, M.M.; Safaei, M.R.; Leon, A.S.; Tlili, I.; Alkanhal, T.A.; Tian, Z.; Goodarzi, M.; Arjomandi, M. Experimental Investigation on Thermal Performance of a PV/T-PCM (Photovoltaic/Thermal) System Cooling with a PCM and Nanofluid. *Energies* **2019**, *12*, 1–16. [CrossRef]
140. Al-Waeli, A.H.A.; Kazem, H.A.; Yousif, J.H.; Chaichan, M.T.; Sopian, K. Mathematical and neural network modeling for predicting and analyzing of nanofluid-nano PCM photovoltaic thermal systems performance. *Renew. Energy* **2020**, *145*, 963–980. [CrossRef]
141. Al-Waeli, A.H.A.; Sopian, K.; Yousif, J.H.; Kazem, H.A.; Boland, J.; Chaichan, M.T. Artificial neural network modeling and analysis of photovoltaic/thermal system based on the experimental study. *Energy Convers. Manag.* **2019**, *186*, 368–379. [CrossRef]
142. Al-Waeli, A.H.A.; Chaichan, M.T.; Sopian, K.; Kazem, H.A.; Mahood, H.B.; Khadom, A.A. Modeling and experimental validation of a PVT system using nanofluid coolant and nano-PCM. *Sol. Energy* **2019**, *177*, 178–191. [CrossRef]
143. Sardarabadi, M.; Passandideh-Fard, M.; Heris, S.Z. Experimental investigation of the effects of silica/water nanofluid on PV/T (photovoltaic thermal units). *Energy* **2014**, *66*, 264–272. [CrossRef]
144. Al-Waeli, A.H.A.; Chaichan, M.T.; Kazem, H.A.; Sopian, K.; Safaei, J. Numerical study on the effect of operating nanofluids of photovoltaic thermal system (PV/T) on the convective heat transfer. *Case Stud. Therm. Eng.* **2018**, *12*, 405–413. [CrossRef]
145. Naghdbishi, A.; Yazdi, M.E.; Akbari, G. Experimental Investigation of the Effect of Multi-Wall Carbon Nanotube—Water/Glycol Based Nanofluids on a PVT System Integrated with PCM-Covered Collector. *Appl. Therm. Eng.* **2020**, *178*, 115556. [CrossRef]
146. Salari, A.; Kazemian, A.; Ma, T.; Hakkaki-Fard, A.; Peng, J. Nanofluid based photovoltaic thermal systems integrated with PCMs: Numerical simulation and thermodynamic analysis. *Energy Convers. Manag.* **2020**, *205*, 112384. [CrossRef]

147. Abdelrahman, H.; Wahba, M.; Refaey, H.; Moawad, M.; Berbish, N. Performance enhancement of photovoltaic cells by changing configuration and using PCM (RT35HC) with nanoparticles Al_2O_3. *Sol. Energy* **2019**, *177*, 665–671. [CrossRef]
148. Abdollahi, N.; Rahimi, M. Potential of Water Natural Circulation Coupled with Nano-Enhanced PCM for PV Module Cooling. *Renew. Energy* **2020**, *147*, 302–309. [CrossRef]
149. Lari, M.O.; Sahin, A.Z. Effect of retrofitting a silver/water nanofluid-based photovoltaic/thermal (PV/T) system with a PCM-thermal battery for residential applications. *Renew. Energy* **2018**, *122*, 98–107. [CrossRef]

photonics

MDPI

Article

The Light-Trapping Character of Pit Arrays on the Surface of Solar Cells

Baohua Zhu [1,2], Le Chen [1,*], Song Ye [1,*] and Wei Luo [1]

[1] Guangxi Key Laboratory of Optoelectronic Information Processing, School of Optoelectronic Engineering, Guilin University of Electronic Technology, Guilin 541004, China
[2] School of Materials Science and Engineering, Guilin University of Electronic Technology, Guilin 541004, China
* Correspondence: chenle11@126.com (L.C.); yesong@guet.edu.cn (S.Y.)

Abstract: Surfaces with light-trapping structures are widely used in solar cells to enhance light capturing and to transform efficiency. The study of light-trapping character is important for light-trapping structures in solar cells. In the present study, the light-trapping character for the regular hemisphere pit arrays (RHPAs) in solar cells was intensively investigated in terms of reducing light reflection, suppressing light escape, and increasing the length of the optical path. Results show that the RHPAs can decrease surface reflectivity by ~54% compared with the plane structure, and can reflect ~33% of the light that has not been absorbed back into the absorption layer of the solar cell. The total optical path of the cell with the RHPAs structure remarkably increased from 2ω to 4ω. To verify the theoretical research conclusions, we produced the glass structure samples with different aspect ratios by using micro/nanometer-processing technology. The reflection ratios for silicon wafers covered by plane and RHPAs glass samples were tested. The test results were compared with the theoretical calculation results, which showed consistency.

Keywords: light-trapping character; regular hemisphere pit arrays (RHPAs); anti-reflection; optical path length

Citation: Zhu, B.; Chen, L.; Ye, S.; Luo, W. The Light-Trapping Character of Pit Arrays on the Surface of Solar Cells. *Photonics* **2023**, *10*, 855. https://doi.org/10.3390/photonics10070855

Received: 31 May 2023
Revised: 28 June 2023
Accepted: 29 June 2023
Published: 24 July 2023

1. Introduction

Photovoltaic energy has broad prospects for green energy development. However, compared with traditional energy resources, solar energy experiences deficiencies such as high manufacturing cost and unideal transform efficiency, which seriously limit its wide application. To improve the conversion efficiency, researchers have developed light-trapping structures for various solar cells, such as nano/micro-pyramids [1–4], nanowires [5–7], nanocone [8–12], nanosphere [13–16], and porous silicon structures [17,18]. Light-trapping structures possess broadband optical absorption abilities, which can effectively lower the loss of light caused by reflection and raise the valid length of optical paths in the active layer of solar cells [19]. The fabrication of the trapping structures often relies on the relatively expensive micro-nano processing technology, such as lithographic development, dry and wet etching, and reactive ion etching. Considering that the processes are often manufactured internally within the solar cells, it is easy to damage the primitive structure, which may reduce the lifetime of the devices. Therefore, the prospect of scaled production of the light-trapping structures prepared inside the cells remains uncertain.

Surface light-trapping technology has been proposed for the significant improvement with two distinct advantages of high reliability and non-disruption of the devices. Quarter-wavelength anti-reflection (AR) coating is among the most common surface light-trapping technologies, which has been widely used for different types of photovoltaic devices or modules [20–23]. However, its conversion effectiveness is dependent on wavelength and the angle of incidence. The preparation of efficient multistory AR coatings is also inseparable from many complex chemical and physical processes, such as sol-gel and magnetron sputtering. Alternative technology is based on a low-cost, flexible, polydimethylsiloxane

(PDMS) thin film, which is processed into 3D nanopit and covered in the outer face of the solar cell without a binding agent [24,25]. Experimental results show that the PDMS thin membrane has superb anti-reflection performance and can be applied to many types of solar cell to promote the transform efficiency. However, the absorption of incident light by the PDMS and the air stored between the PDMS and solar cell may result in wasted light energy. In addition, SiO_2 microparticles with distinct optical characteristics are also attractive [26–29]. It can effectively increase the light absorption through proper arrangement on the cell surface. SiO_2 hemisphere texture surface shows a perfect function and prospect in the application field of high-efficiency photovoltaic cells. However, the preparation of large-area and uniform SiO_2 microsphere surfaces on the surface of the solar cell is difficult.

A novel and efficient surface-trapping technology has attracted the attention of researchers. Wang et al. used an effective light-coupling method, in which a periodic quasi-hemisphere micro-nanostructure was prepared on the front glass surface to improve the property in thin-membrane solar cells [30–32]. This process does not require a change in the processing technology of thin-film solar cells with a flat glass substrate. Experimental results show that the pit arrays have good light-trapping properties in the solar cell, particularly for Si thin-membrane solar cells, including single and tandem solar cells. Shen et al. have reported that, compared with the classical plane surface solar cell, the efficiency of the surface of an a-Si:H/μc-Si:H tandem solar cell of periodic pit array increased from 11.67% to 12.23%, and the short-circuit current density (*Jsc*) increased by 4.6%. In the present study, compared with the flat cell, the trapping structure of hemispherical pit arrays can increase the transformation efficiency by ~6% and the short-circuit current by ~7% for a single a-Si:H thin-film solar cell with a textured surface. The surface with periodic micro-structured arrays has shown good light-capture properties in solar cells. Although there have been many reports on the trapping mechanism of light-trapping structures in solar cells [33–35], there is still a lack of in-depth theoretical research on the surface-trapping mechanism, especially on the surface of the light-capture mechanism. The light-capture mechanism is important for further understanding the nature of the light capture and ameliorating the high-efficiency light-capture structures for solar cells. In this work, numerical analyses and geometrical optics simulations were carried out to investigate the light-capture mechanism of surface pit arrays with different aspect ratios in silicon solar cells, including anti-reflection characteristics, overall reflection ratio, and overall length of light path. Moreover, the theoretical research conclusions were verified by manufacturing the glass structure samples by using micro/nanometer-processing technology. The reflection ratios of silicon wafers covered by the plane and RHPAs glass samples were measured for comparison. Lastly, theoretical calculations were carried out for comparison with the experimental measurements.

2. Fabrication and Simulation of the RHPA Structures

2.1. Fabrication Process and Characterization

Figure 1 shows a thin-film silicon solar cell with the RHPAs on its surface, and a quasi RHPA structure was fabricated on smooth quartz glass. First, a 3-inch optical glass was washed with ethanol and acetone, and then boiled at 180 °C in the admixture of dense H_2SO_4 and H_2O_2 with bulk ratio of 3:1 for the removal of various residual organic matter. Second, a semispherical pit array structure was fabricated on an optical glass by using various types of micro-nanometer processing techniques, including UV lithography, the sputtering of metal seed mask, HF (buffered hydrofluorocarbon acid solution) etching, and ion beam etching. Figure 2 schematically shows the key fabrication process. The pit array's depth and period (the distance between two adjacent pits) could be improved by altering the isotropical etching time in the HF solution and the mask size. A mask with a period of 10 μm and diameter of 2.5 μm was used, because a 10 μm feature size structure can obtain a better aspect ratio of the pit arrays and meet the spectral range of the solar cell to absorb light.

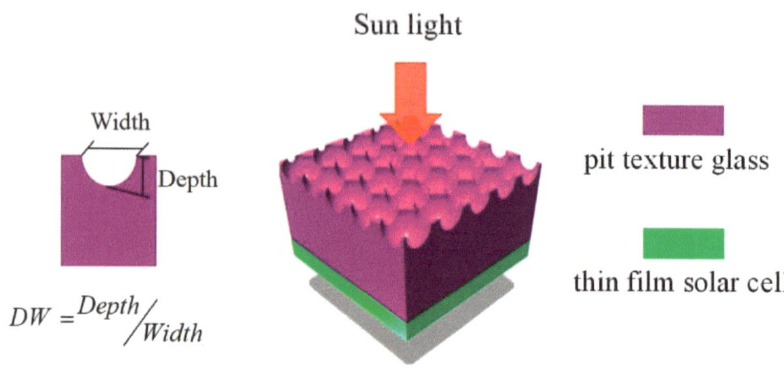

Figure 1. Schematic diagram of the thin-film Silicon solar cell with regular pit arrays.

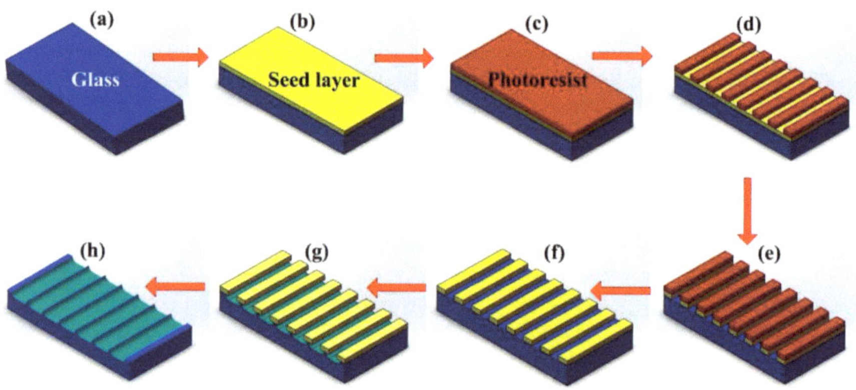

Figure 2. Schematic of key fabrication process of regular pit arrays on glass substrate: (**a**) washing glass substrate; (**b**) sputtering metal seed layer; (**c**) spinning photoresist layer; (**d**) lithography and developing; (**e**) etching metal layer; (**f**) moving photoresist; (**g**) etching glass with HF solution; (**h**) removing metal seed layer.

Notably, the aspect ratio (DW = Depth/Width) of the pit array is equal to or below 0.5 ($DW \leq 0.5$) owing to the isotropic properties of glass materials. Thus, the ultimate RHPA prepared from the above-mentioned approach is a quasi-hemisphere pit array ($DW \approx 0.45$). If the pit structure with high aspect ratio ($DW \geq 0.5$) needs to be prepared on the surface of the glass, the combined processing of HF etching and reactive-ion etching (RIE) needs to be carried out. First, RIE was used to prepare the high-aspect-ratio pit arrays, and then HF etching was carried out to obtain the smooth surface. Other processes are similar to those as described above.

The morphology of the RHPAs' glass experimental products was observed using a field emission scanning electron microscope. The transmission spectrum and light reflection ratio were observed using a UV-vis-NIR spectrophotometer.

2.2. Simulation of Optical Transmission Property

The optical transmission property for the structure of RHPAs in solar cells was studied by using the particle tracing module of COMSOL Multiphysics. For the convenience of research, during the simulation process, the refractive indices of Si, air, and glass were set to 3.5, 1, and 1.5, respectively. At the same time, the imaginary part of their refractive indices was ignored, and the incident optical wavelength was set to 650 nm. The process of light propagation was visualized using a hemispherical pit model (radius = 1, dimensionless) for simulating the propagation of light in the RHPAs. Furthermore, the reflectivity transmittance, scattering rate, and light path length were calculated using the Visual Basic (VB) program.

3. Light-Trapping Mechanisms and Discussions

Figure 3 shows the light transmission properties in a solar cell model with the RHPAs' surface, in which the aspect ratio was $DW = 0.5$. It shows that the hemisphere of the RHPAs can achieve partial incident light re-absorption via the total reflection inside pit arrays. It also depicts that the light incident from region I ($0 < x < 0.5$), where x is the distance from the center of the pit, cannot be trapped in the pit, as shown in Figure 3a. However, the light from the II ($0.5 \leq x < 0.81$) or III ($0.81 \leq x \leq 1$) area can be adjusted using the refraction and total internal reflection of the side of the pit to achieve a two- and three-fold increase in light injection, respectively. The repeated injection of the incoming optical evolved the total reflectance (R) of the solar cells to R^k, where k is a multiple of the injection. The PHRAs' structure can scatter the incident light and enter the solar cells at a certain angle, as shown in Figure 3b. It is beneficial for improving the light path in the cells. The other significant property of RHPAs is that they may return the light that has not been absorbed to the absorbing layer once more. As shown in Figure 3c, when the internal optical ray is transmitted from the glass to the air, if the ray does not meet the requirements of total reflection ($\theta < \theta_C$), the glass would stop it, whereas when $\theta > \theta_C$, the optical ray would be reflected by the RHPAs and would return to the active layer, where θ is the incident angle, and θ_C is the critical angle of total reflection. Therefore, the light-trapping mechanism of the RHPAs can be summarized in terms of reducing light reflection, suppressing light escape, and increasing optical path length, as shown in Figure 3d. Each aspect of the mechanisms was investigated as detailed in the following sections.

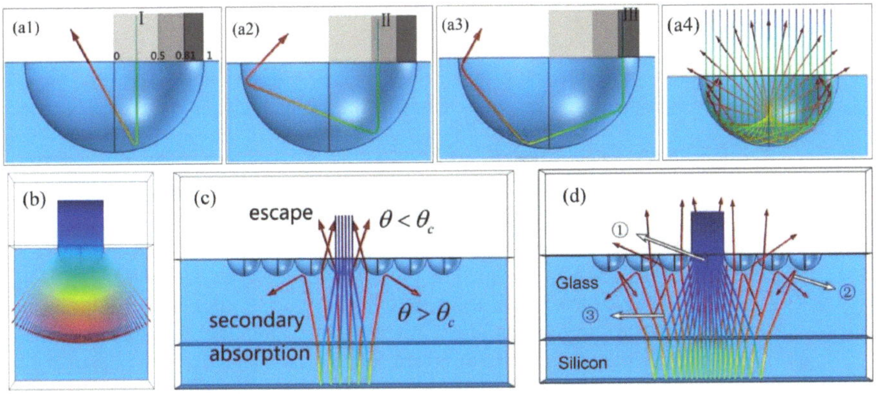

Figure 3. Transmission properties of light in a solar cell model with RHPA surface. The reflection path of incoming rays in area I (**a1**), II (**a2**), and III (**a3**), and a view of the whole reflection (**a4**). (**b**) Scattering light of the hemisphere pit. (**c**) Transmission or reflection path of rays hitting from itself to the RHPAs. (**d**) PHRAs' total light capture mechanism in the solar cell.

(a) Reducing Light Reflection

Generally, the surface reflectivity of glass is ca. 4%. However, if the light can be reflected twice, the reflectivity will reach as low as 0.16%. The higher the time of reflection, the lower the reflectivity is. The hemispherical pit's radius was set to 1. Accordingly, the critical conditions for multiple incidence could be determined with programming calculations, and the schematic diagram is shown in Figure 4, where r_I, r_{II}, and r are the critical conditions when light is injected into the cell the first, second, and greater than or equal to the third time with values of 0.5, 0.81, and 1, respectively. The corresponding areas of $S_I = \pi r_I^2$, $S_{II} = \pi(r_{II}^2 - r_I^2)$, and $S_{III} = \pi(r^2 - r_{II}^2)$ are r_I, r_{II}, and r, respectively. Therefore, for the hemisphere pit arrays, the surface reflectance is as follows:

$$R_{hemisphere} = (\frac{S_0 - S}{S_0} + \frac{S_I}{S})R^1 + \frac{S_{II}}{S}R^2 + \frac{S_{III}}{S}R^3 = 0.4625R + 0.406R^2 + 0.3439R^3 \quad (1)$$

where R is the reflection coefficient of the glass, $S = \pi r^2$ is the total area of the hemisphere pit, and $S_0 = (2r)^2$ is the area of the square cell. For glass with a refractive index $n = 1.5$, $R \approx 0.04$, which is much less than 1. At $r = 1$, $R_{hemisphere} \approx 0.4625R = 0.0185$, indicating that the hemispherical pit structure model can reduce surface reflectance by more than half (~54%) compared with the flat structure. In the actual structure, its surface reflectance may be lower. Thus, the hemisphere pit arrays can reduce the light reflection and capture the majority of the optics into solar cells.

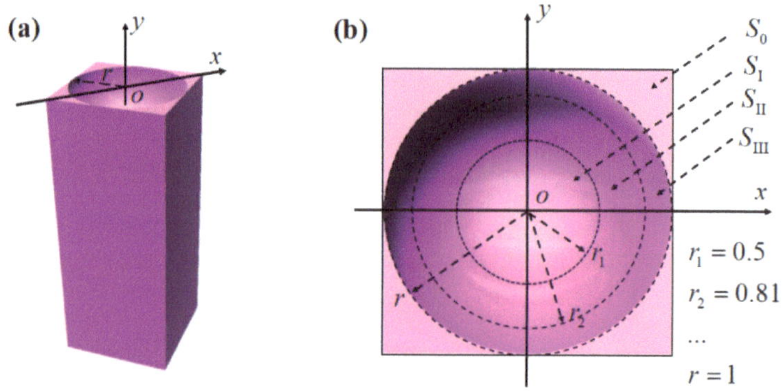

Figure 4. (**a**) Mathematical and physical model of the hemispheric pit. (**b**) The schematic diagram of the interaction between the incident light in different regions and the hemispherical pit structure.

(b) Suppression of Escaped Light

In solar cells, escaping light is another crucial problem. The suppression of escaped light will further enhance the absorption of incoming light. When light arrives at the active layer of solar cells through the glass, it interacts with the active layer or becomes reflected back into the air. The more light that is not absorbed or reflected, the lower the conversion efficiency of the cell. Otherwise, the efficiency will be improved. Therefore, surface-trapping technology is expected to play an active role in this aspect. The RHPA structure plays such an important role because it can return part of the unabsorbed light back to the active layer once more. When internal light is transmitted from the glass to the air, if the light does not reach the requirements of total reflection, it will stop at the glass. Otherwise, it will be reflected into the solar cell again. The critical angle is $\theta_C = 41^\circ$, when the indexes of refraction of air and RHPAs were $n_{air} = 1$ and $n_{RHPAs} = 1.5$, respectively.

Figure 5a shows the diagrammatic sketch of the optical ray in the RHPAs with $DW = 0.5$. It is assumed that the entire optical ray will be totally reflected at the bottom of the RHPAs. When the internal optical ray is transmitted from the RHPAs to the air, it will stop the RHPAs if $\theta < \theta_C$. By contrast, the optical ray would be reflected by the RHPAs and returned to the active layer of the cells again. Based on the total reflection theorem, by using the VB programming method, the quantitative relationship among incoming, transmission, and reflection light was calculated, and the results are shown in Figure 5b. In the figure, p_{tra} and p_{ref} represent the percentage of transmission light and reflection light of each position in all incoming light, respectively. By using the integral method, the probability of total reflection and total transmission was obtained as follows:

$$S_{ref} = \int_{x=0}^{\infty} P_{ref}dx = 64.66 \text{ and } S_{tra} = \int_{x=0}^{\infty} P_{tra}dx = 129.51 \quad (2)$$

$$P_{total-ref} = \frac{S_{ref}}{S_{tra} + S_{ref}} = 33.21\% \text{ and } P_{total-ref} = \frac{S_{tra}}{S_{tra} + S_{ref}} = 66.79\% \quad (3)$$

where S_{ref} and S_{tra} are the weighted integral of the area of p_{ref} and p_{tra}, and $P_{total-ref}$ and $P_{total-tra}$ are the probability of total reflection and total transmission, respectively. Therefore, the hemisphere RHPAs can return 33.21% of the light that has not been absorbed to the absorbing layer once more. This unabsorbed light will be absorbed for a second time. The probability of light that escapes directly without being absorbed is 66.79%, and this part of light escapes directly. These important data show that about 30% of light recycling can be obtained by the hemisphere RHPAs. Therefore, RHPAs can impede the escaping light and enhance the utilization of incoming light as the second aspect of the light-trapping mechanism in this investigation.

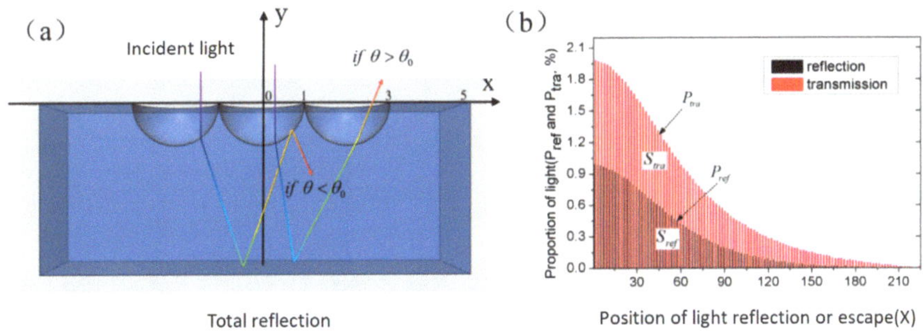

Figure 5. (**a**) (Total reflection.) The diagrammatic sketch of optical rays in RHPAs. (**b**) (Position of light reflection or escape (X).) The diagrammatic sketch of the percentage of transmitted light and reflected light of each position for all incoming light.

(c) Increasing Light Path Length

In addition, RHPAs can scatter light and raise the propagation length of optical rays in the active layer. Reference [36] studied the absorption layer conditions in an ideal suede structure [36], and referring to the physical ideas within it, we will discuss the degree of increase in optical path of silicon thin-film solar cells caused by the hemisphere RHPAs' structure relative to plane structure. Figure 6a shows the propagation of the optical ray in the cell model with RHPAs. Figure 6b shows the comparison of light propagation in the cell model of the plane structure and the cell model of RHPAs, where x_0 represents the point of incidence of light, θ_1 is the incident angle in glass, θ_2 is the refracted angle in the cell, and θ_3 is the scattering angle in Si. The l_{opt} represents the propagation length of the incident optical ray in the silicon layer, ω represents the silicon thickness, and d represents the glass thickness. The values $n_1 = 1$, $n_2 = 1.5$, and $n_3 = 3.5$ were chosen as the refractive index of air, glass, and silicon, respectively. Figure 6c shows the mathematical relationship between x_0, ω, and l_{opt}. As shown in Figure 6d, the relationship between θ and x_0 and the relationship between l_{opt} and x_0 were observed.

In summary, and combined with statistical methods, the average optical scattering angle and optical path length can be obtained using the following formula:

$$< \theta >= \frac{1}{j}\sum_{i=1}^{j} \theta_3 = 6.67^{\circ} \text{ and } < l_{opt} >= \frac{1}{j}\sum_{i=1}^{j} l_{opt} = 1.0069\omega \quad (4)$$

The scattering angle θ increased with the change in incident position x_0. However, considering that Si has a much higher refractive index than glass, the scattering angle varied from 0 to 20°. Accordingly, an average scattering angle of 6.67° was obtained. Similarly, the optical path of scattered light l_{opt} will also change with the location of incoming light. However, it is relatively limited in increasing the propagation length of light according to the incoming position of light, and the increase will be in evidence only near the pit edge. Therefore, the average optical path of scattering light is about 1.0069ω.

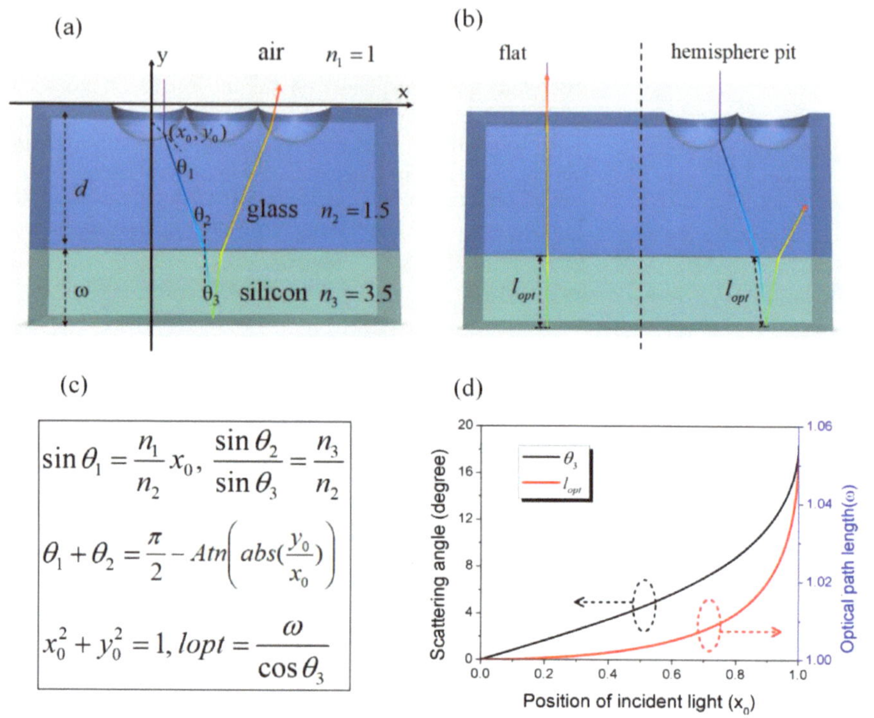

Figure 6. (**a**) The diagram of propagation of the optical ray in the pit texture cell. (**b**) The comparison diagram of light propagation in the plane structure cell and the pit texture structure cell. (**c**) The conversion relationship among light incident positions, cell thickness, and light path length. (**d**) The relationship between scattering angle and incident light position, and the relationship between the optical path and the incident light position.

The above analysis about the optical path and the probability of transmission and reflection was synthesized, and the Lambert surface method can be used to estimate the total optical path length of light in solar cells, which is also a relatively reasonable method [37,38].

$$L_{opt} = \sum dl_{opt} = 2l_{opt} + p_{tra}l_{opt} + 3p_{tra}p_{ref}l_{opt} + 5p_{tra}p_{ref}^2 l_{opt} + \cdots \quad (5)$$

For hemisphere pit arrays, $P_{ref} = 33.21\%$ and $l_{opt} = < l_{opt} > = 1.0069\omega$, Equation (5) is convergent:

$$L_{opt} \approx 4\omega \quad (6)$$

The above calculation results indicate that the total optical path length is increased from initial 2ω to 4ω by the hemisphere pit arrays. Therefore, the light transmission length in RHPAs has risen twice compared with the plane structure glass. Compared with the textured structure inside the solar cell, the limited raise in light propagation length is not obvious for improving the transform efficiency of solar cells, but it has provided rare opportunities and favorable space for reducing the active layer thickness and saving manufacturing expense for cells.

4. Experimental Verification

The light-trapping mechanism of RHPAs is mainly characterized via reduce surface reflectivity for solar cells, because RHPAs can refract and reflect incoming light multiple times before it reaches the active layer of the solar cell. Experimental methods were employed to compare the total reflection capability of solar cells with flat panel and RHPA structures. To verify the theoretical research results, the above-mentioned manufacturing method was used to obtain a PDMS film with flat and RHPA structures.

Figure 7a1,a2 shows the photograph of the quasi-hemispherical pit array textured glass and its diffraction pattern with the laser beam (500 nm) passing through. The diffraction pattern illustrates outstanding regular surface and superior diffraction result of the textured glass sample. Figure 7 shows the scanning electron microscope image of RHPA glass (inclination angle, 30). The pit array has a period of about 20 μm and an aspect ratio of about 0.5. Figure 7c shows the SEM images of the pits at different aspect ratios. Figure 7d provides a comparison of the reflectivity spectrum of flat and textured glass samples. The transmissivity for the quasi-PHRA glass has altered greatly compared with the plane glass. When the light passed through the surface structure, the transmissivity increased by several percent. However, when the light came into the surface structure from the inverse side, the transmissivity decreased to 70–75%. Figure 7e illustrates the light reflection feature of a silicon wafer-covering glass with a quasi-hemisphere pit array structure, and shows a comparison with plane glass. Based on the diagram, the total reflectance of the Si wafer covered by the PDMS film is as follows:

$$R_{total} = R_1 + R_2 \tag{7}$$

For the flat glass sample (without the RHPA structure),

$$R_{1-flat} = \left(\frac{n_{glass} - n_{air}}{n_{glass} + n_{air}}\right)^2, R_{2-flat} = \left(\frac{n_{Si} - n_{glass}}{n_{Si} + n_{glass}}\right)^2 \tag{8}$$

Meanwhile, for the textured glass sample (with the RHPA structure),

$$R_{1-RPAs} = 0.4625R_{1-flat}, R_{2-RPAs} = 0.3321R_{2-flat} \tag{9}$$

where n_{air} represents the air refractive index, n_{glass} represents the glass refractive index, and n_{Si} represents the Si refractive index.

Figure 7f expresses the silicon wafer theory and experiment reflectivity for samples separately covered with plane and hemisphere-pit-array glass when light with a wavelength of 350–800 nm is vertically incident. In order to reduce the impact of the air layer between the silicon and the structured glass on the test results, we introduced a refractive-index-matching solution between the silicon and the structured glass. In order to eliminate the influence of the air layer between the silicon and the structured glass on the test results, we introduced refractive index matching between the silicon and the structured glass, which was prepared with ethanol C_2H_5OH, liquid paraffin, and sodium bromide $C_{10}H_7Br$ in a ratio of 1:1:1. The refractive index was equivalent to that of glass, about 1.5. The small image inserted in the figure is the refractive index curve of the silicon wafer used in the theoretical calculations of this section.

It shows that the theoretical and experimental results are in agreement. Notably, in the theoretical calculation, a strict hemisphere model was obtained, but in the experiment, only a quasi-hemispherical model was used. The aspect ratio of the quasi-hemisphere is ca. 0.45. In the theoretical calculations, the refractive indexes of Si were related to wavelength, which is the famous optical database refractive index [39,40]. In the theoretical calculations, the refractive index of the matching solution is taken as 1.5.

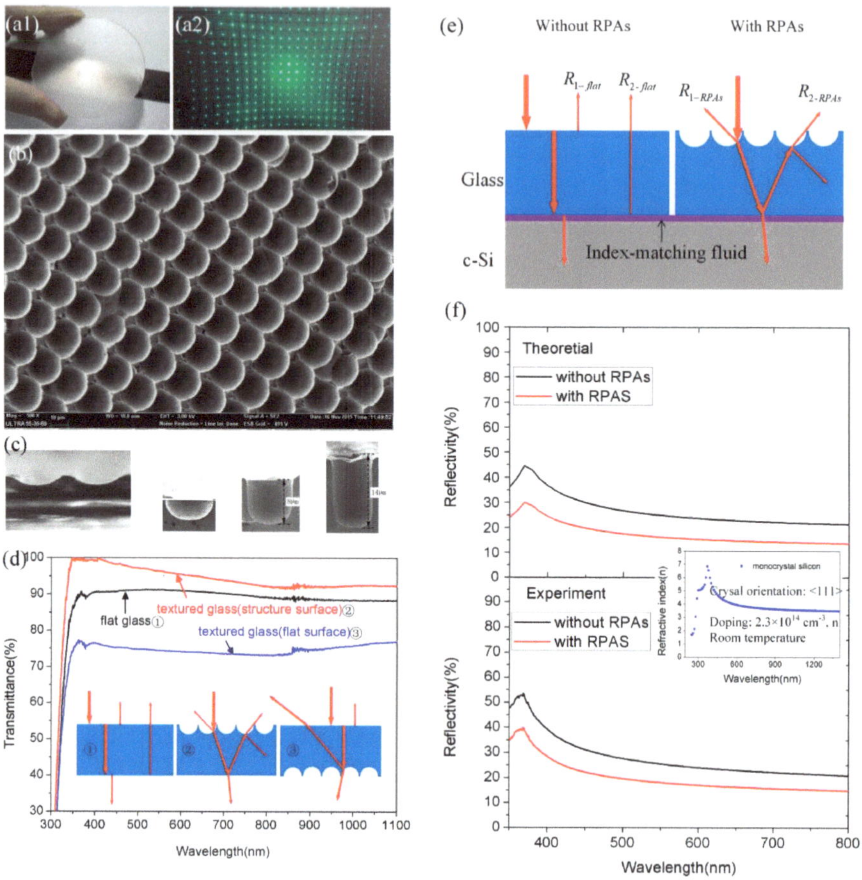

Figure 7. (**a1**,**a2**) Photograph of quasi-hemispherical pit array textured glass and its diffraction pattern with the laser beam (500 nm) passing through. (**b**) SEM image with RHPA array structure with 30-degree inclination. (**c**) SEM images of the pits with different aspect ratios. (**d**) Comparison of reflectivity spectrum of flat and textured glass samples. (**e**) Diagrammatic sketch for light reflectivity of the silicon wafer covered by the plane glass and the pit-array-structured glass. (**f**) The reflectivity curves of a silicon wafer with or without RHPA structure and RHPA-structured glass. The inset is the silicon reflectance graph, which is used in the theoretical calculations.

The theoretical and experimental results were compared digitally by using the following formulas to calculate the total reflectance $< R >$ and relatively reduction ΔR:

$$R = \frac{1}{\lambda_2 - \lambda_1} \int_{\lambda_1}^{\lambda_2} r(\lambda) d\lambda \text{ and } \Delta R = \frac{R_{flat\ glass} - R_{RPAs-glass}}{R_{flat\ glass}} \tag{10}$$

where $r(\lambda)$ represents the theoretical or experimental values as a function of λ, $\lambda_1 = 350$ nm, and $\lambda_2 = 800$ nm. Table 1 shows the comparison results between the theory and the experiment. The theoretical and experimental results of ΔR are 32.95% and 29.64%, respectively, indicating that the RHPA structure can effectively decrease the reflectance of the silicon wafer by about 30%. Although the experimental values are lower than the theoretical values, these discrepancies are caused by the glass used in the experiment. In theoretical calculations, silicon is an ideal semiconductor material, and the designed glass sample is regarded as an ideal shape with a smooth plane and perfect hemispherical pits. However, in the experimental measurement, the surface of the prepared glass sample was not completely smooth, and the cut pits had a hemispherical-like structure. In addition, the refractive index of silicon may not be consistent with the theoretical refractive index. Therefore, the method for analyzing the light-trapping characteristics of RHPAs in this study is reasonable.

Table 1. The values of <*R*> and ΔR under the conditions of theory and experiment.

	<*R*>		ΔR
	Flat	**RHPAs**	
Theory	26.28%	17.62%	32.95%
Experiment	28.51%	20.06%	29.64%

Both the experimental and theoretical results effectively express the pit array texture with good light-trapping properties. The decrease in surface reflectivity indicates an increase in light energy in the solar cell. In other words, the utilization of sunlight is improved. The photoelectric transformation efficiency for solar cell needs to be enhanced, especially for silicon thin-film solar cells.

5. Conclusions

The light-trapping characteristics of regular pit arrays with different aspect ratios in Si solar cells were studied. Numerical analysis and geometrical optics simulation were used to study the antireflection characteristics of the cell surface, the total light reflectivity, and the path length. The theoretical research results were verified using the micro/nano manufacturing method to obtain the flat and the quasi-hemisphere pit array glass samples. For the hemisphere pit arrays, theoretical derivation results indicate that it can reduce surface reflectance by more than half and can cause one third of the unabsorbed light to return to the active layer of the solar cell. Furthermore, the total optical path length for the cells with RHPAs increased from 2ω to 4ω, thus providing a favorable space for reducing the active layer thickness and saving manufacturing expense for cells. Simultaneously, regular pit arrays inhibited the inside light from escaping from the solar cells and returned the unabsorbed light back to the active layer. Thus, regular pit arrays have outstanding AR capability, which could decrease the total reflectivity for solar cells. The research in this article not only helps us to delve deeper into the mechanism of light-trapping, but also provides a reference for improving and researching more effective light-trapping structures. At the same time, it also has significant implications for the design of trapping structures for various optoelectronic devices.

Author Contributions: B.Z. and L.C. conceived the idea and supervised the project. B.Z. and S.Y. designed the simulation experiments, put forward the theoretical model, and contributed to the theoretical calculations. W.L. analyzed the data. B.Z. and L.C. wrote the manuscript. All authors have read and agreed to the published version of the manuscript.

Funding: This research is supported by the Guangxi Science and Technology Base and Talent Project (Department of Science and Technology of Gaungxi, Grant No.AD20297041), the Director's Fund of Guangxi Key Laboratory of Automatic Testing Technology and Instruments (Guangxi Key Laboratory of Automatic Detection Technology and Instruments, Grant No.YQ21106), the National Natural Science Foundation of China (National Natural Science Foundation of China, Grant No.61804097), and the Guangxi Special Expert Team Project.

Data Availability Statement: The raw/processed data cannot be shared at this time as the data also forms part of an ongoing study.

Conflicts of Interest: The authors declare no conflict of interest.

References

1. Liu, H.; Du, Y.; Yin, X.; Bai, M.; Liu, W. Micro/Nanostructures for Light Trapping in Monocrystalline Silicon Solar Cells. J. Nanomater. **2022**, 2022, 8139174. [CrossRef]
2. Tang, Q.; Yao, H.; Xu, B.; Ge, J. Enhanced energy conversion efficiency of Al-BSF c-Si solar cell by a novel hierarchical structure composed of inverted pyramids with different sizes. *Sol. Energy* **2020**, *208*, 1–9. [CrossRef]
3. Tang, Q.; Shen, H.; Yao, H.; Gao, K.; Ge, J.; Liu, Y. Investigation of optical and mechanical performance of inverted pyramid based ultrathin flexible c-Si solar cell for potential application on curved surface. *Appl. Surf. Sci.* **2020**, *504*, 144588. [CrossRef]
4. Lee, Y.; Woo, Y.; Lee, D.-K.; Kim, I. Fabrication of quasi-hexagonal Si nanostructures and its application for flexible crystalline ultrathin Si solar cells. *Sol. Energy* **2020**, *208*, 957–965. [CrossRef]
5. Liu, X.; Ji, Y.; Lu, Z.; Sun, Y.; Yang, H.; Liu, J.; Zhang, Y.; Li, D.; Cao, Y.; Li, W.; et al. Enhanced device performance of Si nanowires/Si nanocrystals heterojunction solar cells with ultrathin Al_2O_3 passivation. *Phys. E Low-Dimens. Syst. Nanostruct.* **2020**, *120*, 114048. [CrossRef]
6. Kordrostami, Z.; Sheikholeslami, H. Optimization of light trapping in square and hexagonal grid inclined silicon nanowire solar cells. *Opt. Commun.* **2020**, *459*, 124980. [CrossRef]
7. Seo, M.; Yoon, S.; Cho, H.; Lee, S.; Kim, K.; Kong, B.D.; Meyyappan, M.; Baek, C.-K. Solar Cell Using Hourglass-Shaped Silicon Nanowires for Increased Light-Trapping Path. *IEEE J. Photovolt.* **2020**, *10*, 475–479. [CrossRef]
8. Yamada, Y.; Iizuka, H.; Mizoshita, N. Silicon Nanocone Arrays via Pattern Transfer of Mushroomlike SiO_2 Nanopillars for Broadband Antireflective Surfaces. *ACS Appl. Nano Mater.* **2020**, *3*, 4231–4240. [CrossRef]
9. Le, C.; Bowen, F.; Qinglin, K.; Wangyang, P.; Hu, K.; Zhang, W. Quasi-hemispherical pit array textured surface for increasing the efficiency of thinfilm solar cells. *AIP Adv.* **2022**, *12*, 015111.
10. Xing, Y.P.; Zhang, K.L.; Zhao, J.S.; Han, P.D.; Yang, Z.C.; Yuan, Y.J.; Ding, Q. Antireflection and absorption properties of silicon parabolic-shaped nanocone arrays. *Optik* **2017**, *128*, 133–138. [CrossRef]
11. Chen, L.; Luo, W.; Fang, B.; Zhu, B.; Zhang, W. Study on light absorption of $CH_3NH_3PbI_3$ perovskite solar cells enhanced by gold nanobipyramids. *Opt. Laser Technol.* **2023**, *159*, 108924. [CrossRef]
12. Xu, Z.; Huangfu, H.; Li, X.; Qiao, H.; Guo, W.; Guo, J.; Wang, H. Role of nanocone and nanohemisphere arrays in improving light trapping of thin film solar cells. *Opt. Commun.* **2016**, *377*, 104–109. [CrossRef]
13. Wang, C.; Zhao, S.; Bian, F.; Du, D.; Wang, C.; Xu, Z. Absorption enhancement of ultrathin crystalline silicon solar cells with frequency upconversion nanosphere arrays. *Commun. Theor. Phys.* **2020**, *72*, 015501. [CrossRef]
14. Chang, Y.-C.; Pollard, M.E.; Payne, D.N.R.; Sprafke, A.; Pillai, S.; Bagnall, D.M. Large-area nanosphere gratings for light trapping and reduced surface losses in thin solar cells. *IEEE J. Photovolt.* **2019**, *9*, 1012–1019. [CrossRef]
15. Chen, L.; Wang, Q.; Chen, W.; Liu, D.; Zhao, Z.; Wang, D. Light trapping mechanism of hemisphere cone arrays for silicon solar cells. *Sol. Energy* **2018**, *163*, 519–525. [CrossRef]
16. Zhang, C.; Guney, D.O.; Pearce, J.M. Plasmonic enhancement of amorphous silicon solar photovoltaic cells with hexagonal silver arrays made with nanosphere lithography. *Mater. Res. Express* **2016**, *3*, 105034. [CrossRef]
17. Khezami, L.; Al Megbel, A.O.; Jemai, A.B.; Ben Rabha, M. Theoretical and experimental analysis on effect of porous silicon surface treatment in multicrystalline silicon solar cells. *Appl. Surf. Sci.* **2015**, *353*, 106–111. [CrossRef]
18. Ben Rabha, M.; Mohamed, S.B.; Dimassi, W.; Gaidi, M.; Ezzaouia, H.; Bessais, B. Optoelectronic enhancement of monocrystalline silicon solar cells by porous silicon-assisted mechanical grooving. *Phys. Status Solidi C* **2011**, *8*, 887–890. [CrossRef]
19. Eduardo, C.A.; Hannah, J.J. Transparent Quasi-Random Structures for Multimodal Light Trapping in Ultrathin Solar Cells with Broad Engineering Tolerance. *ACS Photonics* **2022**, *9*, 2724–2735.
20. Senthilkumar, N.; Arulraj, A.; Nandhakumar, E.; Ganapathy, M.; Vimalan, M.; Potheher, I.V. Green mediated synthesis of plasmonic nanoparticle (Ag) for antireflection coating in bare mono silicon solar cell. *J Mater. Sci. Mater. Electron.* **2018**, *29*, 12744–12753. [CrossRef]
21. Voroshilov, P.M.; Simovski, C.R.; Belov, P.A.; Shalin, A.S. Light-trapping and antireflective coatings for amorphous Si-based thin film solar cells. *J. Appl. Phys.* **2015**, *117*, 203101. [CrossRef]
22. Elshorbagy, M.H.; Abdel-Hady, K.; Kamal, H.; Alda, J. Broadband anti-reflection coating using dielectric Si3N4 nanostructures. Application to amorphous-Si-H solar cells. *Opt. Commun.* **2017**, *390*, 130–136. [CrossRef]
23. Kephart, J.M.; Geisthardt, R.M.; Sampath, W.S. Optimization of CdTe thin-film solar cell efficiency using a sputtered, oxygenated CdS window layer. *Prog. Photovolt. Res. Appl.* **2015**, *23*, 1484–1492. [CrossRef]
24. Rosell, A.; Martin, I.; Garin, M.; Lopez, G.; Alcubilla, R. Textured PDMS Films Applied to Thin Crystalline Silicon Solar Cells. *IEEE J. Photovolt.* **2019**, *10*, 351–357. [CrossRef]
25. Gao, Z.; Lin, G.; Chen, Y.; Zheng, Y.; Sang, N.; Li, Y.; Chen, L.; Li, M. Moth-eye nanostructure PDMS films for reducing reflection and retaining flexibility in ultra-thin c-Si solar cells. *Sol. Energy* **2020**, *205*, 275–281. [CrossRef]
26. Chen, G.; Hu, D.Q.; Li, C.; Wang, W.W.; Zhang, J.Q.; Wu, L.L.; Li, W. Process study about silica anti-reflection coatings prepared by sol-gel method for cadmium telluride solar cells. *J. Mater. Sci.* **2018**, *53*, 15588–15599. [CrossRef]
27. Zhang, C.; Song, Y.; Wang, M.; Yin, M.; Zhu, X.F.; Tian, L.; Wang, H.; Chen, X.Y.; Fan, Z.Y.; Lu, L.F.; et al. Efficient and Flexible Thin Film Amorphous Silicon Solar Cells on Nanotextured Polymer Substrate Using Sol-gel Based Nanoimprinting Method. *Adv. Funct. Mater.* **2017**, *27*, 1604720. [CrossRef]

28. Jannat, A.; Lee, W.; Akhtar, M.S.; Li, Z.Y.; Yang, O.B. Low cost sol-gel derived SiC-SiO_2 nanocomposite as anti reflection layer for enhanced performance of crystalline silicon solar cells. *Appl. Surf. Sci.* **2016**, *369*, 545–551. [CrossRef]
29. Kim, K.; Kim, S.; An, S.; Lee, G.H.; Kim, D.; Han, S. Anti-reflection porous SiO_2 thin film deposited using reactive high-power impulse magnetron sputtering at high working pressure for use in a-Si:H solar cells. *Sol. Energy Mater. Sol. C* **2014**, *130*, 582–586. [CrossRef]
30. Liu, D.; Wang, Q. Light-trapping surface coating with concave arrays for efficiency enhancement in amorphous silicon thin-film solar cells. *Opt. Commun.* **2018**, *420*, 84–89. [CrossRef]
31. Ashish, P.; Jordi, L.; Patrícia, C.S. Broadband and Omnidirectional Antireflection Surfaces Based on Deep Subwavelength Features for Harvesting of the Solar Energy. *Sol. RRL* **2021**, *5*, 2100548.
32. Shen, X.Q.; Wang, Q.K.; Wangyang, P.H.; Huang, K.; Chen, L.; Liu, D.M. Performance enhancement in a-Si:H/μc-Si:H tandem solar cells with periodic microstructured surfaces. *Opt. Lett.* **2015**, *40*, 1290–1293. [CrossRef] [PubMed]
33. Jovanov, V.; Palanchoke, U.; Magnus, P.; Stiebig, H.; Hüpkes, J.; Sichanugrist, P.; Konagai, M.; Wiesendanger, S.; Rockstuhl, C.; Knipp, D. Light trapping in periodically textured amorphous silicon thin film solar cells using realistic interface morphologies. *Opt. Express* **2013**, *21*, A595–A606. [CrossRef]
34. Tamang, A.; Pathirane, M.; Parsons, R.; Schwarz, M.M.; Iheanacho, B.; Jovanov, V.; Wagner, V.; Wong, W.S.; Knipp, D. Zinc oxide nanowire arrays for silicon core/shell solar cells. *Opt. Express* **2014**, *22*, A622–A632. [CrossRef]
35. Preinfalk, J.B.; Donie, Y.J.; Egel, A.; Hecht, M.; Hüpkes, J.; Bittkau, K.; Lemmer, U.; Gomard, G. On the fabrication of disordered nanostructures for light extraction in corrugated OLEDs. In *Solid-State Lighting;* Optica Publishing Group: Washington, DC, USA, 2017; p. JW5A.20.
36. Yablonovitch, E. Statistical ray optics. *JOSA* **1982**, *72*, 899–907. [CrossRef]
37. Deckman, H.W.; Roslo, C.B.; Yablonovitch, E. Maximum statistical increase of optical absorption in textured semiconductor films. *Opt. Lett.* **1983**, *8*, 491–493. [CrossRef]
38. Nelson, J. *The Physics of Solar Cells;* Imperial College Press: London, UK, 2003; p. 223.
39. Available online: https://refractiveindex.info/?shelf=main&book=Si&page=Aspnes (accessed on 3 April 2022).
40. Aspnes, D.E.; Studna, A.A. Dielectric functions and optical parameters of Si, Ge, GaP, GaAs, GaSb, InP, InAs, and InSb from 1.5 to 6.0 eV. *Phys. Rev. B* **1983**, *27*, 959–1009. [CrossRef]

energies

Article

Techno-Economic Assessment of CPVT Spectral Splitting Technology: A Case Study on Saudi Arabia

Cesar Lucio *, Omar Behar and Bassam Dally

CCRC, King Abdullah University of Science and Technology (KAUST), Thuwal 23955, Saudi Arabia; omar.behar@kaust.edu.sa (O.B.); bassam.dally@kaust.edu.sa (B.D.)
* Correspondence: cesar.cifuentes@kaust.edu.sa

Abstract: Concentrating PV thermal (CPVT) collector with spectral splitting technology is a promising solution for heat and electricity production. To extend the use of this technology, a novel and cost-effective CPVT collector for harsh environments, such as those in Saudi Arabia, is presented and evaluated using theoretical energy, economy, and environmental analysis. Two questions are answered in this study, namely: which is the best operation strategy, and which is the best energy storage technology for CPVT. The potential of using a CPVT under the climate conditions of six cities in Saudi Arabia is also evaluated. It is found that a heat/electricity production strategy and a thermal energy storage are the most suitable for the CPVT technology. The economic assessment shows a levelized cost of electricity (LCOE) of $0.0847/kWh and a levelized cost of heat (LCOH) of $0.0536/kWh when water is used as a spectral filter, and a LCOE of $0.0906/kWh and a LCOH of $0.0462/kWh when ZnO nanoparticles are added. The CO_2-equivalent emissions in a 20 MW CPVT plant are cut from 5675 tonnes to 7822 tonnes per year for Saudi Arabian weather and present power generation conditions.

Keywords: CPVT; spectral filtering; solar energy

Citation: Lucio, C.; Behar, O.; Dally, B. Techno-Economic Assessment of CPVT Spectral Splitting Technology: A Case Study on Saudi Arabia. *Energies* **2023**, *16*, 5392. https://doi.org/10.3390/en16145392

Academic Editor: Armando Oliveira

Received: 28 May 2023
Revised: 4 July 2023
Accepted: 12 July 2023
Published: 14 July 2023

1. Introduction

Solar energy can play an important role in a sustainable worldwide energy supply to address carbon emission and climate change. The range and the applications of solar energy conversion devices have expanded dramatically in recent years, with the objective of reducing reliance on fossil fuels. Solar energy can be converted into useful energy using thermal and photovoltaic (PV) collectors.

Traditional PV collectors convert part of the solar spectrum into electricity (typical efficiency of traditional PV panels is about 20%). The rest of the energy received by the PV panel is converted into heat, decreasing its performance. This constraint has led research groups all around the world to seek ways to use the solar radiation that cannot be converted by the PV cells into electricity, in other words, to be able to exploit the entire solar spectrum while preventing photovoltaic cell from overheating.

In this regard, spectrum beam splitting (SBS) has been the technique that has undergone the greatest progress in recent years. It employs filters that split the incoming solar radiation into different wavelengths. The solar radiation within the spectral window, useful for the photovoltaic effect, is directed to PV panels, while the unutilized energy by the PV panels is directed and absorbed by a heat transfer fluid (HTF) to generate heat. The PV panel and the solar thermal collector is combined into a single unit, which is known as, concentrating solar photovoltaic thermal (CPVT) collector.

There are three main methods to split solar radiation into different ranges of wavelengths: interference filtering, use of semi-transparent PV panels, and selective absorption. The challenge of using interference filters is their complicated manufacturability and high cost [1,2]. Some limitations of the semi-transparent PV panels include development of semi-transparent electrodes [3], insulation issues [4] and that some materials are made semi-transparent by reducing the semiconductor's layer thickness; however, doing so results in a reduction in performance [5]. Alternatively, selective absorbers employing HTFs could be a more affordable approach. An HTF that is transparent to the desired wavelengths for PV cells is located in front of them, letting those wavelengths be transmitted to the cells. The HTF is highly absorbing in the rest of the spectrum [2]. From the economic point of view, selective absorption is a cost-effective technique since the working fluid can be water [6].

Several researchers have been working to develop the CPVT technology. One of the first studies was performed by Soule [7], who proposed, in 1987, a CPVT using dome-shaped linear Fresnel lenses as the concentrator with a dielectric-Au-dielectric multilayer filter. The system produced electricity, low-temperature thermal energy (50–70 °C), and high-temperature thermal energy (150–250 °C). The corresponding efficiencies are 9.5%, 41.9%, and 17.8%, respectively. A CPVT with SBS and a parabolic trough collector (PTC) has been proposed by Zhang et al. [8]. The system achieved a maximum electrical efficiency of 22.64%. Some studies showed that a CPVT with PTC can reach an overall thermal efficiency of 70% and an overall electrical efficiency of 25%, while a system with a Linear Fresnel Collector (LFC) can achieve a thermal efficiency of more than 60% and an electrical efficiency of more than 20% [9].

Ling et al. [10] investigated a CPVT with LFC and a selective filter and found a levelized cost of electricity (LCOE) of $0.20/kWh. Recently, Liew et al. [11] proposed a photovoltaic/concentrated solar power hybrid plant to increase the performance of a concentrated solar power plant currently operating in California, USA. The proposed hybrid system performed 9% better than the actual one and was also 4% more efficient than the virtual photovoltaic-alone scenario.

Instead of using solid filters for SBS, liquid absorptive filters can be used and have several advantages [12]. The absorptive liquid is often inexpensive and can perform numerous functions: it absorbs the unused spectral solar irradiance by PV cells; thermal energy can be transported and stored by absorptive fluids; and it could be used as the coolant of PV modules to extract the dissipated heat from the solar cells. Sabry et al. [13] theoretically demonstrated that an ideal liquid filter, which matches the spectral response of silicon solar cells, significantly reduces the solar cells' operating temperature and increases their efficiency by 30%. The performance of a combined liquid and solid absorptive filter on a compact CPVT receiver for an LFC was investigated by Manfred et al. [14]. They found that, for Seville (Spain), the receiver can achieve an electrical efficiency of up to 6.2% and a thermal efficiency of up to 61.2%.

Advances in nanotechnology have resulted in nanoparticles that can selectively filter solar radiation and can be added to a base fluid to modify its optical characteristics. Meraje et al. [15] designed and validated a CPVT based on LFC and a nanofluid spectrum splitting filter. They evaluated several volume concentrations of ZnO nanoparticles. The closest spectrum match with a silicon solar cell was determined to be 0.00089 vol%. Recently, Barthwal et al. [16] examined the utilization of deionized water and ZnO nanoparticles as optical filters in a compound parabolic-concentrate-based CPVT. They evaluated it for conditions in New Delhi (India) and concluded that the cell temperature was kept near the standard test. Wang et al. [17] studied a CPVT with compact LFC and Ag/$CoSO_4$-PG nanofluids. The performance estimation showed that the PV module has a photoelectric efficiency of 30.2%, and the receiver has a thermal efficiency up to 49.3%.

In terms of the applications for CPVT, Su et al. [18] investigated the feasibility of applying CPVT to boost biomethane generation in anaerobic digestion via biogas upgrading. They also proposed the use of CPVT for trigeneration (heat, cooling, and electricity) [19]. At Tucson (United States), Fernandes et al. [20] carried out a simulation for a small-scale nanofluid spectral filtering CPVT for domestic applications. The possibility of using CPVT for water desalination has also been investigated by several authors as reviewed by Anand. et al. [21]. Another recent application of a CPVT was proposed by Youssef et al. [22].

While many of the previous studies have investigated different types of CPVT collectors and highlighted their thermal performance, very few publications have reported on the operation strategy, the optimum heat versus electricity storage, or evaluated the benefits under harsh weather conditions, such as extremely high ambient temperatures and high levels of aerosols prevalent in places like Saudi Arabia. The objective of this paper is to address these shortcomings using Saudi Arabia as a case study.

To do so, a detailed techno-economic theoretical assessment is carried out. A CPVT with a novel receiver design, suitable for the harsh conditions, is investigated under the climate of six cities in Saudi Arabia. To provide a comprehensive analysis, a mathematical model is developed to investigate the optical and thermal performance of the proposed CVPT. For each location studied, a year-round performance assessment considering the hourly variation of solar radiation, sun position, ambient temperature, and wind speed is conducted. A comparison is then made for all cities and under all operating and storage scenarios.

2. Materials and Methods

2.1. Description of the CPVT

A Linear Fresnel Collector (LFC) with a hybrid receiver fitted at its focal axis is proposed in this study. As illustrated in Figure 1, the proposed system consists of mirrors, a thermal receiver with cooling channel, a heat transfer fluid that also plays the role of a filter, and a silicon bifacial PV module with a 22% nominal efficiency at 25 °C. The mirrors focus direct normal irradiance on the receiver's front surface. The fluid is used as spectral filter, absorbing low and high-energy photons and converting them into useful heat. As a result, a suitable solar radiation spectrum for silicon PV cells reaches the PV module, which is placed above the nanofluid. Due to the bifaciality factor of the solar cell, the side with the highest efficiency faces the concentrated solar radiation to maximize energy production. The cooling channel is used to reduce the PV module temperature. The design values of the proposed system are presented in Table 1.

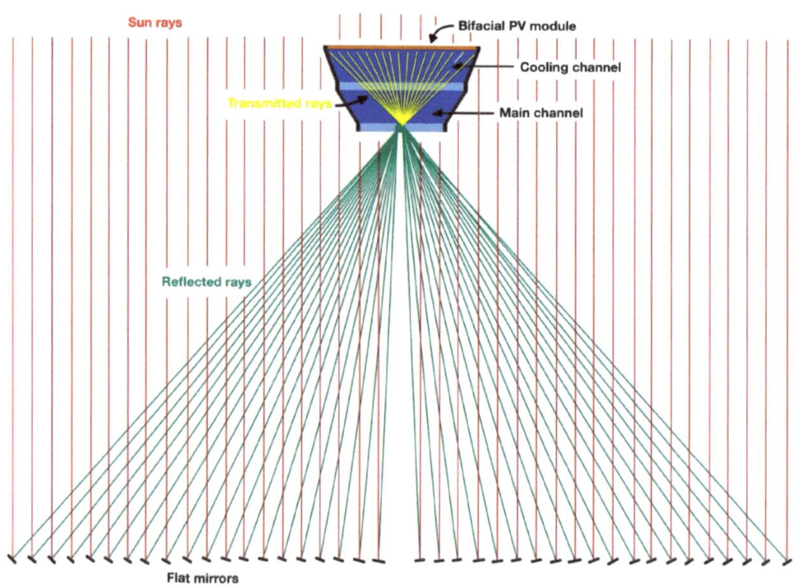

Figure 1. Basic design of the CPVT.

Table 1. Design data of the CPVT.

Component	Parameter	Value	Units
Linear Fresnel collector	Length	10	m
Receiver	Receiver focal length	1.5	m
	Height of the receiver	0.08	m
	Wide of the mirrors	0.1	m
	Wide of the receiver front surface	0.2	m
	Wide of the receiver back surface	0.33	m

A detailed design of the receiver is presented in Figure 2a. It consists of the main liquid channel and the cooling channel together with the PV module. As illustrated in Figure 2b, these two channels are linked by a U-shaped pipe to enhance thermal efficiency [22]. The liquid initially flows at room temperature through the cooling channel to cool down the PV panel. As a result, the panel's temperature drops, its efficiency increases, and the HTF is preheated before entering the main receiver channel. Figure 2c highlights the main parts of the receiver.

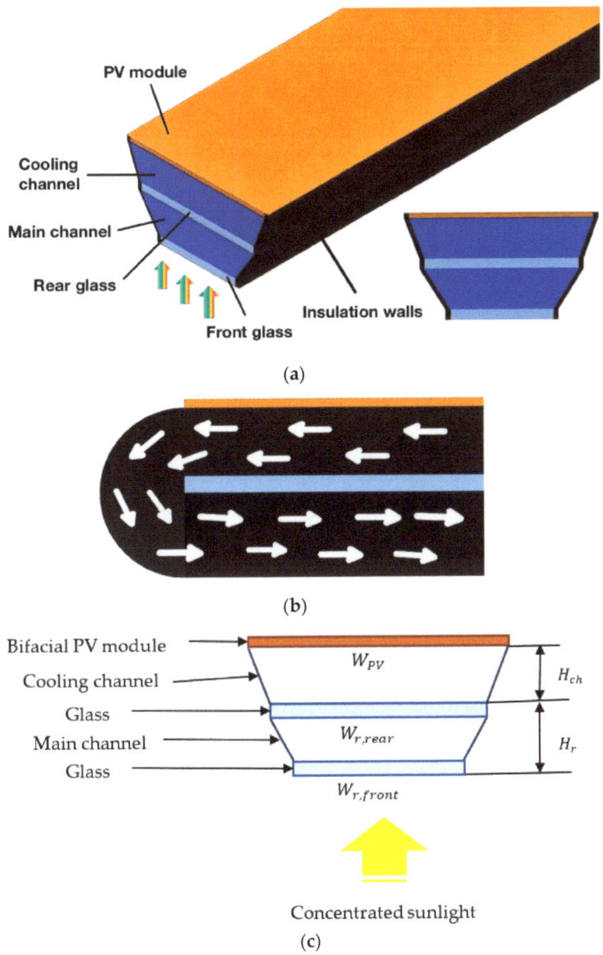

Figure 2. (**a**) CAD design of the receiver (cross-section). (**b**) U-shape pipe linking the cooling channel to the main channel. (**c**) Simplified design of the receiver.

As highlighted in Figure 2c, concentrated light passes through the highly transparent glass and across the working fluid. The working fluid acts as a spectral filter, absorbing solar radiation with wavelengths less than 700 nm or greater than 1100 nm. As a result, only solar radiation within the spectral window of between 700 nm and 1100 nm reaches the PV module. The receiver's side walls are painted with selective, highly absorbent materials.

In this study, two different working fluids, namely water and a water-based ZnO nanofluid (0.01 wt%), were examined. The introduction of nanoparticles into the water resulted in alterations within the thermophysical and spectral characteristics of the fluid, as documented in Tables 2 and 3, respectively. The evaluation of the thermophysical properties was carried out under atmospheric pressure and at an approximate average fluid temperature of 62.5 °C, representing an average working fluid temperature of our system.

Table 2. Thermo-physical properties of water and ZnO water-based nanofluid.

Symbol	Fluid Properties	Water	ZnO (0.01 wt%)
μ	Dynamic viscosity, mPas	0.47 [23]	0.47 *
k	Thermal conductivity, $Wm^{-1}K^{-1}$	0.65 [23]	0.86 [24]
Cp	Specific heat capacity, $Jkg^{-1}K^{-1}$	4185 [23]	4148 [25]
ρ	Density, kgm^{-3}	983.7 [23]	976.9 [25]

* Due to a lack of data, it is presumed that the dynamic viscosity does not change due to the low concentration of ZnO particles.

Table 3. Average spectral transmittance of water and ZnO water-based nanofluid for specific spectral windows.

Spectral Window	200–700 nm	700–1100 nm	1100–2400 nm
Water [2]	97.1	88.1	11.7
ZnO (0.01 wt%) [24]	64.3	79.8	5.1

The present investigation focuses on the photovoltaic active range of 700 nm to 1100 nm for silicon solar cells, in accordance with prior research [26]. Notably, the study does not encompass the photovoltaic active spectrum spanning 400–700 nm, where energy states surpass the bandgap energy of silicon, resulting in the thermal relaxation of excess photon energy. Nevertheless, the examined fluids exhibit a notable degree of radiation transmission within the 400–700 nm range, as demonstrated in Table 3, thus signifying their potential efficacy in capturing solar energy from this specific region.

2.2. Design of the CPVT

The CPVT collector is north–south orientated and rotates along the east–west horizontal axis to increase the overall optical performance and reduce variation in energy delivery during the day [27].

Three parameters are important in the design of the LFC (see Figure 3): location (M_n), tilt angle (δ_n), and distance of adjacent mirrors (S_n). These may be obtained using elementary geometrical optics by using the following formulas [28]:

$$\delta_n = \frac{\text{atan}\left(\frac{M_n}{fc_r}\right)}{2} \tag{1}$$

$$S_n = \frac{W_{mirror}}{2 \times \left[(\sin(\delta_n) + \sin(\delta_{n-1})) \times \tan(2\delta_n) + \cos(\delta_n) + \cos(\delta_{n-1})\right]} \tag{2}$$

$$M_n = M_{n-1} + S_n \tag{3}$$

where fc_r is the focal length of the receiver, W_{mirror} the width of the primary mirrors, and the subscript n is the number of the primary mirror.

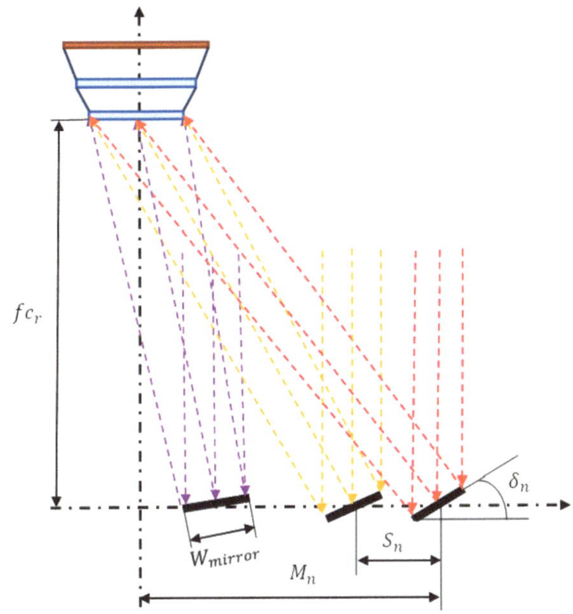

Figure 3. Schematic of the CPVT collector.

2.3. Optical and Thermal Modelling

2.3.1. Optical Efficiency

The following expression is used to estimate the optical efficiency of the LFC [29]:

$$\eta_{opt} = \eta_{opt,nom} K_T(\theta_T) K_L(\theta_L) \tag{4}$$

where $\eta_{opt,nom}$ is the nominal optical efficiency measured at solar noon, $K_T(\theta_T)$ is the transversal incidence angle modifier, θ_T is the transversal incidence angle in degree, $K_L(\theta_L)$ is the longitudinal incidence angle modifier, and θ_L is the longitudinal incidence angle in degree.

For a collector aligned along the north–south axis, the transversal and longitudinal angles are calculated as follows [27]:

$$\theta_T = \tan^{-1}(\sin(Az) \times \tan(Z)) \tag{5}$$

$$\theta_L = \tan^{-1}(\cos(Az) \times \tan(Z)) \tag{6}$$

where Az and Z are the Azimuth and Zenith angles, respectively.

In addition, the transversal and the longitudinal incidence angle modifiers are calculated using the following expressions, respectively [27]:

$$K_T(\theta_T) = \cos\left(\frac{\theta_T}{2}\right) - \frac{\frac{W_{field}}{4}}{fc_r + \sqrt{fc_r^{\,2} + (\frac{W_{field}}{4})^2}} \times \sin(\frac{\theta_T}{2}) \tag{7}$$

$$K_L(\theta_L) = \cos(\theta_L) - \frac{fc_r}{L_r} \times \sqrt{1 + \left(\frac{W_{field}}{4fc_r}\right)^2} \times \sin(\theta_L) \tag{8}$$

where L_r is the receiver length, and W_{field} is the field width.

2.3.2. Heat Transfer Model

To examine the heat flow inside the receiver, a heat transfer model is developed. The flowchart outlining the model's structure and methodology can be found in Appendix A. The model takes into account the following set of assumptions:

- Steady state heat transfer model
- Thin PV module
- Side walls of the receiver are adiabatic
- Uniform temperature distribution
- The nanofluid flow is uniform

Furthermore, considering the phenomenon of self-absorption exhibited by the fluid and the similarity in emissivity between the fluid and the glass window, it is assumed that the heat radiation losses can be directly attributed to the glass window.

Heat Transfer in the Receiver

According to Newton's law of cooling, the convection heat transfer from the absorber's interior surface to the HTF is:

$$Q_{conv,r-fl} = h_{fl} \times A_{r,in} \times \left(T_{r,in} - T_{r,fl,mean}\right) \quad (9)$$

where $A_{r,in}$ is the inside surface of the thermal receiver, $T_{r,in}$ is the temperature of the inside surface of the thermal receiver, $T_{r,fl,mean}$ is the mean temperature of the fluid in the receiver and h_{fl} is the fluid heat transfer coefficient defined in the following way:

$$h_{fl} = \frac{Nu_{fl} \times k_{fl}}{Dh_r} \quad (10)$$

where Nu_{fl} is the fluid Nusselt number, k_{fl} is the fluid thermal conductivity, and Dh_r is the hydraulic diameter of the receiver. For the case of laminar flow, the Nusselt number is considered constant:

$$Nu_{fl\_laminar} = 4.36 \quad (11)$$

For the case of turbulent flow, the following Nusselt number correlation is used:

$$Nu_{fl\_turbulent} = 0.023 \times {Re_{fl}}^{3/4} \times {Pr_{fl}}^{0.3} \quad (12)$$

where Re_{fl} is the Reynolds number and Pr_{fl} is the fluid Prandtl number.

Conduction through the front and rear glass of the receiver can be represented as follows:

$$Q_{cond,r} = \left(\frac{k_{glass} \times A_{r,glass} \times (T_{r,in} - T_{r,out})}{t_{glass}}\right) \quad (13)$$

where k_{glass} is the glass thermal conductivity, $A_{r,glass}$ is the area of the front and rear glass, $T_{r,out}$ is the temperature of the outside surface of the thermal receiver and t_{glass} the glass thickness.

The rear glass surface of the receiver is connected to the cooling channel, and the walls are insulated, so convective heat exchange with the ambient air is only considered on the front glass surface of the receiver. Consequently, following Newton's law of cooling, the convection heat transfer from the receiver's outside surface to the atmosphere is:

$$Q_{conv,r-amb} = h_{air} \times A_{r,out,front} \times (T_{r,out} - T_{amb}) \quad (14)$$

where $A_{r,out,front}$ is the front glass surface of the thermal receiver, T_{amb} is the ambient temperature during sun hours, and h_{air} is the air heat transfer coefficient defined in the following way:

$$h_{air} = \frac{Nu_{air} \times k_{air}}{Dh_r} \quad (15)$$

where Nu_{air} is the air Nusselt number, and k_{air} is the air thermal conductivity. For laminar flow over a flat plate, the Nusselt number is expressed as follows:

$$Nu_{air\_laminar} = 0.664 \times {Re_{air}}^{0.5} \times {Pr_{air}}^{1/3} \quad (16)$$

For turbulent flow over a flat plate, the Nusselt number is expressed as follows:

$$Nu_{air\_turbulent} = 0.037 \times {Re_{air}}^{0.8} \times {Pr_{air}}^{1/3} \quad (17)$$

Because the receiver's front glass surface is in contact with the ambient air and the sidewalls are insulated, convective heat exchange with the cooling channel is only evaluated on the receiver's rear glass surface. As a result, according to Newton's law of cooling, the convection heat transfer from the outer surface of the receiver to the cooling channel is:

$$Q_{conv,r-ch} = h_{fl} \times A_{r,out,rear} \times \left(T_{r,out} - T_{ch,fl,mean}\right) \quad (18)$$

where $A_{r,in}$ is the inside surface of the thermal receiver, $T_{r,in}$ is the temperature of the inside surface of the thermal receiver, $T_{ch,fl,mean}$ is the mean temperature of the fluid in the cooling channel and h_{fl} the fluid heat transfer coefficient.

According to the Stefan–Boltzmann law of radiation, the radiation heat transfer from the external surface of the receiver to the atmosphere is:

$$Q_{rad,r-atm} = \sigma \times \varepsilon_{glass} \times A_{r,out,front} \times \left({T_{r,out}}^4 - {T_{sky}}^4\right) \quad (19)$$

where σ is the Stefan-Boltzmann constant $\left(5.67 \times 10^{-8}\ \text{Wm}^{-2}\text{K}^{-4}\right)$, ε_{glass} is the glass emissivity, and T_{sky} is the sky temperature estimated using the following expression [30]:

$$T_{sky} = 0.0522 \times {T_{amb}}^{1.5} \quad (20)$$

Radiation heat exchange with the *PV* panel is only evaluated on the receiver's rear glass surface. As a result, the expression that estimates the radiation heat transfer between two parallel plates is used:

$$Q_{rad,r-PV} = \left(\frac{\sigma \times A_{r,out,rear} \times \left({T_{r,out}}^4 - {T_{PV}}^4\right)}{\frac{1}{\varepsilon_{glass}} + \frac{1}{\varepsilon_{PV}} - 1}\right) \quad (21)$$

where T_{PV} is the temperature of the *PV* panel, and ε_{PV} is its emissivity.

Heat Transfer in the PV Panel

The solar radiation on the rear surface of PV cell follows the Stefan–Boltzmann law of radiation:

$$Q_{rad,PV-atm} = \sigma \times \varepsilon_{PV} \times A_{PV,rear} \times ({T_{PV}}^4 - {T_{sky}}^4) \quad (22)$$

where $A_{PV,rear}$ is the area of the *PV* panel rear surface.

Newton's law of cooling states that the convective heat transfer from the *PV* panel to the cooling channel is:

$$Q_{conv,PV-ch} = h_{fl} \times A_{PV,front} \times (T_{PV} - T_{ch,fl,mean}) \quad (23)$$

where $A_{PV,front}$ is the front surface of the *PV* panel, and h_{fl} is the fluid heat transfer coefficient.

The convective heat transfer from the *PV* panel to the ambient air is:

$$Q_{conv,PV-amb} = h_{air} \times A_{PV,rear} \times (T_{PV} - T_{amb}) \quad (24)$$

where h_{air} is the air heat transfer coefficient.

Power, Efficiency and Energy

The efficiency of bifacial crystalline silicon PV cells can be estimated using the following expression, which considers a temperature coefficient of −0.45%/°C:

$$\eta_{PV} = \eta_{PV,nom}[1 - (0.0045 \times (T_{PV} - T_{PV,ref}))] \quad (25)$$

where $\eta_{PV,nom}$ is the nominal efficiency of the PV panel at the reference temperature $T_{PV,ref}$. The electric energy produced by the PV panel can be calculated using the following equation:

$$Q_{u,PV,el} = DNI \times A_{ap} \times \eta_{opt} \times (1 - f_{opt}) \times tr_{fl,700-1100nm} \times \eta_{PV} + GHI \times A_{PV,rear} \times \eta_{PV} \quad (26)$$

where DNI is the direct normal irradiance, f_{opt} is the fraction of optical loss in the receiver, $tr_{fl,700-1100\text{nm}}$ is the average spectral transmittance of the fluid filter between the 700–1100 nm spectral window, A_{ap} is the aperture area of the primary mirrors, and GHI is the global horizontal irradiance.

The power absorbed by the receiver is calculated using the following equation:

$$Q_{abs,r} = DNI \times A_{ap} \times \eta_{opt} \times f_r \quad (27)$$

where f_r is the fraction of radiation absorbed by the receiver.

The useful thermal power absorbed by the fluid in the receiver is:

$$Q_{u,fl,th,r} = \dot{m}_{fl} \times Cp_{fl} \times (T_{r,fl,out} - T_{ch,fl,in}) \quad (28)$$

where Cp_{fl} is the specific heat capacity of the fluid, $\dot{m}_{fl}$ is the fluid mass flow rate, $T_{r,fl,out}$ is the temperature of the fluid in the outlet of the receiver, and $T_{ch,fl,in}$ is the temperature of the fluid in the inlet of the cooling channel. As a result, the thermal efficiency of the receiver may be calculated as follows:

$$\eta_r = \frac{Q_{u,fl,th,r}}{Q_{abs,r}} \quad (29)$$

The organic Rankine cycle (ORC) has received a great deal of attention as a well-accepted technology because it can make effective use of low-grade thermal energy sources, such as solar thermal [31]. In the present study, one of the scenarios examined considers that the thermal energy stored in the fluid is converted to electrical energy through an ORC. Therefore, the overall electrical efficiency of the system is defined as follows:

$$\eta_{total,el} = \eta_{PV} + [\eta_r \times \eta_{heat-Carnot} \times \left(1 - \frac{T_{amb}}{T_{r,fl,out}}\right)] \quad (30)$$

where $\eta_{heat-Carnot}$ is the thermodynamic efficiency of heat engine to Carnot efficiency [32].

Lastly, the net solar-to-electric efficiency of the system, which incorporates the total incident solar power as a common denominator, is presented as:

$$\eta_{NSE} = \frac{Q_{u,PV,el} + Q_{u,fl,th,r} \times \eta_{heat-Carnot} \times \left(1 - \frac{T_{amb}}{T_{r,fl,out}}\right)}{\text{DNI} \times A_{ap}} \quad (31)$$

3. Results and Discussion

3.1. Ray Tracing and Optimum Geometric Concentration Ratio of the CPVT

A ray-tracing simulation of the LFC has been carried out using Tonatiuh software to assess the design of the proposed CPVT (see Figure 4). Figure 5 illustrates the heat flux distribution on the front glass of the receiver. As can be noticed, the flux distribution corresponds to that of a typical LFC.

Figure 4. Ray tracing simulation with 250 rays using Tonatiuh software.

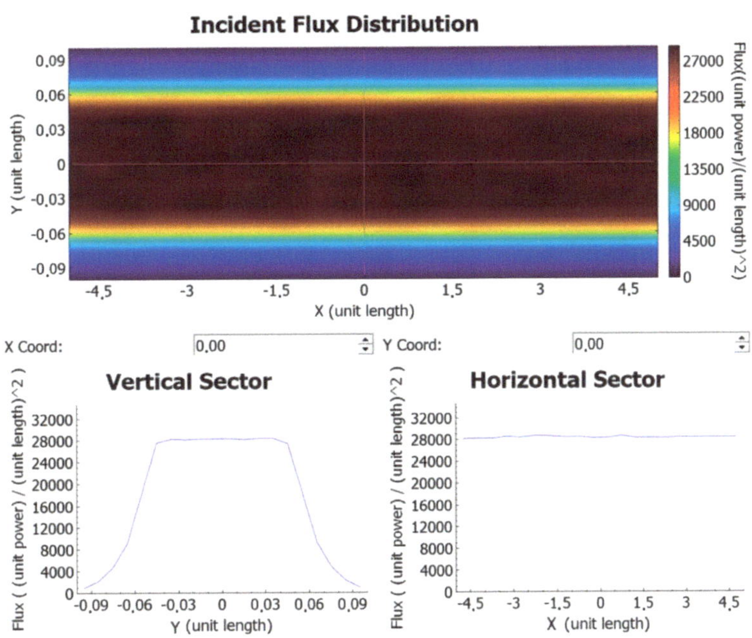

Figure 5. Front glass of the receiver flux distribution, simulation with 1×10^7 rays using Tonatiuh software.

A parametric study is carried out to determine the optimum concentration ratio for the CPVT collector. Average weather data for Tabuk was employed for this optimization process. Figure 6 illustrates the variation of the overall electric efficiency and the temperature of the PV module as a function of the geometric concentration ratio (GCR) of the CPVT collector. As can be seen, the optimum GCR that maximize the overall electric efficiency of the CPVT collector is about 20. At this GCR, the temperature of the PV module is less than 85 degree C (the maximum operating temperature of crystallin PV cells). Therefore, this value is used in this study.

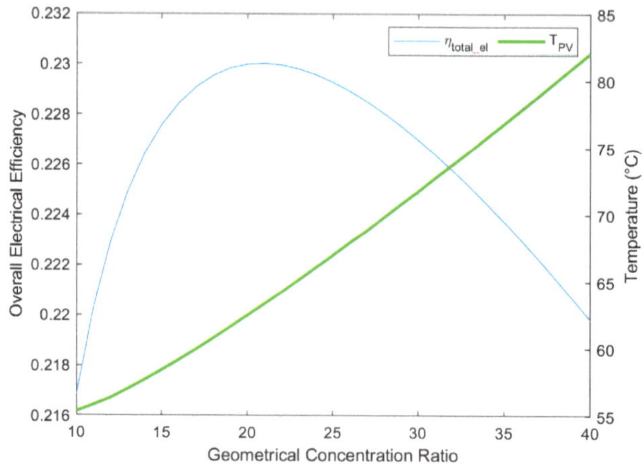

Figure 6. Optimum geometric concentration of the CPVT collector.

3.2. Advantages of the Proposed Receiver Design

To highlight the advantages of the proposed receiver design, an annual performance comparison between a receiver with cooling the PV module (denoted C in this paper) and a receiver without cooling the PV module (denoted NC in this paper) has been conducted. Six different locations and two different HTFs—water (denoted W in this paper) and water with ZnO nanoparticles at 0.01 wt% concentration (denoted W+ZnO in this paper)—are considered.

As can be noticed in Figure 7, the average temperature of the PV module is lower for the case with cooling than for the case without cooling (more than 10 °C difference). This results in higher efficiency of the PV cells. The addition of ZnO nanoparticles to water improves the heat transfer, which further reduces the temperature of the PV module; thus, high electric efficiency is achieved.

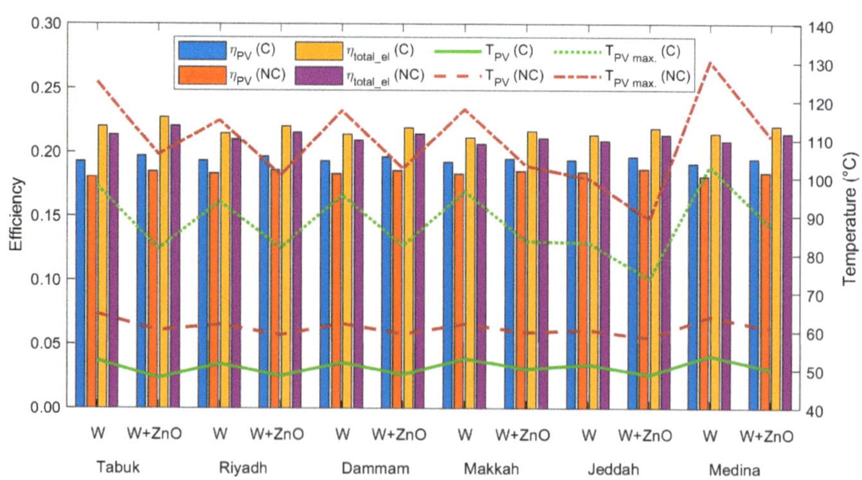

Figure 7. Annual average values of PV module efficiency (η_{PV}), total efficiency of CPVT collector ($\eta_{total,el}$), temperature of PV module (T_{PV}), and maximum temperature of PV module.

Overall, the performance of the CPVT collector at Tabuk is better than other locations because of the low ambient temperature and the high solar irradiance (see Table 4).

Table 4. Annual average values of direct normal irradiance (DNI), global horizontal irradiance (GHI), ambient temperature, wind speed, and optical efficiency of CPVT collector (η_{opt}).

Location	*DNI* (W m^{-2})	*GHI* (W m^{-2})	T_{amb} (°C)	V_{wind} (ms^{-1})	η_{opt} (%)
Tabuk	599	524	27.2	3.5	54.2
Riyadh	452	501	30.6	3.7	55.3
Dammam	441	494	31.1	3.6	55.3
Makkah	427	487	34.6	4.4	55.0
Jeddah	426	493	32.8	4.4	54.9
Medina	516	512	32.5	4.1	54.8

3.3. Thermal Performance of the CPVT

The yearly energy production of the CPVT at different locations is illustrated in Figure 8. The PV electrical energy output and thermal energy output are higher when the CPVT is installed in Tabuk. When ZnO nanoparticles are added to water, the thermal energy increases in all the considered locations, but the electrical energy provided by the PV panel slightly decreases. Although the drop in electrical energy is small compared to the gain in thermal energy, if all the thermal energy is converted to electricity, less energy is obtained compared with the case of using water.

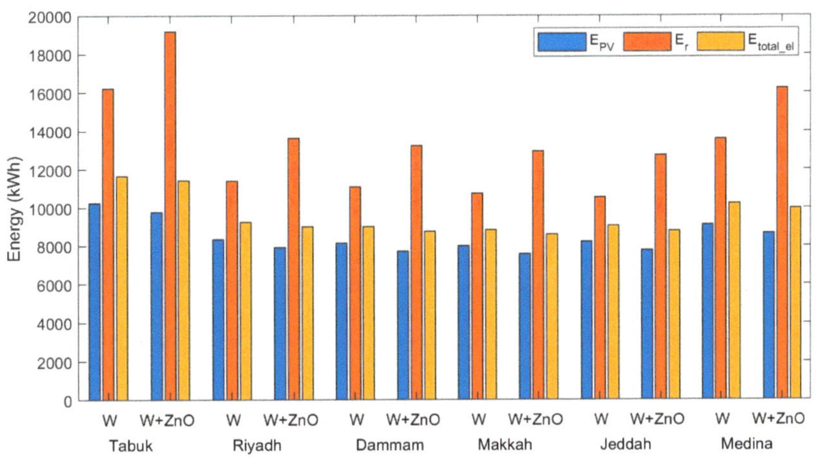

Figure 8. Annual energy production of the CPVT system.

The monthly energy production in Tabuk is shown in Figure 9. Summer months always have the highest energy output. The amount of energy produced varies dramatically throughout the year, with the summer period producing twice as much electrical energy and up to four times more thermal energy compared to winter months. This is because solar radiation is higher in the summer than in the winter, and the optical efficiency of the system is also higher.

Figure 9 also highlights that, when ZnO nanoparticles are added to water, the thermal energy output increases. This is because the working fluid absorbs 6.2% more solar radiation, as shown in Table 5, due to the variation in the spectral transmittance property, when ZnO is added to the water. On the other hand, when the working fluid contains nanoparticles, the electrical production is slightly lower. The scientific reason behind it is that the nanofluid absorbs more solar radiation at wavelengths between 700 nm and 1100 nm; these wavelengths are used to generate energy through the photovoltaic effect for silicon-based PV panels. In this spectral window, water alone has an average spectral transmittance of 88.1% [2], which drops to 79.8% [24] when ZnO nanoparticles are added.

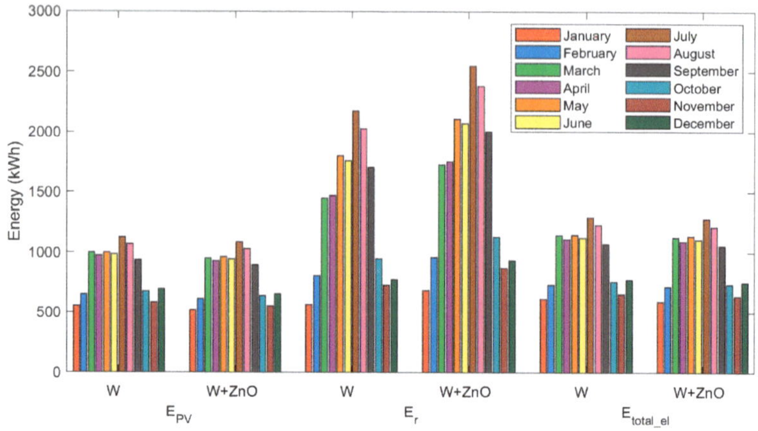

Figure 9. Energy production per month in Tabuk.

Table 5. Percentage of light power absorbed by component using different fluid filters.

Fluid Filter	PV Module (%)	Fluid (%)	Receiver Walls (%)	Thermal Unit (%)	Optical Loss (%)
Water [2]	31.5	23.1	33.2	56.3	12.2
Water-ZnO	25.3	29.3	33.2	62.5	12.2

Values for water-ZnO nanofluid calculated using the spectral transmittance presented by Huaxu et al. [33] at 0.01 wt% concentration.

A monthly analysis of the efficiency for the PV panel and the receiver in Tabuk is illustrated in Figure 10. It is notable that the variation in the efficiency for the PV panel is not significant during the year. The PV efficiency is slightly better in the winter period compared to the summer period due to lower ambient temperature. In contrast, the net solar-to-electric and thermal receiver efficiencies follow the same trend as thermal energy generation, being higher in summer than in winter.

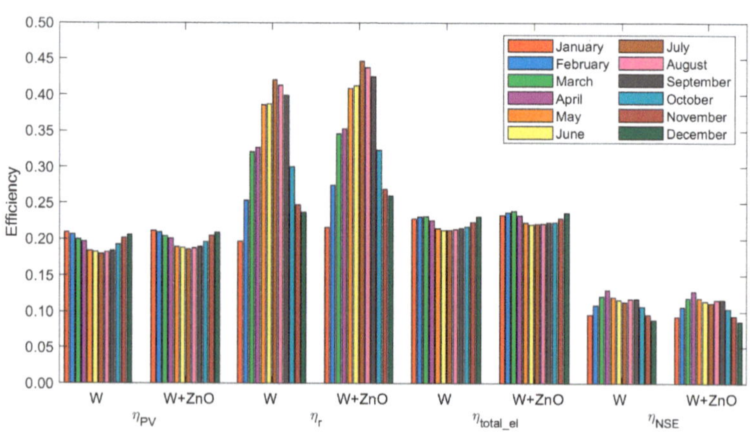

Figure 10. Energy efficiency per month in Tabuk.

3.4. Economic Analysis

3.4.1. CAPEX of the CPVT

The estimation of the cost of the CPVT is based on the Hyperlight Energy project. Table 6 highlights the specific costs as well as the total CAPEX of the CPVT collector considered in this study [34,35]. The cost of bifacial photovoltaic panels has been evaluated based on an average of projects completed in the last few years following the IRENA report [36]. Estimation showed that the CPVT collector with nano-particles costs $8550, while a CPVT collector with water as a HTF costs $200 less.

Table 6. CAPEX of a CPVT collector.

Component	Value	Unit	Cost ($)
Site improvement	5 [34]	$/m²mirror	200
Primary mirrors	110 [34]	$/m²mirror	4400
Thermal receiver (HTF, piping, etc.)	60 [34]	$/m²mirror	2400
Bifacial crystalline silicon cells (Total ins.)	1.5 [36]	$/Wp	1350
ZnO 0.01 wt% (preparation, product)	5 [37]	$/m²mirror	200
TOTAL CPVT COST			8550

3.4.2. The LCOE and LCOH for a 20 MW CPVT Plant

The LCOE and LCOH represent the average of the net present cost of energy production for the plant over its lifetime. The IRENA methodology is used in this paper [36]:

$$LCOE\ or\ LCOH = \frac{\sum_{t=1}^{n} \frac{I_t + OM_t}{(1+r)^t}}{\sum_{t=1}^{n} \frac{E_t}{(1+r)^t}} \quad (32)$$

where I_t are the investment expenditures in the year t, OM_t are the operations and maintenance expenditures in the year t, E_t is the energy generation in the year t, r is the discount rate, and n is the lifetime of the system.

The financial parameters that were used by IRENA are adapted in this study. These include: a 10% discount rate, a lifetime of 25 years, and 3% of the CAPEX were considered for the maintenance and operation costs.

Two operation strategies are considered, namely electricity production strategy and heat/electricity production strategy. In the former, the heat absorbed by the HTF is converted into electricity using an ORC cycle. The electricity may either be fed directly into the power grid or used to power an industry or households in remote places. A 20 MWh thermal energy storage (TES) consisting of a water tank is considered to store thermal energy with 82% round-trip-efficiency. In a heat/electricity production strategy, it is assumed that there is an industry nearby requiring hot water at 95 °C. This is highly feasible given that the industrial sector with low-temperature heat processes accounts for 7.1% of world energy consumption [38].

To calculate the LCOH and the LCOE of a large-scale CVPT power plant, it is important to estimate the CAPEX. Considering that the LFC uses an average of 70% of the total land, the cost of the land required was estimated at $5/m². The cost of a TES system with a capacity of 20,000 kWh has been evaluated taking data from the European Association of Storage of Energy [39], and adding an extra cost of 15$/kWh for each system owing to the cost of building work and additional materials, like pipes. For the situation when the thermal energy is converted into electricity, the cost of the plant required has been determined using a Pratt and Whitney ORC catalogue [40]. Project efforts have also been considered and are estimated at 22.5% of the total cost of the solar plant.

Finally, an additional 5% has been added to the overall expenditures to compensate for any unanticipated occurrences throughout the project's execution phase. The sum of all the expenditures is the capital expenditure (CAPEX), which is given in Tables 7 and 8 for both operation strategies.

Table 7. CAPEX for electricity production strategy, 20 MW CPVT plant with water + ZnO installed in Tabuk.

Component	Value	Unit	Cost ($)	
Design Type			**NC**	**C**
CPVT system	8550	\$/system	5,985,000	
Land costs (60,060 m^2)	5	\$/m^2	300,300	
Water storage system (20,000 kWh)	30 [39]	\$/kWh	600,000	
Power plant unit (all included)	2400 [40]	\$/kW	4,451,089	3,966,506
Project efforts (22.5% of solar plant costs)	22.5%	N/A	2,550,687	2,441,656
Uncertainties (5% of total costs)	5%	N/A	694,353.8	664,673
CAPEX			14,581,431	13,958,136

Table 8. CAPEX for electricity + heat production strategy, 20 MW CPVT plant with water + ZnO installed in Tabuk.

Component	Value	Unit	Cost ($)	
Design Type			**NC**	**C**
CPVT system	8550	\$/system	5,985,000	
Land costs (60,060 m^2)	5	\$/m^2	300,300	
Water storage system (20,000 kWh)	30 [39]	\$/kWh	600,000	
Project efforts (22.5% of solar plant costs)	22.5%	N/A	1,549,192	
Uncertainties (5% of total costs)	5%	N/A	421,724	
CAPEX			8,856,217	

Table 9 illustrates the values of the LCOE, LCOH, and CAPEX in different locations. Overall, the LCOE is lower for the proposed design (C) compared with the traditional design (NC). This proves the advantages of using the novel design proposed in this study. The addition of ZnO nanoparticles to the water increases the LCOE but decreases the LCOH.

An important finding of this study is that the heat/electricity production strategy is much better than the electricity production strategy. For instance, at Tabuk, the LCOE for our proposed design with water as HTF is 0.2232 USD/kWh when the electricity production strategy is selected. However, it is only 0.0847 USD/kWh when heat/electricity production is selected. Indeed, the LCOE when the heat/electricity production strategy is selected is lower than that of CSP (and the heat is produced as a by-product for free).

The most suitable location for installing CPVT technology in Saudi Arabia is at Tabuk. The analysis shows a LCOE of \$0.0847/kWh and a LCOH of \$0.0536/kWh when water is used as a spectral filter, and a LCOE of \$0.0906/kWh and a LCOH of \$0.0462/kWh when ZnO nanoparticles are added.

The LCOE of CPVT systems has been investigated in a limited number of scientific papers. In this study, we compare our results with some previously published studies, as shown in Table 10. Some of these studies have used PTC, which is more expensive than the technology of LFC used in our proposed design. Additionally, Fernandez et al. [41] have employed more expensive materials, such as ITO nanocrystals and Au nanoparticles instead of the ZnO nanoparticles used in our system. Furthermore, Ling et al. [10] have used a solid oxide fuel cell (SOFC) instead of an ORC to transform electrical energy into thermal energy, which requires the purchase of methanol, implying additional expenses. Moreover, according to the NREL database [42], the DNI and GHI in Shiraz are 7% and 10% lower, respectively, compared to Tabuk. Taking into account these differences and the novelty of our design for harsh environments, the proposed CPVT system offers a lower LCOE, making it a more cost-effective solution for the given geographical area.

Table 9. LCOE, LCOH, and CAPEX of the CPVT power plants for different scenarios. Note, system with cooling is denoted as C and without cooling as NC.

Location	Scenario	Fluid Filter	LCOE ($/kWh)		LCOH ($/kWh)		CAPEX ($)	
	Design Type		NC	C	NC	C	NC	C
Tabuk	All electricity	Water	0.2451	0.2232			13,533,991	12,985,307
		Water+ZnO	0.2670	0.2442			14,581,431	13,958,136
	Electricity + heat	Water	0.0916	0.0847	0.0479	0.0536	8,676,142	8,676,142
		Water+ZnO	0.0977	0.0906	0.0416	0.0462	8,856,217	8,856,217
Riyadh	All electricity	Water	0.2697	0.2495			11,993,164	11,555,130
		Water+ZnO	0.2942	0.2727			12,803,027	12,298,728
	Electricity + heat	Water	0.1109	0.1042	0.0669	0.0762	8,676,142	8,676,142
		Water+ZnO	0.1193	0.1120	0.0574	0.0651	8,856,217	8,856,217
Dammam	All electricity	Water	0.2742	0.2539			11,878,377	11,448,747
		Water+ZnO	0.2988	0.2771			12,663,564	12,168,329
	Electricity + heat	Water	0.1138	0.1068	0.0687	0.0784	8,676,142	8,676,142
		Water+ZnO	0.1223	0.1148	0.0590	0.0670	8,856,217	8,856,217
Makkah	All electricity	Water	0.2719	0.2552			11,540,492	11,275,797
		Water+ZnO	0.2976	0.2794			12,301,948	11,985,618
	Electricity + heat	Water	0.1147	0.1085	0.0736	0.0808	8,676,142	8,676,142
		Water+ZnO	0.1238	0.1170	0.0625	0.0685	8,856,217	8,856,217
Jeddah	All electricity	Water	0.2666	0.2500			11,647,527	11,338,953
		Water+ZnO	0.2927	0.2744			12,439,096	12,074,460
	Electricity + heat	Water	0.1116	0.1057	0.0741	0.0824	8,676,142	8,676,142
		Water+ZnO	0.1207	0.1142	0.0627	0.0696	8,856,217	8,856,217
Medina	All electricity	Water	0.2542	0.2359			12,401,737	12,058,328
		Water+ZnO	0.2779	0.2585			13,295,356	12,895,731
	Electricity + heat	Water	0.1014	0.0951	0.0583	0.0638	8,676,142	8,676,142
		Water+ZnO	0.1089	0.1021	0.0500	0.0546	8,856,217	8,856,217

Table 10. Comparison of the LCOE.

References	Location	Technology	LCOE ($/kWh)
Ling et al. [10]	Not available	LFC with solid filter	0.2000
Rodrigeus et al. [41]	Blythe, California	PTC with nanofluid filter	0.1783
Abedanzadeh et al. [43]	Shiraz, Iran	PTC with pieces of mirrors	0.1293
Present study	Tabuk, Saudi Arabia	LFC with fluid filter	0.0847
Present study	Tabuk, Saudi Arabia	LFC with nanofluid filter	0.0906

3.4.3. CO_2 Emission Analysis

According to the Brown to Green 2019 report, the national emissions in Saudi Arabia associated with electricity generation in 2019 were 0.723 kgCO_2-equivalent for each kWh produced [44]. Moreover, according to the Ministry of Spain, emissions from stationary combustion equipment powered by natural gas (such as boilers) are 0.209 kgCO_2-equivalent per kWh generated [45]. These two variables are used as the electricity and thermal emissions factors, respectively, to compute, based on the energy production, the emission savings due to the use of CPVT technology. Figures 11 and 12 show the results for one single CPVT collector and for a 20 MW CPVT plant, respectively.

The implementation of the proposed CPVT can cut off annual emissions by 11.2 tCO_2eq (Tabuk) per system and from 5675 tCO_2eq (Makkah) to 7822 tCO_2eq (Tabuk) per 20 MW plant if the heat/electricity production strategy is selected. However, if the electricity production strategy is selected, this technology can save a total of 8.5 tCO_2eq per collector and 5968 tCO_2eq per 20 MW CPVT plant annually.

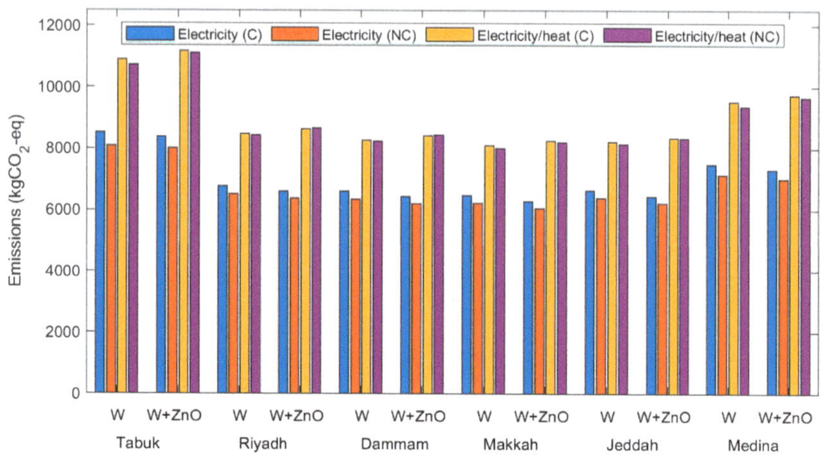

Figure 11. kgCO_2-equivalent emissions saved in KSA with one CPVT system.

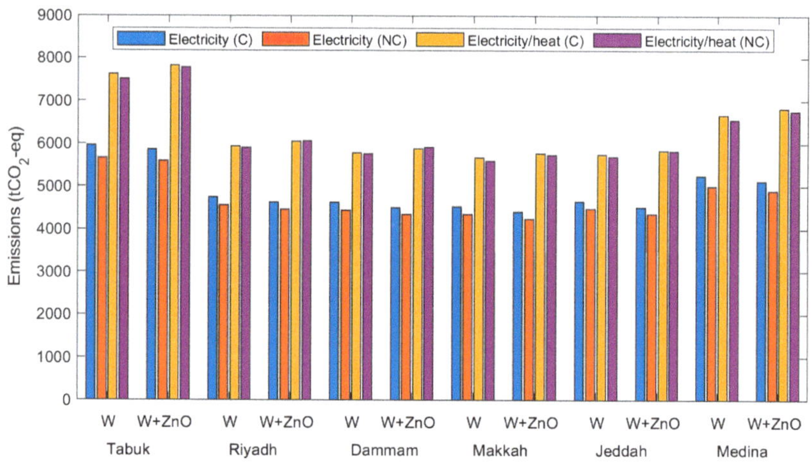

Figure 12. tCO_2-equivalent emissions saved in KSA with a 20 MW plant.

3.4.4. Battery vs. TES

A large increase in battery production is expected in the coming years. However, the materials needed for their production are limited, so it is critical to look for other ways of storing energy. For this reason, a thermal energy storage (TES) system has been considered in this study, and its comparison in economic terms is shown below.

The cost of lithium-ion battery packs has increased for the first time since 2010 because of rising inflation and prices of raw materials and battery components, reaching an average of 151$/kWh in 2022 [46]. Moreover, in the most optimistic scenario, lithium-ion batteries have a lifetime of 15 years, so they would have to be replaced at least once to match the lifespan of the solar power system [47,48]. On the other hand, according to the European Association for Energy Storage, the price for a hot water storage tank is 15$/kWh with an average 30-year working life.

For the battery scenario, the heat is first converted to electricity using an ORC with an efficiency on the order of 10% for the working temperatures considered in this study (see Equation (30)). Afterwards, electricity is stored into a 20 MWh lithium-ion battery which nowadays reach up DC round-trip efficiency values as high as 95% [49]. Alternatively, the MWh TES system with 82% round-trip efficiency previously described in Section 3.4.2 can be used, with no need to convert the heat into electricity for storage. Table 11 compares the two suggested storage systems installed in Tabuk based on the LCOE, LCOH, and CAPEX for a 20 MW large-scale plant, in which an extra cost of 15$/kWh has been estimated for each system owing to the cost of building work and additional materials like cables or pipes.

Table 11. LCOE, LCOH, and CAPEX of the CPVT with cooling for different energy storage types in Tabuk.

Location	Scenario	Fluid Filter	LCOE ($/kWh)		LCOH ($/kWh)		CAPEX ($)	
	Storage Type		**TES**	**Battery**	**TES**	**Battery**	**TES**	**Battery**
Tabuk	All electricity	Water	0.2232	0.3502			12,985,307	20,368,382
		Water+ZnO	0.2442	0.3734			13,958,136	21,341,211
	Electricity + heat	Water	0.0847	0.1568	0.0536	0.0991	8,676,142	16,059,217
		Water+ZnO	0.0906	0.1661	0.0462	0.0848	8,856,217	16,239,292

The use of batteries compared to a TES system based on a water tank represents an additional increase of 5.74 million dollars, considering that the batteries will need to be replaced once during the lifetime of the solar plant. This represents an even greater increase in the CAPEX, which is reflected in the LCOE and LCOH costs, which increase up to 56% and 84% respectively.

3.5. Future Work

In terms of future work, there are several areas of research that could be explored to further enhance the performance of concentrated photovoltaic-thermal (CPVT) systems. Building a prototype to experimentally validate the theoretical outcomes of this study would be a valuable next step. This would provide a more accurate representation of the real-world performance of the CPVT system.

Another possible avenue of research would be to include the capability of joining multiple CPVT systems in series in the mathematical model presented. This would allow for a higher heat transfer fluid output temperature, expanding the range of potential applications beyond just low-temperature heating.

Furthermore, it may be useful to explore the recommendations of An et al. [50] and investigate the effectiveness of using two reflectors at the sides of the solar receiver to minimize the effect of imprecise sun tracking and receiver installation. This could potentially reduce the current optical losses of the proposed solar receiver (12.2%) and improve the overall performance of the CPVT system.

Overall, this study contributes to the body of knowledge on CPVT systems and provides valuable insights for future research in this area. There is still much to be explored in terms of optimizing the performance and applicability of CPVT systems, and the proposed future research directions could help to advance this field.

4. Conclusions

The design and performance evaluation of a novel CPVT with spectral beam splitting technology, a cooling channel, and nanofluid is presented in this paper. A raytracing simulation tool is used to assess optical performance of the proposed CPVT collector, while an optical-thermal model is used to estimate the performance of the system.

The investigation revealed that using fluids as a filter in the CPVT collector has numerous benefits, including a low operating temperature for the PV cells and a high energy output. By adding the cooling channel and ZnO nanoparticles, it is found that a significant decrease in the average and maximum temperature of the PV panel is achieved, where they are lowered by 16.6 °C and 43.4 °C, respectively. This allows conventional silicon photovoltaic panels which have a maximum operating temperature of 85 °C [51] to be used. Therefore, without these design improvements, we would have to resort to special high-temperature PV panels, which are in very limited supply from manufacturers and present lower efficiencies due to the increased temperature.

The calculated yearly average values of the efficiencies, with the addition of the cooling channel and nanofluid, are, for Tabuk, 19.74% for the photovoltaic panel, 35.65% for the thermal collector, and 22.65% for the total conversion to electricity.

The economic assessment showed that the CPVT system has great possibilities to lead the Saudi renewable energy production in the coming years. Under a heat/electricity production strategy, a LCOE of $0.0847/kWh and a LCOH of $0.0536/kWh when only water is used a HTF are obtained. At the same time, a LCOE of $0.0906/kWh and a LCOH of $0.0462/kWh are obtained when ZnO particles are added. The analysis showed that due to the low performance and high costs of converting thermal energy into electricity, the CPVT technology is less competitive when the electricity production strategy is selected. The results showed an LCOE of $0.2232/kWh with water only and $0.2442/kWh with the addition of ZnO nanoparticles.

Furthermore, after comparing battery energy storage against a TES system, a large increase in the CAPEX was observed if batteries are used, reflected in the LCOE and LCOH costs, which increase up to 56% and 84% (compared with the case of TES), respectively. Thus, a CPVT plant with TES operating under a heat/electricity production strategy is better than a CPVT plant with battery operating under an electricity production strategy.

The study showed that a 20 MW CPVT plant cuts CO_2-equivalent emissions up to 7822 tonnes every year under Saudi Arabian conditions. Another benefit in terms of sustainability is the ease of recycling the proposed CPVT technology, taking up less space and requiring less photovoltaic material to capture the same sunlight as non-concentrating PV modules. Thus, the process is less dependent on the silicon supply chain.

Regarding the practicality of the technology presented, it has been demonstrated that the system is technically feasible through a series of rigorous computations and simulations. Specifically, the results indicate that the proposed design offers significant advantages when operating in harsh environments when compared to traditional designs. Additionally, an economic study was conducted which revealed that the system can be constructed at a relatively low cost in comparison to previous publications, resulting in an improved levelized cost of energy (LCOE) for this technology. Overall, these findings support the practicability of the technology, and suggest that it has the potential to be a viable cost-effective solution for a range of real-world energy applications.

Author Contributions: Conceptualization, C.L., O.B. and B.D.; Methodology, C.L., O.B. and B.D.; Software, C.L. and O.B.; Formal analysis, C.L.; Investigation, C.L., O.B. and B.D.; Resources, B.D.; Writing—original draft, C.L.; Writing—review & editing, O.B. and B.D.; Visualization, C.L.; Supervision, O.B. and B.D.; Project administration, B.D.; Funding acquisition, B.D. All authors have read and agreed to the published version of the manuscript.

Funding: Project is supported by Prof Bassam Dally KAUST baseline fund.

Data Availability Statement: Not applicable.

Acknowledgments: The authors gratefully acknowledge the funding of the King Abdullah University of Science and Technology.

Conflicts of Interest: The authors declare no conflict of interest.

Nomenclature

Abbreviations:

PV	Photovoltaic
CPVT	Concentrating solar photovoltaic thermal
LCOE	Levelized cost of electricity
LCOH	Levelized cost of heat
SBS	Spectrum beam splitting
HTF	Heat transfer fluid

PTC	Parabolic trough collector
LFC	Linear Fresnel collector
TE	Thermoelectric generator
NREL	National Renewable Energy Laboratory
DNI	Direct normal irradiance
GHI	Global horizontal irradiance
ORC	Organic Rankine cycle
GCR	Geometric concentration ratio
GMT	Greenwich mean time
CSP	Concentrated solar power
TES	Thermal energy storage
C	CPVT with cooling channel
NC	CPVT without cooling channel
CAPEX	Capital expenditure
DR	Diffuse irradiance
LST	Local solar time
LT	Local time
EoT	Equation of time
TC	Net time correction factor
LSTM	Local standard time meridian
HRA	Hour angle

Nomenclature:

W	Width $[\mathrm{m}]$
H	Height $[\mathrm{m}]$
M	Location of the mirrors $[\mathrm{m}]$
S	Distance of adjacent mirrors $[\mathrm{m}]$
fc	Focal length $[\mathrm{m}]$
K	Incidence angle modifier
Az	Azimuth angle $[°]$
Z	Zenith angle $[°]$
L	Length $[\mathrm{m}]$
A	Area $\left[\mathrm{m}^2\right]$
Q	Heat flux $[\mathrm{W}]$
T	Temperature $[\mathrm{K}]$
h	Heat transfer coefficient $\left[\frac{\mathrm{W}}{\mathrm{m}^2\mathrm{K}}\right]$
Nu	Nusselt number
k	Thermal conductivity $\left[\frac{\mathrm{W}}{\mathrm{mK}}\right]$
Dh	Hydraulic diameter $[\mathrm{m}]$
Re	Reynolds number
Pr	Prandtl number
t	Thickness $[\mathrm{m}]$
tr	Average spectral transmittance
f	Fraction of radiation absorbed
$\dot{m}$	Fluid mass flow rate $\left[\frac{\mathrm{kg}}{\mathrm{s}}\right]$
Cp	Specific heat capacity $\left[\frac{\mathrm{J}}{\mathrm{kgK}}\right]$
wt	Mass fraction
d	Day number of the year, ranging from 1 to 365
V	Velocity $\left[\frac{\mathrm{m}}{\mathrm{s}}\right]$

Greek letters:

δ	Tilt angle $[°]$
η	Efficiency
θ	Incidence angle $[°]$
σ	Stefan–Boltzmann constant $\left[\frac{\mathrm{W}}{\mathrm{m}^2\mathrm{K}^4}\right]$
ε	Emissivity
α	Elevation angle $[°]$
φ	Local latitude $[°]$
β	Declination of the sun $[°]$

Subscripts:

n	Number of primary mirror
ch	Channel
r	Receiver
opt	Optical
T	Transversal
L	Longitudinal
fl	Fluid
amb	Ambient
conv	Convection
cond	Conduction
rad	Radiation
ref	Reference
ap	Aperture
abs	Absorbed
atm	Atmosphere
th	Thermal
el	Electrical
NSE	Net solar-to-electric

Appendix A

A mathematical code is developed in Matlab to simulate the performance of the CPVT. The set of equations presented in the previous sections are solved using an iteration process. The flowchart of the model is illustrated in Figure A1. The output HTF temperature, location information, geometrical parameters, and fluid characteristics are used as input. The model uses the direct normal irradiance (DNI), global horizontal irradiance (GHI), wind speed, and ambient temperature from the NREL database [42]. The optical efficiency of the system is calculated using the location, solar angles, and geometrical data. Next, the variables to be determined are set up, and an iterative procedure based on energy balance is used. This process ends when all the energy and mass balance equations are satisfied. Lastly, the power and energy performance are determined.

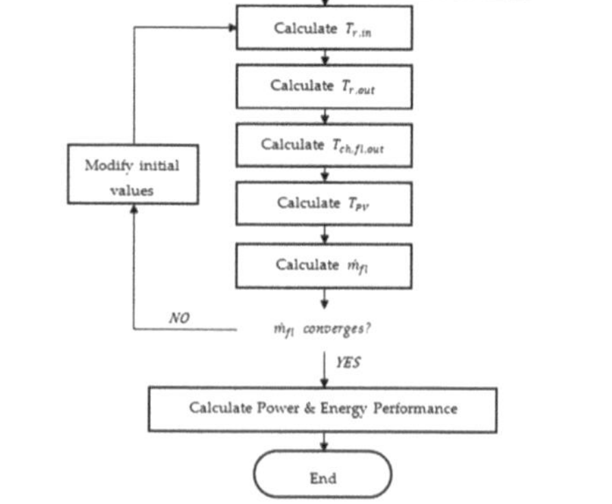

Figure A1. Flowchart of calculation model.

References

1. Han, X.; Tu, L.; Sun, Y. A spectrally splitting concentrating PV/T system using combined absorption optical filter and linear Fresnel reflector concentrator. *Sol. Energy* **2021**, *223*, 168–181. [CrossRef]
2. Mojiri, A.; Stanley, C.; Rosengarten, G. Spectrally Splitting Hybrid Photovoltaic/thermal Receiver Design for a Linear Concentrator. *Energy Procedia* **2014**, *48*, 618–627. [CrossRef]
3. Ramírez Quiroz, C.O.; Levchuk, I.; Bronnbauer, C.; Salvador, M.; Forberich, K.; Heumüller, T.; Hou, Y.; Schweizer, P.; Spiecker, E.; Brabec, C.J. Pushing efficiency limits for semitransparent perovskite solar cells. *J. Mater. Chem. A* **2015**, *3*, 24071–24081. [CrossRef]
4. Jarimi, H.; Lv, Q.; Ramadan, O.; Zhang, S.; Riffat, S. Design, mathematical modelling and experimental investigation of vacuum insulated semi-transparent thin-film photovoltaic (PV) glazing. *J. Build. Eng.* **2020**, *31*, 101430. [CrossRef]
5. Mujahid, M.; Chen, C.; Zhang, J.; Li, C.; Duan, Y. Recent advances in semitransparent perovskite solar cells. *InfoMat* **2021**, *3*, 101–124. [CrossRef]
6. Joshi, S.S.; Dhoble, A.S.; Jiwanapurkar, P.R. Investigations of Different Liquid Based Spectrum Beam Splitters for Combined Solar Photovoltaic Thermal Systems. *J. Sol. Energy Eng.* **2016**, *138*, 021003. [CrossRef]
7. Soule, D.E. Hybrid Solar Energy Generating System. 1987. Available online: https://www.osti.gov/biblio/5984591 (accessed on 28 May 2022).
8. Zhang, J.J.; Qu, Z.G.; Zhang, J.F. MCRT-FDTD investigation of the solar-plasmonic-electrical conversion for uniform irradiation in a spectral splitting CPVT system. *Appl. Energy* **2022**, *315*, 119054. [CrossRef]
9. Daneshazarian, R.; Cuce, E.; Cuce, P.M.; Sher, F. Concentrating photovoltaic thermal (CPVT) collectors and systems: Theory, performance assessment and applications. *Renew. Sustain. Energy Rev.* **2018**, *81*, 473–492. [CrossRef]
10. Ling, Y.; Li, W.; Jin, J.; Yu, Y.; Hao, Y.; Jin, H. A spectral-splitting photovoltaic-thermochemical system for energy storage and solar power generation. *Appl. Energy* **2020**, *260*, 113631. [CrossRef]
11. Liew, N.J.Y.; Yu, Z.; Holman, Z.; Lee, H.-J. Application of spectral beam splitting using Wavelength-Selective filters for Photovoltaic/Concentrated solar power hybrid plants. *Appl. Therm. Eng.* **2022**, *201*, 117823. [CrossRef]
12. Ju, X.; Xu, C.; Han, X.; Du, X.; Wei, G.; Yang, Y. A review of the concentrated photovoltaic/thermal (CPVT) hybrid solar systems based on the spectral beam splitting technology. *Appl. Energy* **2017**, *187*, 534–563. [CrossRef]
13. Sabry, M.; Gottschalg, R.; Betts, T.R.; Shaltout, M.A.M.; Hassan, A.F.; El-Nicklawy, M.M.; Infield, D.G. Optical filtering of solar radiation to increase performance of concentrator systems. In Proceedings of the Conference Record of the IEEE Photovoltaic Specialists Conference, New Orleans, LA, USA, 19–24 May 2002; pp. 1588–1591. Available online: https://www.scopus.com/inward/record.uri?eid=2-s2.0-0036957820&partnerID=40&md5=9aae59a3fe7add1b1f8cc02576b1774e (accessed on 29 May 2022).
14. Hangweirer, M.; Höller, R.; Schneider, H. Design and analysis of a novel concentrated photovoltaic–thermal receiver concept. *Jpn. J. Appl. Phys.* **2015**, *54*, 08KE01. [CrossRef]
15. Meraje, W.C.; Huang, C.-C.; Barman, J.; Huang, C.-Y.; Kuo, C.-F.J. Design and experimental study of a Fresnel lens-based concentrated photovoltaic thermal system integrated with nanofluid spectral splitter. *Energy Convers. Manag.* **2022**, *258*, 115455. [CrossRef]
16. Barthwal, M.; Rakshit, D. Holistic opto-thermo-electrical analysis of a novel spectral beam splitting-based concentrating photovoltaic thermal system. *J. Clean. Prod.* **2022**, *379*, 134545. [CrossRef]
17. Wang, G.; Ge, Z.; Lin, J. Design and performance analysis of a novel solar photovoltaic/thermal system using compact linear Fresnel reflector and nanofluids beam splitting device. *Case Stud. Therm. Eng.* **2022**, *35*, 102167. [CrossRef]
18. Su, B.; Wang, H.; Zhang, X.; He, H.; Zheng, J. Using photovoltaic thermal technology to enhance biomethane generation via biogas upgrading in anaerobic digestion. *Energy Convers. Manag.* **2021**, *235*, 113965. [CrossRef]
19. Su, B.; Han, W.; Qu, W.; Liu, C.; Jin, H. A new hybrid photovoltaic/thermal and liquid desiccant system for trigeneration application. *Appl. Energy* **2018**, *226*, 808–818. [CrossRef]
20. Rodrigues Fernandes, M.; Schaefer, L.A. Long-term environmental impacts of a small-scale spectral filtering concentrated photovoltaic-thermal system. *Energy Convers. Manag.* **2019**, *184*, 350–361. [CrossRef]
21. Anand, B.; Shankar, R.; Murugavelh, S.; Rivera, W.; Midhun Prasad, K.; Nagarajan, R. A review on solar photovoltaic thermal integrated desalination technologies. *Renew. Sustain. Energy Rev.* **2021**, *141*, 110787. [CrossRef]
22. George, M.; Pandey, A.K.; Abd Rahim, N.; Tyagi, V.V.; Shahabuddin, S.; Saidur, R. Concentrated photovoltaic thermal systems: A component-by-component view on the developments in the design, heat transfer medium and applications. *Energy Convers. Manag.* **2019**, *186*, 15–41. [CrossRef]
23. National Institute of Standards and Technology. Thermophysical Properties of Fluid Systems. Available online: https://webbook.nist.gov/chemistry/fluid/ (accessed on 14 May 2022).
24. Han, X.; Lu, L.; Yan, S.; Yang, X.; Tian, R.; Zhao, X. Stability, Thermal Conductivity and Photothermal Conversion Performance of Water-Based ZnO Nanofluids. *J. Therm. Sci.* **2021**, *30*, 1581–1595. [CrossRef]
25. Farzanehnia, A.; Sardarabadi, M. Exergy in Photovoltaic/Thermal Nanofluid-Based Collector Systems. In *Exergy and Its Application: Toward Green Energy Production and Sustainable Environment*; IntechOpen: Rijeka, Croatia, 2019. Available online: https://www.intechopen.com/chapters/66854 (accessed on 3 June 2022).
26. Hassani, S.; Taylor, R.A.; Mekhilef, S.; Saidur, R. A cascade nanofluid-based PV/T system with optimized optical and thermal properties. *Energy* **2016**, *112*, 963–975. [CrossRef]

27. Rungasamy, A.E.; Craig, K.J.; Meyer, J.P. A review of linear Fresnel primary optical design methodologies. *Sol. Energy* **2021**, *224*, 833–854. [CrossRef]
28. Zhu, Y.; Shi, J.; Li, Y.; Wang, L.; Huang, Q.; Xu, G. Design and thermal performances of a scalable linear Fresnel reflector solar system. *Energy Convers. Manag.* **2017**, *146*, 174–181. [CrossRef]
29. Said, Z.; Ghodbane, M.; Hachicha, A.A.; Boumeddane, B. Optical performance assessment of a small experimental prototype of linear Fresnel reflector. *Case Stud. Therm. Eng.* **2019**, *16*, 100541. [CrossRef]
30. Radwan, A.; Emam, M.; Ahmed, M. Chapter 2.15—Comparative Study of Active and Passive Cooling Techniques for Concentrated Photovoltaic Systems. In *Exergetic, Energetic and Environmental Dimensions*; Dincer, I., Colpan, C.O., Kizilkan, O., Eds.; Academic Press: Cambridge, MA, USA, 2018; pp. 475–505.
31. Wang, H.; Li, H.; Wang, L.; Bu, X. Thermodynamic Analysis of Organic Rankine Cycle with Hydrofluoroethers as Working Fluids. *Energy Procedia* **2017**, *105*, 1889–1894. [CrossRef]
32. An, W.; Li, J.; Ni, J.; Taylor, R.A.; Zhu, T. Analysis of a temperature dependent optical window for nanofluid-based spectral splitting in PV/T power generation applications. *Energy Convers. Manag.* **2017**, *151*, 23–31. [CrossRef]
33. Huaxu, L.; Fuqiang, W.; Dong, L.; Jie, Z.; Jianyu, T. Optical properties and transmittances of ZnO-containing nanofluids in spectral splitting photovoltaic/thermal systems. *Int. J. Heat Mass Transf.* **2019**, *128*, 668–678. [CrossRef]
34. Kincaid, N.; Mungas, G.; Kramer, N.; Wagner, M.; Zhu, G. An optical performance comparison of three concentrating solar power collector designs in linear Fresnel, parabolic trough, and central receiver. *Appl. Energy* **2018**, *231*, 1109–1121. [CrossRef]
35. U.S. Department of Energy. PROJECT PROFILE: Hyperlight Energy (CSP: COLLECTS). Available online: https://www.energy.gov/eere/solar/project-profile-hyperlight-energy-csp-collects (accessed on 30 May 2022).
36. IRENA. *Renewable Power Generation Costs in 2020*; International Renewable Energy Agency: Abu Dhabi, UAE, 2021. Available online: https://www.irena.org/publications/2021/Jun/Renewable-Power-Costs-in-2020 (accessed on 2 May 2022).
37. Liang, H.; Wang, F.; Yang, L.; Cheng, Z.; Shuai, Y.; Tan, H. Progress in full spectrum solar energy utilization by spectral beam splitting hybrid PV/T system. *Renew. Sustain. Energy Rev.* **2021**, *141*, 110785. [CrossRef]
38. Abarca, V. Energía Solar Térmica en la Industria: Potencial y Aplicaciones. Fundación de la Energía de la Comunidad de Madrid. Available online: https://www.fenercom.com/wp-content/uploads/2020/06/1_Energia_solar_termica_en_la_industria_potencial_y_aplicaciones_ASIT_fenercom-2020.pdf (accessed on 3 December 2022).
39. Thermal Hot Water Storage. Available online: https://ease-storage.eu/wp-content/uploads/2016/03/EASE_TD_HotWater.pdf (accessed on 5 December 2022).
40. Pratt & Whitney. Organic Rankine Cycle Power Generation. Available online: https://aeenewengland.org/images/downloads/AEE_ASHRAE_JointMeeting_March2013/orc_power_generation_pratt_and_whitney.pdf (accessed on 30 May 2022).
41. Rodrigues Fernandes, M.; Schaefer, L.A. Levelized cost of energy of hybrid concentrating photovoltaic-thermal systems based on nanofluid spectral filtering. *Sol. Energy* **2021**, *227*, 126–136. [CrossRef]
42. National Renewable Energy Laboratory (NREL). NSRDB: National Solar Radiation Database. Available online: https://nsrdb.nrel.gov/data-viewer (accessed on 27 November 2022).
43. Abedanzadeh, A.; Borgheipour, H.; Fakouriyan, S.; Torabi, F. Economic Evaluation of Supplying Commercial Thermal Load by a New CPVT System: A Case Study in Iran. *J. Appl. Comput. Mech.* 2021. [CrossRef]
44. Saudi Arabia, Country Profile 2019. Climate Transparency. Available online: https://www.climate-transparency.org/media/saudi-arabia-country-profile-2019#:~:text=This%20country%20profile%20is%20part,several%20studies%20by%20renowned%20institutions (accessed on 20 November 2022).
45. Ministerio de España. Factores de Emisión. Available online: https://www.miteco.gob.es/ca/cambio-climatico/temas/mitigacion-politicas-y-medidas/factores_emision_tcm34-446710.pdf (accessed on 30 May 2022).
46. Bloomberg. Lithium-Ion Battery Pack Prices Rise for First Time to an Average of $151/kWh. Available online: https://about.bnef.com/blog/lithium-ion-battery-pack-prices-rise-for-first-time-to-an-average-of-151-kwh (accessed on 10 December 2022).
47. National Renewable Energy Laboratory. Battery Lifespan. Available online: https://www.nrel.gov/transportation/battery-lifespan.html (accessed on 11 December 2022).
48. Sunrun. What Is the Life Expectancy of a Solar Battery? Available online: https://www.sunrun.com/go-solar-center/solar-articles/what-is-the-life-expectancy-of-a-solar-battery (accessed on 11 December 2022).
49. Six Years and 26 Batteries Later. Available online: https://itpau.com.au/knowledge/itps-battery-test-centre/#:~:text=A%20typical%20lithium%20ion%20battery,losses%20for%20lead%2Dacid%20systems (accessed on 11 December 2022).
50. An, W.; Zhang, J.; Zhu, T.; Gao, N. Investigation on a spectral splitting photovoltaic/thermal hybrid system based on polypyrrole nanofluid: Preliminary test. *Renew. Energy* **2016**, *86*, 633–642. [CrossRef]
51. How Hot Do Solar Panels Get and How Does It Affect My System? Solar Reviews. Available online: https://www.solarreviews.com/blog/how-hot-can-solar-panels-get (accessed on 13 December 2022).

technologies

Article

An Improved Photovoltaic Module Array Global Maximum Power Tracker Combining a Genetic Algorithm and Ant Colony Optimization

Kuo-Hua Huang, Kuei-Hsiang Chao * and Ting-Wei Lee

Department of Electrical Engineering, National Chin-Yi University of Technology, Taichung 41170, Taiwan; huangkh@ncut.edu.tw (K.-H.H.); ylps60849@gmail.com (T.-W.L.)
* Correspondence: chaokh@ncut.edu.tw; Tel.: +886-4-2392-4505 (ext. 7272); Fax: +886-4-2392-2156

Abstract: In this paper, a hybrid optimization controller that combines a genetic algorithm (GA) and ant colony optimization (ACO) called GA-ACO algorithm is proposed. It is applied to a photovoltaic module array (PVMA) to carry out maximum power point tracking (MPPT). This way, under the condition that the PVMA is partially shaded and that multiple peaks are produced in the power-voltage (P-V) characteristic curve, the system can still operate at the global maximum power point (GMPP). This solves the problem seen in general traditional MPPT controllers where the PVMA works at the local maximum power point (LMPP). The improved MPPT controller that combines GA and ACO uses the slope of the P-V characteristic curve at the PVMA work point to dynamically adjust the iteration parameters of ACO. The simulation results prove that the improved GA-ACO MPPT controller is able to quickly track GMPP when the output P-V characteristic curve of PVMA shows the phenomenon of multiple peaks. Comparing the time required for tracking to MPP with different MPPT approaches for the PVMA under five different shading levels, it was observed that the improved GA-ACO algorithm requires 19.5~35.9% (average 29.2%) fewer iterations to complete tracking than the mentioned GA-ACO algorithm. Compared with the ACO algorithm, it requires 74.9~79.7% (average 78.2%) fewer iterations, and 75.0~92.5% (average 81.0%) fewer than the conventional P&O method. Therefore, it is proved that by selecting properly adjusted values of the Pheromone evaporation rate and the Gaussian standard deviation of the proposed GA-ACO algorithm based on the slope scope of the P-V characteristic curves, a better response performance of MPPT is obtained.

Keywords: genetic algorithm (GA); ant colony optimization (ACO); photovoltaic module array (PVMA); maximum power point tracking (MPPT); global maximum power tracking (GMPP); local maximum power point (LMPP); P-V characteristic curve

Citation: Huang, K.-H.; Chao, K.-H.; Lee, T.-W. An Improved Photovoltaic Module Array Global Maximum Power Tracker Combining a Genetic Algorithm and Ant Colony Optimization. *Technologies* **2023**, *11*, 61. https://doi.org/10.3390/technologies11020061

Academic Editors: Dongran Song, Valeri Mladenov, Jayanta Deb Mondol, Annamaria Buonomano and Biplab Das

Received: 17 January 2023
Revised: 20 March 2023
Accepted: 17 April 2023
Published: 20 April 2023

1. Introduction

In recent years, with the rise of environmental protection awareness and the exhaustion of petroleum, natural gas, coal mines and other forms of energy, scientists have begun searching for environmentally-friendly and sustainable alternative energy. For scientists, solar power has become one of the ideal forms of alternative energy as it is not bound by geographical conditions and is easily installed. In order to realize energy independence and reduce carbon emissions to alleviate global warming, photovoltaic power generation that does not rely on imports has become one of the renewable energies actively developed by the world's governments. The governments around the world have also set their own capacity goals. In order to achieve these goals, an appropriate MPPT controller has been designed to increase the power generation utilization of photovoltaic power generation systems.

The conventional algorithms generally applied to track the maximum power of PVMAs include perturbation and observation (P&O) [1–3] and incremental conductance

(INC) [4,5]. These two conventional approaches, while being able to effectively track the MPPs when PVMAs operate under normal working conditions, when PVMAs are shaded, these conventional approaches may only track the LMPPs rather than the GMPP. Thus, the power generation efficiency of PVMAs is reduced. Therefore, the problem of tracking the maximum power point when shading occurs in the photovoltaic module cannot be solved [6].

To address the aforesaid issue, in recent years, some solutions for determining the optimal values have been proposed which reduce the probability that local optimal solutions are obtained during the solution process. For instance, an efficient hybrid starling murmuration optimizer that combines dynamic opposition, a Taylor-based optimal neighborhood strategy, and a crossover operator (DTCSMO) [7], an efficient enhanced modified chameleon swarm algorithm termed MCSA [8] and an enhanced hierarchical guided slime mould algorithm called HG-SMA [9] etc. have been developed. While these optimization algorithms may effectively address the issue of optimal solutions, there is no practical case for tracking the GMPP of PVMAs when multiple peak values appear in the P-V characteristic curves.

In order to solve the GMPP tracking problem under the condition that the PVMA module is partially shaded, many practical smart maximum power tracking controllers have been proposed and applied to solve the problem [10–17]. This is because the smart maximum power tracking controller can search the GMPP generated in the nonlinear multi-peak P-V characteristic curve under the condition that the PVMA module is partially shaded. The more commonly used smart methods include particle swarm optimization (PSO) [11–13], ant colony optimization (ACO) [14–16], genetic algorithms (GA) [17–20], and cuckoo search algorithms (CSA) [21,22], etc. The literature indicates that the smart algorithm-based MPPT controllers have a better steady-state response and tracking response compared with traditional methods. Additionally, when the photovoltaic module is partially shaded, the GMPP can be accurately and quickly tracked, unlike traditional method-based MPPT controllers that can only track the LMPP. However, these smart maximum power tracking controllers adopt fixed parameter values in the iterative formulas adopted, so there is still room for improvement in terms of the speed of dynamic tracking responses and the performance of steady-state tracking.

The improved artificial bee colony (I-ABC algorithm) [23] combining the artificial bee colony algorithm and the perturb and observe (P&O) method has the advantage that the GMPP is searched via the bee colony algorithm, and the correct direction for the next tracking is determined by the P&O method. While this approach reduces the issue of tendency where local optimal solutions are obtained during the solution process and addresses the issue that the P&O method is unable to track the MPPs if the PVMAs are abnormal, the computation is more complex, and the tracking responses are not fast enough. Additionally, the conventional cuckoo search (CS) can be improved by adjusting the step factors of CS depending on the slopes and iterations of the PVMAs' P-V characteristic curves [24]. While the GMPPs may be tracked faster and more precisely when partial modules are shaded in a PVMA and multiple peak values are generated for the P-V characteristic curve, the improved CS is only applicable in the simulation phase. Practical testing results for PVMAs under different connection configurations and shading ratios may enable the improved CS to track the GMPP in less time and improve the power generation efficiency of the photovoltaic power generation system.

Although Chao and Rizal [25] proposed a MPPT controller with a new GA and ACO hybrid algorithm, the proposed MPPT controller also demonstrates the characteristics of GA and ACO algorithms. In particular, the GA has excellent features when searching for the best solution and enabling the system to slowly converge. When used independently, more computation time is needed, possibly because there are more populations, resulting in the disadvantage of a longer tracking time [19]. On the other hand, ACO features the ability to quickly search the subspace and converge to the best non-global solution in advance. Hence, the incorporation of GA can complement the ACO algorithm, thereby enhancing the speed of maximum power tracking and enabling the PVMA to output the global maximum power. Based on the above reasons, an improved GA-ACO MPPT controller is proposed in this paper. The same circuit structure and tracking steps as [25] were also employed. The optimization of the GA-ACO parameters differs depending on the P-V characteristic curves generated under different shading conditions of PVMAs, and thus, no principle can be found for the parameter optimization. Therefore, it is learnt in tests that when the tracking approaches the MPP and as the slope of the P-V characteristic curve declines, the Pheromone evaporation rate ρ and the Gaussian standard deviation x increase; and the ρ and x parameters are required to be greater as MPP is approached. In contrast, the farther the MPP is, the more ρ and x must decrease as the slope of the P-V characteristic curve increases. Therefore, the optimal adjusted value of the Pheromone evaporation rate, $\Delta\rho$, and the optimal adjusted value of Gaussian standard deviation, Δx, may be obtained via multiple simulations based on the slope values of P-V characteristic curves of PVMAs. Through the location of the work point, the slope of the P-V characteristic curve was calculated to automatically adjust the Pheromone evaporation rate ρ and the Gaussian standard deviation x in the iterative formula. The global maximum power tracking time was reduced to obtain better steady-state responses.

2. The Shading Characteristics of a PVMA

The output power of a photovoltaic module changes with the environment, weather and temperature. In the PVMA, any shaded module will affect the total output power, because each part of the module in the PVMA is connected in series [26]. Therefore, even if a photovoltaic module is shaded in series, the output current of the entire PVMA is also affected. Using MATLAB software and under the standard test conditions (STC) (solar irradiance: 1000 W/m^2, air mass (AM): 1.5 and temperature: 25 °C), the I-V and P-V characteristic curves of the photovoltaic module array were simulated. Figure 1 illustrates the P-V and I-V characteristic curves of a four photovoltaic module array with one module under 50% shade [26]. Because the photovoltaic module array consists of four photovoltaic modules in series, one of which is shaded by 50%, with the rest unshaded, two peaks appear in the P-V characteristic curve of the PV module array, and there is a considerable decrease in the maximum power output, as shown in the P-V characteristic curve. A similar pattern is observed in other situations. For any shading occurring on a photovoltaic module array, there will be more than one maximum power point (MPP) observed in the power-voltage (P-V) characteristic curve of the photovoltaic module array. However, only the local maximum power point (LMPP) can be tracked by the traditional maximum power point tracker, but not the global maximum power point (GMPP). Therefore, in this paper, an intelligent maximum power point tracker based on an improved GA-ACO algorithm is presented to overcome this problem.

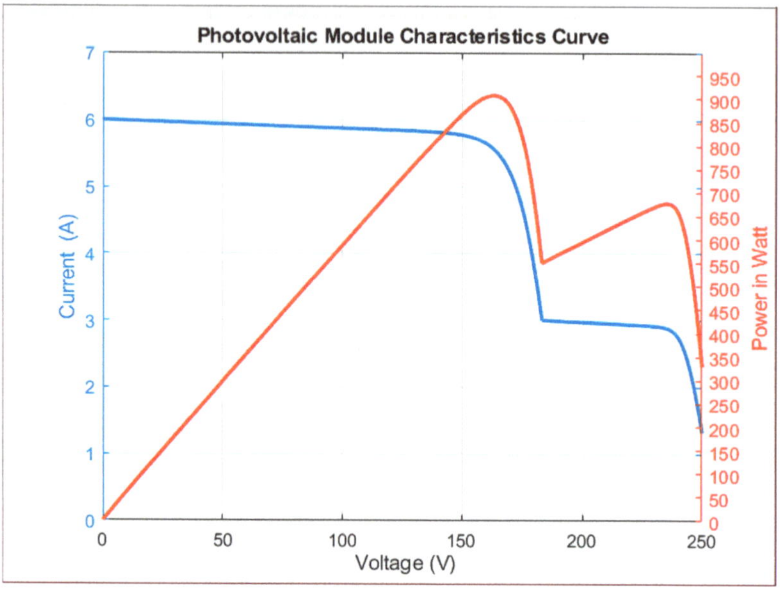

Figure 1. P-V and I-V characteristic curves of the photovoltaic module array with four series and one parallel structure and with one module under 50% shade [26].

3. The Proposed Improved MPPT Methods

In order to improve the tracking performance of the MPPT controller in a PVMA, an improved GA-ACO MPPT controller is proposed in this paper. The principle involves automatically adjusting the Pheromone evaporation rate ρ and Gaussian standard deviation x of the traditional GA-ACO MPPT controller, thereby shortening the tracking time. The MATLAB software was used to simulate maximum power tracking of the GA-ACO MPPT controller under different shading ratios in order to verify the excellent performance of the tracking methods proposed.

3.1. Genetic Algorithm

The theoretical basis of genetic algorithms originates from "On the Origin of Species" written by Charles Darwin in 1859. A genetic algorithm is a search-type algorithm based on natural selection and genetic mechanisms in the field of biology. It simulates natural selection among organisms in nature, as well as the phenomena of breeding, crossover and mutation. Moreover, in each iteration, several candidate populations (solutions) are retained and superior individuals are selected from the candidate populations. Through genetic factors (crossover and mutation), a new generation of candidate populations is produced until the best individuals are found. A genetic algorithm features multi-point search in order to prevent becoming caught in the local best solution. However, if the population quantity is too large, considerable time may be required for calculation, leading to low search efficiency. Thus, when the need to search bulk data arises, a genetic algorithm may take a long time to compute before a search is completed [19].

3.2. Ant Colony Optimization

Ant colony optimization (ACO) is a type of algorithm for searching the best path. It can also mimic ant behaviors [14]. In nature, ants leave behind pheromones secreted along their foraging trails in order to mark their trails. When ants behind reach the location previous ants have reached, they choose trails with higher pheromone values and leave more pheromones to strengthen the likelihood of ants behind taking the trials. Therefore as long as the trials with the highest pheromone values exist, the trails have a higher chance of attracting ants to move toward foraging. The ACO algorithm has many advantages, including robustness, the ability to search for a better solution, and good feedback, etc. However, the ACO algorithm may cause the search to slow down in the initial phase due to inadequate information obtained.

3.3. Traditional GA-ACO MPPT Controller

First, the traditional GA-ACO algorithm is applied in the PVMA to explain the MPPT steps. The traditional GA-ACO MPPT controller implementation steps are described below [25].

Step 1. First initialize the GA and ACO parameters. The GA parameter settings include: number of iterations (*Itmax*), the number of solution (k), the number of populations (*nPop*), crossover percentage (*pc*), factor for crossover (γ), mutation percentage (*pm*), mutation rate (mu), tournament size (*ts*), etc. The ACO parameter settings include: number of ants (*Ant*), Pheromone evaporation rate (ρ), etc. The populations are subsequently initialized. Each population has k solutions. In order to initialize the populations, the solution of each initial population randomly selects the output voltage of PVMA and substitutes it into the iterative formula.

Step 2. For all the populations, calculate the fitness of each population through the fitness function.

Step 3. In *nPop*, randomly select several populations (i.e., The tournament size (*ts*) value). After comparing the randomly selected populations, the best population is selected as the father, and the mother is chosen through the same approach. The parents go through crossover to create offspring. The quantity of offspring produced is determined by the crossover percentage (*pc*) value. Through the same crossover method, population mutation occurs. The number of mutated populations is determined by the mutation percentage (*pm*) value. The cost function should be calculated for all the offspring and mutated populations produced. The new populations produced replace inferior populations and the next generation is added. Better populations are retained and selected as the ACO initial conditions.

Step 4. In order to initialize ACO, the fitness of all the solutions in the retained populations should be calculated. It can be observed from Step 1 that all the solutions retained are output voltages (V_{pv}) of the PVMA. The fitness of these solutions refers to the output power (P_{pv}) corresponding to each voltage (V_{pv}). Then, all the solutions of better populations retained from the GA undergo pheromone initialization. The initialization steps are as follows:

Step 4.1. Calculate the distance ΔV_n ($n = 1, \ldots, k$) between each voltage value (V_n) and the best solution (V_{best}) in the population retained from the GA. In particular, V_{best} refers to the voltage solution of the maximum power value in a population.

$$\Delta V_n = \|V_n - V_{best}\| \tag{1}$$

Step 4.2. In order to calculate the pheromone value (τ_n) of each solution, the Gaussian normal distribution in Equation (2) should be used to obtain φ_n and each solution is computed. Through the Gaussian normal distribution, the normal distribution distance of all the solutions can be calculated. The shortest distance represents the best solution, the Gaussian value approximates zero and the farthest distance is the worst solution, with the Gaussian value approximating 1.

$$\varphi_n = \frac{1}{x\sqrt{2\pi}} \exp\left(-\frac{(\Delta V_n)^2}{2x^2}\right) \tag{2}$$

Here, x is the Gaussian standard deviation (usually set as $x = 0.5$).

Step 4.3. Use Equation (3) to calculate the pheromone value (τ_n) of all the solutions.

$$\tau_n = (1-\rho)\cdot\frac{\varphi_n}{\sum_{i=1}^{K}\varphi_i} \tag{3}$$

In particular, the ant path is determined by the pheromone value (τ_n) calculated from each solution in the previously retained population of the GA. The higher the pheromone value of a solution, the more likely it is to attract ants to move toward foraging. The pheromone evaporation rate (ρ) balances the Pheromone value of each solution in a population. Ants not are not only attracted to the solution with the highest pheromone value, but there is also a chance that they are attracted by Pheromone values generated from other computed solutions. However, the trail that attracts the highest number of ants is regarded as the maximum power point, and this solution is selected as the ACO tracking result.

Step 5. Repeat Step 3 and Step 4 until the number of iterations has reached the preset maximum iterations (*Itmax*) at which point the iterations end.

3.4. Improved GA-ACO MPPT Controller

The improved GA-ACO algorithm proposed in this paper implements adjustments, mainly targeting the Pheromone evaporation rate (ρ) in the ACO algorithm and the Gaussian standard deviation (x) and based on the slope in the P-V characteristic curve. Equations (2) and (3) show that when the two parameters of ρ and x are adjusted, the Pheromone value (τ_n) can be changed. When the Pheromone value increases, the rate of ant colony convergence to the maximum power point can be accelerated, which in turn, enhances the tracking response performance of the algorithm at the maximum power point. Thus, based on the slope (m) of the P-V characteristic curve in the PVMA in this paper, the Pheromone evaporation rate (ρ) and Gaussian standard deviation (x) are adjusted. In particular, slope (m) is defined in Equation (4):

$$m = \frac{P_{(it)} - P_{(it-1)}}{V_{(it)} - V_{(it-1)}} \tag{4}$$

where it represents the current number of iterations, $it-1$ represents the previous number of iterations, and $P_{(it)} - P_{(it-1)}$ represents the difference in the output power of the PVMA in the two iterations.

In this paper, based on the changes in the slope of the P-V characteristic curve, the changed Pheromone evaporation rate (ρ) and the Gaussian standard deviation (x) are as shown in Equations (5) and (6):

$$\rho = \left|\sqrt{m}\right| \times \rho + \Delta\rho \tag{5}$$

$$x = |0.5m| \times x + \Delta x \tag{6}$$

where $\Delta\rho$ is the adjustment value of ρ under different m, adjusted as shown in Table 1; and Δx is the adjustment value of x under different m, adjusted as shown in Table 2.

Table 1. The adjustment value $\Delta\rho$ of ρ under different slopes of the P-V characteristic curve.

$m = \frac{P_{(it)}-P_{(it-1)}}{V_{(it)}-V_{(it-1)}}$	$\Delta\rho$
$m > 2$	−0.2
$2 \geq m \geq 1.5$	−0.15
$1.5 \geq m \geq 1$	−0.09
$1 \geq m \geq 0.5$	+0.07
$0.5 \geq m \geq 0$	+0.17
$m = 0$	0
$0 \leq m \leq -0.5$	+0.17
$-0.5 \leq m \leq -1$	+0.07
$-1 \leq m \leq -1.5$	−0.09
$-1.5 \leq m \leq -2$	−0.15
$m < -2$	−0.2

Table 2. The adjustment value Δx of x under different slopes of the P-V characteristic curve.

$m = \frac{P_{(it)} - P_{(it-1)}}{V_{(it)} - V_{(it-1)}}$	Δx
$m > 2$	−0.285
$2 \geq m \geq 1.5$	−0.14
$1.5 \geq m \geq 1$	−0.02
$1 \geq m \geq 0.5$	+0.1
$0.5 \geq m \geq 0$	+0.2
$m = 0$	0
$0 \leq m \leq -0.5$	+0.2
$-0.5 \leq m \leq -1$	+0.1
$-1 \leq m \leq -1.5$	−0.02
$-1.5 \leq m \leq -2$	−0.14
$m < -2$	−0.285

Since the optimized value of the GA-ACO parameter differs due to the P-V characteristic curve generated under different shading conditions, the optimal rules for parameter adjustment cannot be identified. However, we learned from the test that the slope of the P-V characteristic curve reduced when the tracking reached within proximity of the maximum power point. Therefore, the Pheromone evaporation rate ρ and Gaussian standard deviation x should increase; the closer it gets to the MPP, the greater the increase in parameter values ρ and x that are required. Conversely, the further it is away from the MPP, the slope of P-V characteristic curve becomes greater, where ρ and x should reduce. Based on this, $\Delta\rho$ and Δx can apply the slope of the P-V characteristic curve for the PVMA accordingly to derive more optimized experience values for the $\Delta\rho$ and Δx adjustments (as shown in Tables 1 and 2) through multiple simulations.

3.5. The Maximum Power Tracking Processes and Architecture of the Proposed Improved GA-ACO

Figure 2 shows the flowchart of the maximum power tracking controller based on the improved GA-ACO proposed in this paper. The iterations in the last block of the flowchart in Figure 2 are indeed the set maximum iterations *Itmax*. Figure 3 shows the system structural diagram of the proposed maximum power controller. It mainly consists of the PVMA, boost DC-DC converter, improved GA-ACO maximum power tracking controller and voltage and current detectors. Table 3 shows the component specifications of a boost DC-DC converter.

Table 3. The specifications of the main components of a boost DC-DC converter.

Items	Specifications
Energy storage inductance (L_m)	250 uH, 10 A
Filter capacitor (C_{in})	390 uF, 450 V
Filter capacitor (C_{out})	330 uF, 450 V
Fast diode (D) IQBD60E60A1	withstand voltage V_{RRM} = 600 V, withstand current I_{FAV} = 60 A
Power semiconductor (S) MOSFET IRFP460	withstand voltage V_{DSS} = 500 V, withstand current I_D = 20 A

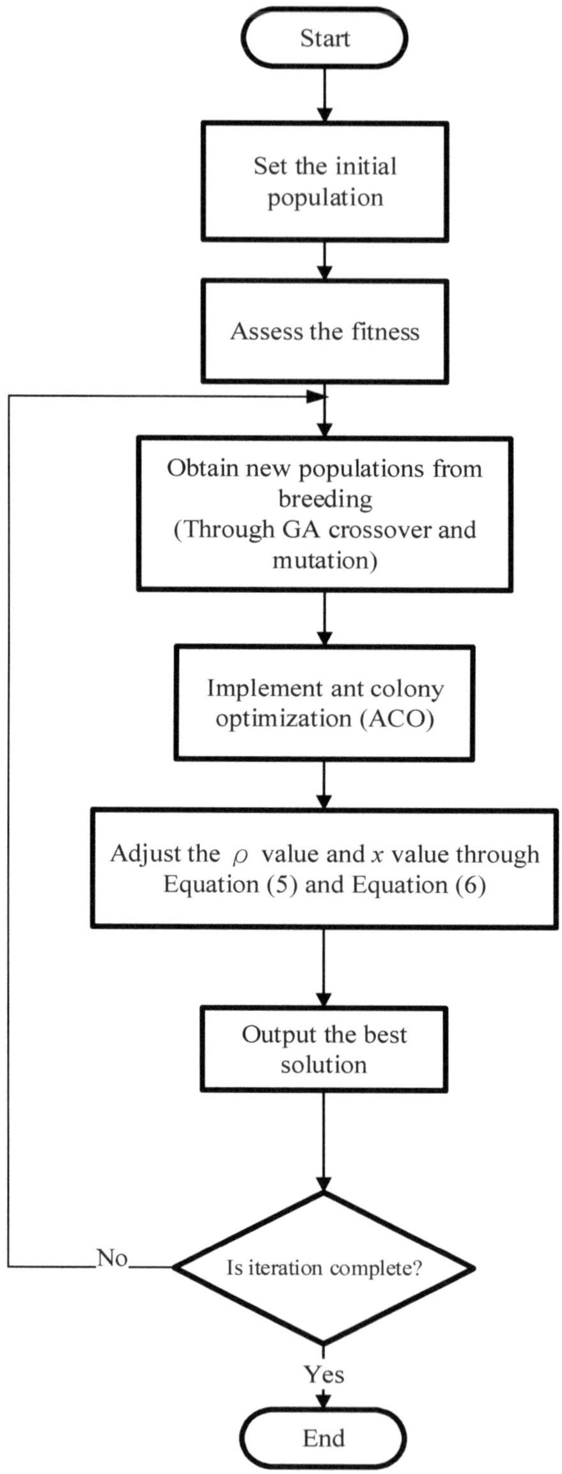

Figure 2. The flowchart of the proposed improved GA-ACO maximum power tracking.

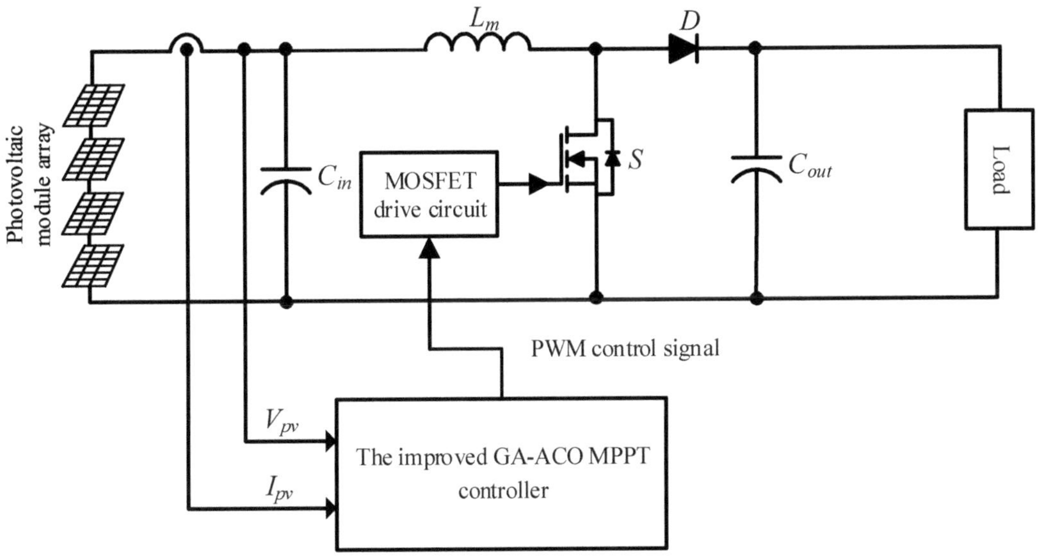

Figure 3. The structural diagram of the proposed improved GA-ACO MPPT controller.

4. Simulation Results

First, the MATLAB software was adopted to carry out maximum power tracking simulation by applying the improved GA-ACO algorithm to the photovoltaic module array (PVMA). The simulation results obtained from traditional GA-ACO, ACO and P&O MPPT controllers were compared for performance. The electrical parameter specifications of the photovoltaic module in this paper are shown in Table 4. As shown in Table 4, four photovoltaic modules were configured as four-series/one-parallel arrays and a two-series/two-parallel array. Under the same temperature condition, maximum power tracking simulation under five different shading conditions was carried out. It can be observed in Table 5 that under the five different shading conditions, the P-V characteristic curves obtained from the simulation showed different numbers of peaks. Then, through simulation, the proposed improved GA-ACO MPPT method under five different shading conditions was verified to be superior to the other traditional methods.

Table 4. The electric parameter specifications of the photovoltaic module adopted.

Parameters	Specifications
Rated maximum output power (P_{mp})	40.75 W
Current at maximum output power point (I_{mp})	1.74 A
Voltage at maximum output power point (V_{mp})	23.42 V
Short-circuit current (I_{sc})	2 A
Open-circuit voltage (V_{oc})	36 V

Table 5. The test cases of five shading ratios under different parallel series configurations.

Case	Series-Parallel Configuration and Shading Ratio	The Number of Peaks in the P-V Curve
1	0% shading + 0% shading + 0% shading + 0% shading	Single-peak
2	0% shading + 35% shading + 35% shading + 35% shading	Double-peak
3	0% shading + 25% shading + 40% shading + 40% shading	Triple-peak
4	0% shading + 25% shading + 35% shading + 50% shading	Quadruple-peak
5	(0% shading + 35% shading) // (35% shading + 35% shading)	Double-peak

Note: 0% shading means no shading; "+" means "series"; "//" means "parallel".

The parameter setting values of the improved GA-ACO, traditional GA-ACO, ACO and P&O MPPT methods adopted for the simulations in this paper are shown in Table 6.

Table 6. The parameter setting values of the improved GA-ACO, traditional GA-ACO, ACO, and P&O MPPT methods adopted for the simulation.

Parameter Name	Value
Maximum number of iterations (*Itmax*)	50
Number of solutions (*k*)	3
Number of populations (*nPop*)	5
Crossover percentage (*pc*)	0.7
Factor for crossover (γ)	0.4
Mutation percentage (*pm*)	0.4
Mutation rate (*mu*)	0.3
Tournament size (*ts*)	3
Ant count (*Ant*)	5
Pheromone evaporation rate (ρ)	0.37
Gaussian standard deviation (x)	0.5
Duty cycle disturbance (Δd)	0.02

4.1. Case 1: 0% Shading + 0% Shading + 0% Shading + 0% Shading

Figure 4 shows the four modules in series adopted to simulate the P-V and I-V characteristic curves of the photovoltaic module array under the condition of no shading through the MATLAB software. Since the photovoltaic module is in series, the voltages and powers are added. Therefore, it can be observed from Figure 4 that the voltage of the maximum power point and the maximum power point value are four times those of a single photovoltaic module. The simulation results in Figure 5 show that the improved GA-ACO managed to track the GMPP through just one iteration. On the other hand, the traditional GA-ACO, ACO, and P&O methods required 3, 10 and 16 iterations to track the GMPP. In addition, the P&O method continued to oscillate near the maximum power point.

4.2. Case 2: 0% Shading + 35% Shading + 35% Shading + 35% Shading

Figure 6 shows the simulation results when four photovoltaic modules in series are adopted and the shading ratio of the three photovoltaic modules is 35%. When one photovoltaic module is completely unshaded, the P-V and I-V characteristic curves are simulated through MATLAB software. It can be observed in Figure 6 that double-peak values appeared with a GMPP of 121.6 W and a GMPP voltage of 104.2 V. It can be observed from the simulation results in Figure 7 that the improved GA-ACO managed to track the GMPP with just one iteration. On the other hand, the traditional GA-ACO and ACO methods required 3 and 17 iterations to track the GMPP. As for the P&O method, the GMPP could not even be tracked.

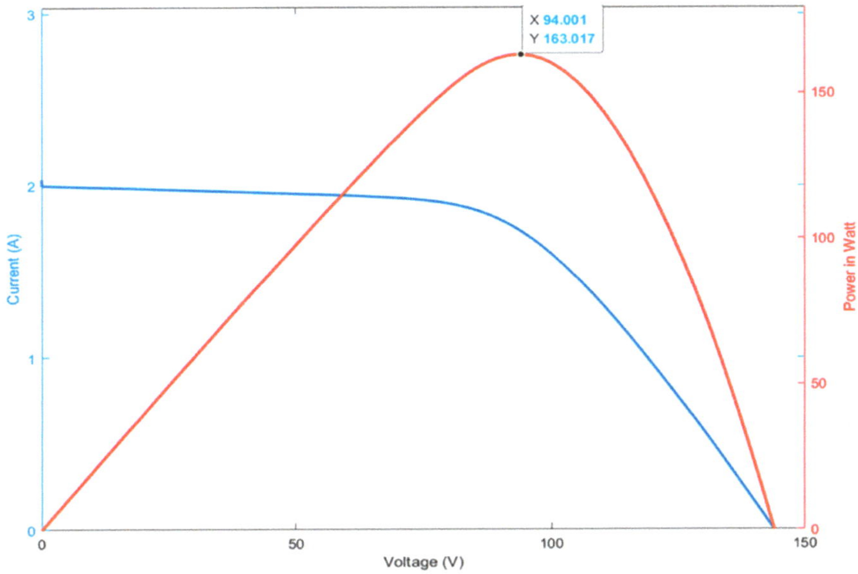

Figure 4. The P-V characteristic curve simulation (Red represents the P-V characteristic curve; blue represents the I-V characteristic curves) for Case 1.

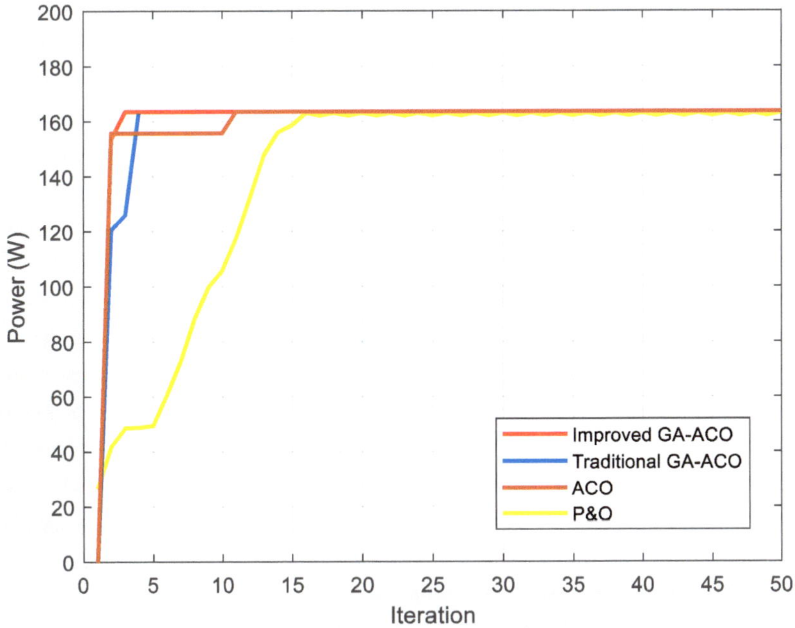

Figure 5. Simulation results of performance comparison among the improved GA-ACO, traditional GA-ACO, ACO and P&O MPPT methods for Case 1.

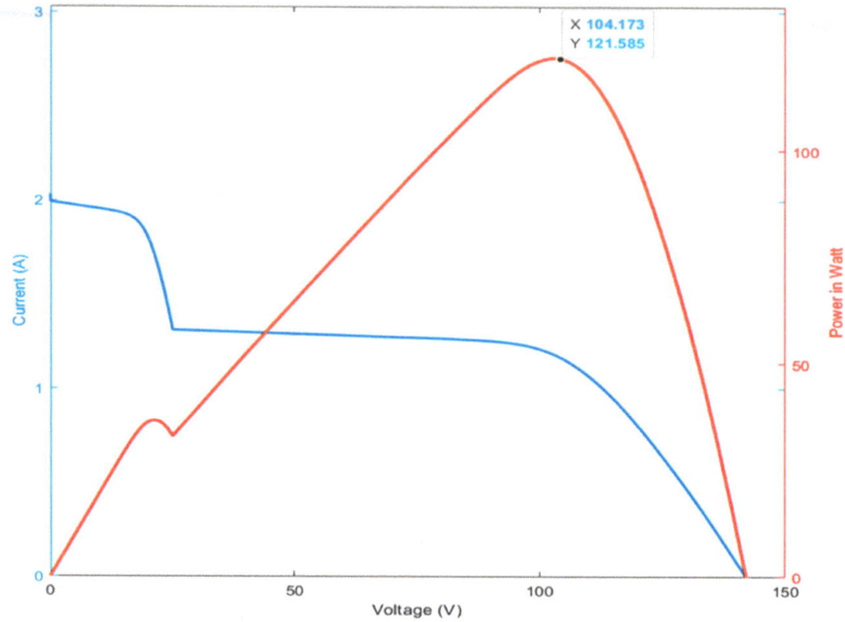

Figure 6. The P-V and I-V characteristic curves (Red represents the P-V characteristic curve; blue represents the I-V characteristic curve) for Case 2.

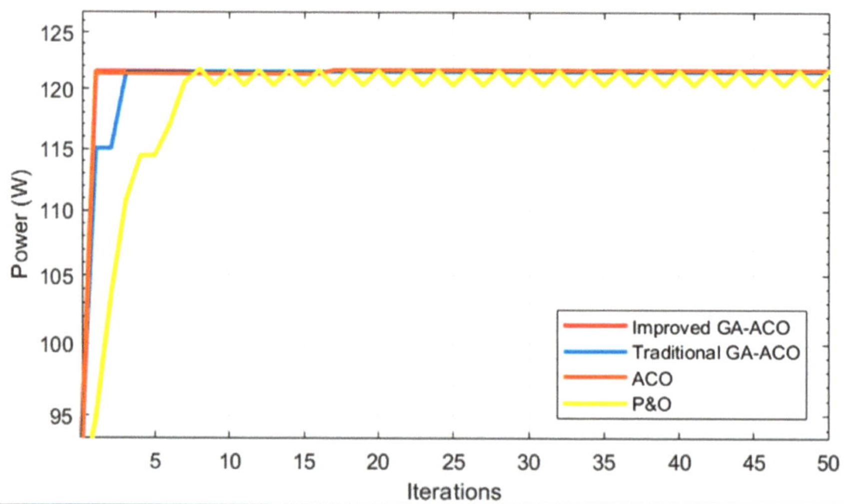

Figure 7. Simulation results of performance comparison among the improved GA-ACO, traditional GA-ACO, ACO and P&O MPPT methods for Case 2.

4.3. Case 3: 0% Shading + 25% Shading + 40% Shading + 40% Shading

When four photovoltaic modules in series are adopted, the shading ratios of the three photovoltaic modules are 40%, 40% and 25%, respectively. When one photovoltaic module is completely unshaded, the P-V and I-V characteristic curves simulated through MATLAB software are shown in Figure 8. It can be observed from Figure 8 that three-peak values appeared in the P-V characteristic curve, with the GMPP of 117.8 W and the GMPP voltage of 106.1 V. The simulation results in Figure 9 show that the improved GA-ACO method managed to track the GMPP after just two iterations. On the other hand, the traditional GA-ACO and ACO methods required 4 and 27 iterations to track the GMPP. As for the P&O method, the GMPP still failed to be tracked.

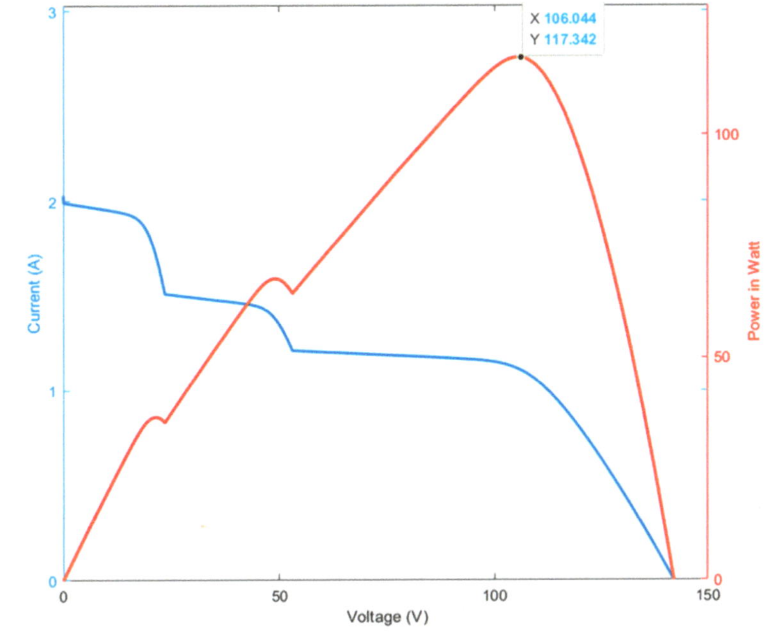

Figure 8. The P-V and I-V characteristic curves (Red represents the P-V characteristic curve; blue represents the I-V characteristic curve) for Case 3.

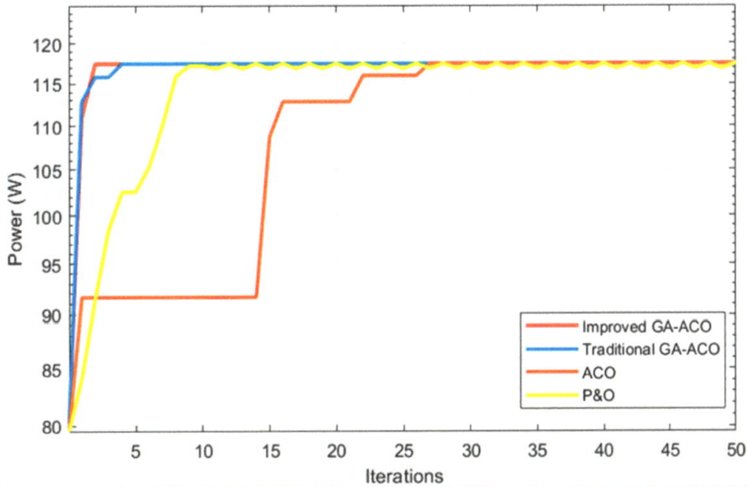

Figure 9. Simulation results of performance comparison among the improved GA-ACO, traditional GA-ACO, ACO, and P&O MPPT methods for Case 3.

4.4. Case 4: 0% Shading + 20% Shading + 35% Shading + 50% Shading

The shading ratios of the three modules set in this case are 20%, 35% and 50%, respectively. One module is without shading. Figure 10 shows the Case 4 simulation results of the P-V and I-V characteristic curves. It can be observed from the simulation results that four peaks appeared in the P-V characteristic curve, while the GMPP occurred at 105.8 W Figure 11 shows the simulation results of the improved GA-ACO, traditional GA-ACO, ACO and P&O MPPT methods. It can be observed from Figure 11 that even though four peaks appeared in the P-V characteristic curve, the improved GA-ACO method managed to track the GMPP with just three iterations. On the other hand, the traditional GA-ACO and ACO methods required 7 and 34 iterations to track the GMPP. As for the P&O method, the GMPP still failed to be tracked with a limited set of iterations.

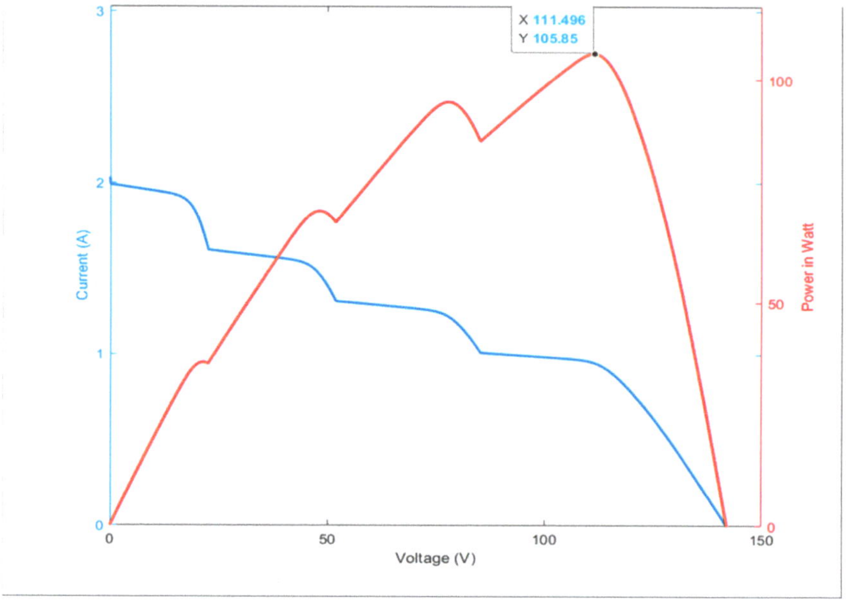

Figure 10. The P-V and I-V characteristic curves (Red represents the P-V characteristic curve; blue represents the I-V characteristic curve) for Case 4.

4.5. Case 5: (0% Shading + 35% Shading) // (35% Shading + 35% Shading)

Figure 12 shows Case 5 P-V and I-V characteristic curves obtained from the simulation. The module array is configured as a two-series/two-parallel array. One of the modules is under the condition of 0% shading, while the rest of the modules are under the condition of 35% shading. Since the photovoltaic module is connected as a two-series/two-parallel array, only two peaks are produced in the P-V characteristic curve. The power values of two of the peaks are 69.29 W and 121.75 W, respectively. Figure 13 shows that the improved GA-ACO method managed to track the GMPP with just one iteration. On the other hand, the traditional GA-ACO and ACO methods required 3 and 13 iterations, respectively, to track the GMPP. On the other hand, the P&O method still failed to successfully track the GMPP. It can be observed from the simulation results in Figure 13 that even though the PVMA in Case 5 was changed, it was also confirmed that the tracking speed using the improved GA-ACO method proposed in this paper was not affected by changes in the connection method.

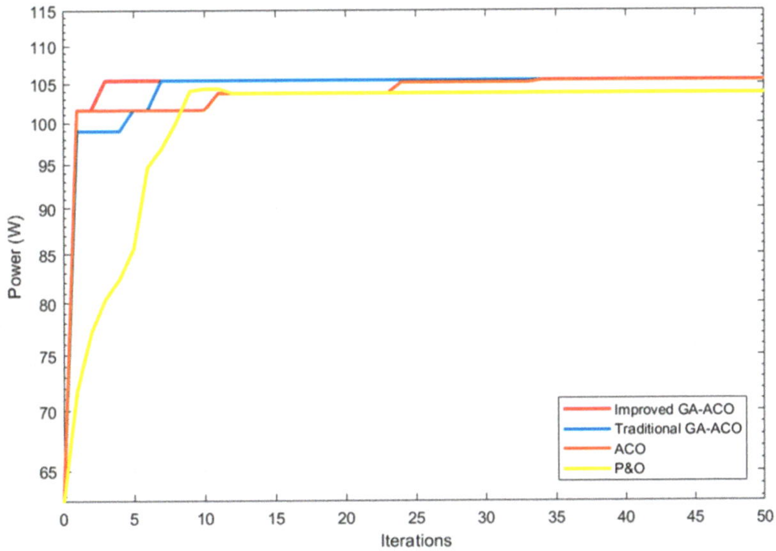

Figure 11. Simulation results of performance comparison among the improved GA-ACO, traditional GA-ACO, ACO and P&O MPPT methods for Case 4.

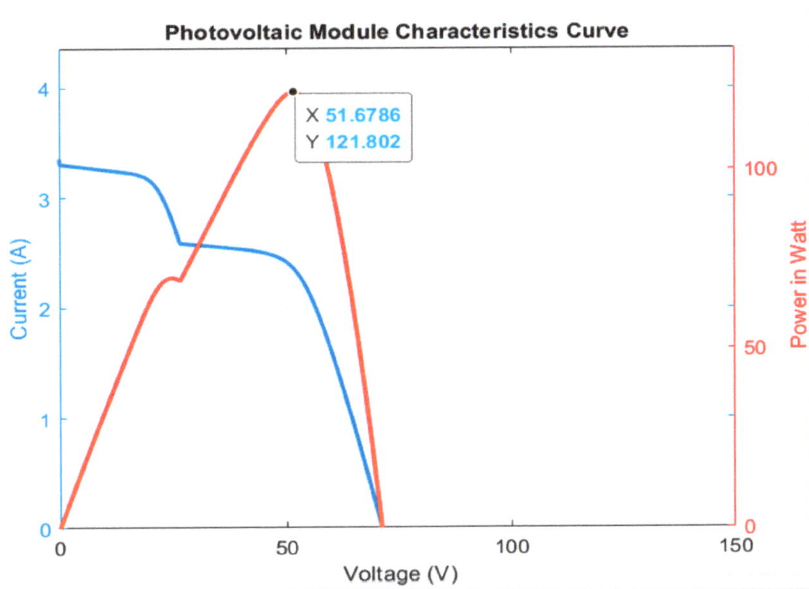

Figure 12. The P-V and IV characteristic curves (Red represents the P-V characteristic curve; blue represents the I-V characteristic curve) for Case 5.

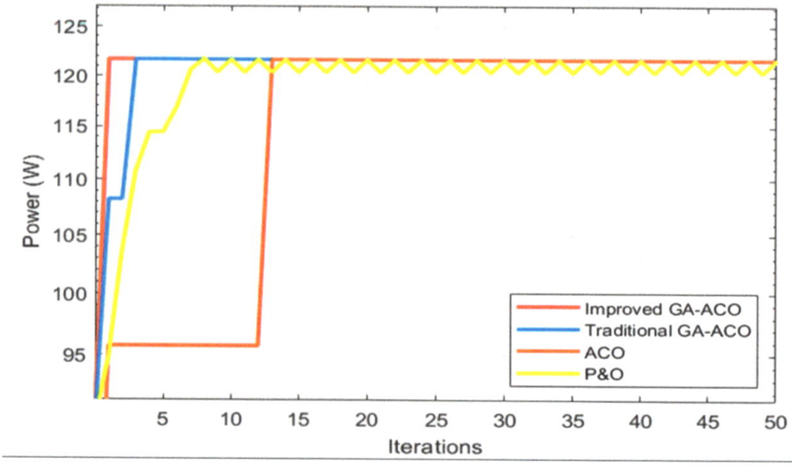

Figure 13. Simulation results of performance comparison among the improved GA-ACO, traditional GA-ACO, ACO and P&O MPPT methods for Case 5.

The simulation results of the five cases above show that the improved GA-ACO method is superior to the traditional GA-ACO, ACO and P&O methods in terms of performance. In addition, under each case, the improved GA-ACO algorithm managed to track the GMPP with fewer iterations. On the other hand, the traditional GA-ACO and ACO algorithms required more iterations to track the GMPP. As for Cases 1, 2, 3 and 5, although the P&O method managed to track GMPP, it oscillated to and fro near the maximum power point, while in Case 4, it was unable to track GMPP with a limited set of iterations. In addition, based on the five cases selected, maximum power tracking was conducted 50 times using the improved GA-ACO, traditional GA-ACO, ACO and P&O MPPT methods. The numbers of iterations of GMPP tracked each time were added up and averaged, as shown in Table 7. Table 7 shows that although the three algorithms of the improved GA-ACO, traditional GA-ACO and ACO can track the GMPP among the five cases, the improved GA-ACO method required on average fewer iterations compared with those of the traditional GA-ACO and ACO algorithms. This demonstrates that the proposed improved GA-ACO MPPT method has better tracking performance. In particular the higher the number of peaks in the P-V characteristic curve, the greater the differences in the tracking performance.

Table 7. The comparison of the five cases tracking the average number of iterations of GMPP using different algorithms.

Case	ACO MPPT [15]		P&O MPPT [3]		GA-ACO MPPT [25]		Proposed GA-ACO MPPT	
	Iter GMPP	**Total Time**	**Iter GMPP**	**Total Time**	**Iter GMPP**	**Total Time**	**Iter GMPP**	**Total Time**
1	15.64	3.1 ms	17.85	3.0 ms	4.85	1.2 ms	3.24	0.8 ms
2	18.16	3.6 ms	18.25	3.1 ms	6.74	1.7 ms	4.56	1.2 ms
3	30.96	6.2 ms	32.35	5.5 ms	8.42	2.1 ms	6.78	1.8 ms
4	36.87	7.4 ms	Stuck in LMPP	–	11.68	2.9 ms	7.49	1.9 ms
5	25.38	5.1 ms	23.15	3.9 ms	7.14	1.8 ms	5.36	1.4 ms

Note: Iter GMPP signifies the average number of iterations to obtain GMPP, Total Time signifies the average total time to reach MPP, and LMPP signifies the local maximum power point.

In fact, the more complex the algorithm is, the slower the calculation will be. We can conclude from the calculation time comparison that the proposed improved GA-ACO

MPPT has the slowest calculation time. Even though the proposed improved GA-ACO MPPT is the slowest to calculate an iteration, this algorithm is the fastest to reach the GMPP because it requires fewer iterations. The iterations of the proposed improved GA-ACO are the least compared with the other three algorithms. The detailed comparison is shown in Table 7 which demonstrates that the proposed GA-ACO MPPT is an improvement on the conventional MPPT algorithm.

In addition, the proposed hybrid method was compared with different hybrid MPPT controllers including the improved artificial bee colony (I-ABC) algorithm [23] and the improved cuckoo search (I-CS) algorithm [24]. The comparison between the proposed hybrid MPPT and existing hybrid MPPTs [23,24] is shown in the Table 8. The partial shading conditions that were tested in [23,24] were one peak, two peaks, three peaks, and four peaks of P-V curve peaks, yet they used different PV specifications and different exact irradiances. Their results are compared with the proposed hybrid MPPT in Table 8. The proposed method is better than the I-ABC and I-CS MPPT methods [23,24] in all cases.

Table 8. The hybrid MPPT comparison.

Case	Number of Iterations		
	I-ABC MPPT [23]	**I-CA MPPT [24]**	**Proposed GA-ACO MPPT**
1	4.56	4.21	3.24
2	6.45	5.68	4.56
3	7.39	7.21	6.78
4	10.18	8.32	7.49
5	10.57	7.98	5.36

Based on the PVMA in Table 7, the performance was compared in terms of the time response when using different MPPT methods under five different shades for tracking the maximum power point. The results show that the method proposed in this paper indeed provided a better tracking speed response. Therefore, it was verified that the $\Delta\rho$ and Δx adjustment values selected from the slope range for the P-V characteristic curve in Tables 1 and 2 led to performance of the MPPT response.

For each of the test cases, although the shading conditions for the selected simulation were set at fixed values, they could be seen as the change in equivalent shading ratio since all the shading conditions of the different test cases differed from each other. Furthermore, it can be observed from the simulation results that the proposed MPPT method could obtain better tracking performance under all the changes in shading conditions. Therefore, the five different test conditions listed in Table 5 could be treated as the tests of changing between different shading conditions.

Since the slow change of actual irradiance seemed unable to reveal the superior performance of the MPPT methods proposed, irradiance levels with greater step changes were adopted in this paper for conducting the test and verifying the tracking response of the proposed methods. Under the condition of slow changes in irradiance, the MPPT methods proposed could also produce the same superior tracking performance, only not significant enough.

In this paper, the PVMA went through MPPT tests under five different shading conditions. From Figures 4, 6, 8, 10 and 12, it can be observed that under such different shading conditions, the P-V characteristic curves show different local peak values, and the curve types differ accordingly. Moreover, from the simulation results in Figures 5, 7, 9, 11 and 13, it can also be observed that with the proposed MPPT method, at any point in time during the tracking process, the power value tracked produces a minimum difference between all the compared MPPT methods and the global maximum power point (GMPP). Therefore, it can be determined that the integral of squared error (ISE), integral of time-squared error (ITSE), integral of absolute error (IAE) and integral of time-absolute error (ITAE), which are calculated according to references [27–29], would be minimum throughout the simulation.

In this paper, the maximum power tracking test was conducted on the PVMA under five different shading conditions, where each shading condition was equivalent to certain changes in temperature and irradiance parameters. Therefore, the test was the same as the robustness test that considers the MPPT of parametric uncertainties [30,31]. As indicated by the test results in Table 7, all the MPPT methods herein produced better tracking response performances compared with other methods, which demonstrates that the MPPT methods proposed did indeed show robustness.

5. Discussion

The proposed improved GA-ACO algorithm combining the ant colony optimization (ACO) and the genetic algorithm (GA) referred to in reference [25] determine the initial value of the iterative parameters of the ant colony algorithm. To shorten the number of iterations needed to obtain the optimal value, it is necessary to address the issue that the conventional ACO tends to track the local maximum power point (LMPP) when the optimal value is applied to search the global maximum power point (GMPP) if the photovoltaic module arrays (PVMAs) are abnormal. However, the optimization of the GA-ACO parameters differs depending on the P-V characteristic curves generated from different shading conditions. Thus, no principle is to be found for parameter optimization. Provided that it is learned in tests that when the tracking approaches the MPP and as the slope of the P-V characteristic curve declines, the Pheromone evaporation rate ρ and the Gaussian standard deviation x increase, and the ρ and x parameters are required to be greater when approaching the MPP. In contrast, the farther the MPP is, the ρ and x must be decreased as the slope of the P-V characteristic curve increases. Therefore, the optimal adjusted value of the Pheromone evaporation rate, $\Delta\rho$, and the optimal adjusted value of Gaussian standard deviation, Δx, may be obtained via multiple simulations based on the slope values of the P-V characteristic curves of PVMAs, as indicated in Tables 1 and 2. Comparing the responses of the time tracking to MPP with different MPPT approaches for the PVMA in Table 7, Section 4, under five different shading levels for their performances, it is observed that the improved GA-ACO algorithm proposed in this paper indeed has better tracking speed response. When five different peak values are found in the P-V characteristic curve in Table 7, the proposed improved GA-ACO algorithm has 19.5~35.9% (average 29.2%) fewer iterations when tracking than the GA-ACO algorithm mentioned in [25]. Compared with the ACO algorithm [15], it has 74.9~79.7% (average 78.2%) fewer, and 75.0~92.5% (average 81.0%) fewer than the conventional P&O method [3].

6. Conclusions

In this paper, an improved GA-ACO algorithm was proposed for application to photovoltaic module arrays to carry out MPPT. The simulation results have validated that its trackability is significantly superior to those of traditional GA-ACO, ACO and P&O MPPT controllers. The MPPT method proposed combines the superior characteristics of GA and ACO. In addition, based on the slope of the P-V characteristic curve in the location of the photovoltaic module array work point, the Pheromone evaporation rate ρ and the Gaussian standard deviation x in the ACO iterative formula are automatically adjusted. The ACO algorithm can then more speedily search the subspace and output the local best solution. The simulation results prove that the improved GA-ACO MPPT controller is superior to traditional GA-ACO, ACO and P&O MPPT controllers in terms of tracking response performance under different connection configurations and shading ratios. The proposed improved GA-ACO MPPT controller even managed to track the global maximum power point during the first iteration. On the other hand, the traditional GA-ACO and ACO MPTT controllers required more iterations to track the GMPP. As for the P&O method, other than in Case 1 (0% shading ratio) when it managed to successfully track the GMPP and generate oscillation near its maximum power point, in the rest of the cases, it was unable to track the GMPP with a limited set of iterations. Therefore, since the improved GA-ACO MPPT required fewer iterations to accurately track the GMPP, the power generation utilization rate of the photovoltaic module array was enhanced.

Author Contributions: The conceptualization was proposed by K.-H.C., who was also responsible for the writing, review and editing of this paper. K.-H.H. completed the formal analysis of the improved GA-ACO algorithm optimization smart algorithm. T.-W.L. was responsible for the data curation, software program, and simulation validation. K.-H.C. was in charge of project administration. All authors have read and agreed to the published version of the manuscript.

Funding: The authors gratefully acknowledge the support and funding of this project by Ministry of Science and Technology, Taiwan, under the Grant Number MOST 110-2221-E-167-007-MY2.

Institutional Review Board Statement: Not applicable.

Informed Consent Statement: Not applicable.

Data Availability Statement: This study did not report any data.

Conflicts of Interest: The authors declare no conflict of interest.

Nomenclature

Acronyms	
MPPT	maximum power point tracking
AM	air mass
MPP	maximum power point
P-V	power-voltage
I-V	current-voltage
GA	genetic algorithm
ACO	ant colony optimization
P&O	perturbation and observation
INC	incremental conductance
LMPP	local maximum power point
GMPP	global maximum power point
PVMA	photovoltaic
STC	standard test condition
ISE	integral of square error
ITSE	integral of time-square error
IAE	integral of absolute error
ITAE	integral of time- absolute error
Symbols	
$Itmax$	number of iterations
k	number of solutions
$nPop$	number of populations
pc	crossover percentage
pm	mutation percentage
m	slope of the P-V characteristic curve in the PVMA
mu	mutation
γ	factor for crossover
t_s	tournament size
Ant	number of ants
dx	length of a jump
ρ	pheromone evaporation rate
$\Delta\rho$	adjustment value of ρ
V_{pv}	output voltage of PVMA
P_{pv}	output power of PVMA corresponding to each voltage V_{pv}
ΔV_n	distances between each voltage V_n and the best solution (n = 1 . . . k)
V_n	solution from the archive (n = 1 . . . k)
V_{best}	best solution in the population retained from GA
Φ_n	Gaussian normal distribution value

x	Gaussian standard deviation
τ_n	Pheromone value
Δx	adjustment of value of x
Δd	duty cycle disturbance
L_m	energy storage inductor
C_{in}	input filter capacitor
C_{out}	output filter capacitor

References

1. Femia, N.; Granozio, D.; Petrone, G.; Spagnuolo, G.; Vitelli, M. Predictive and Adaptive MPPT Perturb and Observe Method. *IEEE Trans. Aerosp. Electron. Syst.* **2007**, *43*, 934–950. [CrossRef]
2. Salman, S.; Xin, A.I.; Wu, Z. Design of a P-&-O Algorithm Based MPPT Charge Controller for a Stand-Alone 200W PV System. *Prot. Control Mod. Power Syst.* **2018**, *3*, 25.
3. Sahu, R.K.; Ghosh, A. Maximum Power Generation from Solar Panel by Using P&O MPPT. In Proceedings of the International Conference on Intelligent Controller and Computing for Smart Power (ICICCSP), Hyderabad, India, 21–23 July 2022; pp. 21–23.
4. Hang, L.; Guo, H.; Zhu, W. An improved MPPT Control Strategy Based on Incremental Conductance Algorithm. *Prot. Control Mod. Power Syst.* **2020**, *5*, 14.
5. Bhattacharyya, S.; Kumar, D.S.; Samanta, S.; Mishra, S. Steady Output and Fast Tracking MPPT (SOFT-MPPT) for P&O and INC Algorithms. *IEEE Trans. Sustain. Energy* **2021**, *12*, 293–302.
6. Aquib, M.; Jain, S. A fast global maximum power point tracking technique for partially shaded PV arrays. In Proceedings of the IEEE International Students' Conference on Electrical, Electronics and Computer Science (SCEECS), Bhopal, India, 24–25 February 2018; pp. 1–5.
7. Hu, G.; Zhong, J.; Wei, G.; Chang, C.T. DTCSMO: An Efficient Hybrid Starling Murmuration Optimizer for Engineering Applications. *Comp. Methods Appl. Mech. Eng.* **2023**, *405*, 115878. [CrossRef]
8. Hu, G.; Yang, R.; Qin, X.; Wei, G. MCSA: Multi-strategy Boosted Chameleon-Inspired Optimization Algorithm for Engineering Applications. *Comp. Methods Appl. Mech. Eng.* **2023**, *403*, 115676. [CrossRef]
9. Hu, G.; Du, B.; Wei, G. HG-SMA: Hierarchical Guided Slime Mould Algorithm for Smooth Path Planning. *Artif. Intell. Rev.* **2023**, *56*, 1–61. [CrossRef]
10. Ma, Y.; Zhou, X.; Gao, Z.; Bai, T. Summary of the novel MPPT (maximum power point tracking) algorithm based on few intelligent algorithms specialized on tracking the GMPP (global maximum power point) for photovoltaic systems under partially shaded conditions. In Proceedings of the IEEE International Conference on Mechatronics and Automation (ICMA), Takamatsu, Japan, 6–9 August 2017; pp. 311–315.
11. Luong, X.T.; Bui, V.H.; Do, D.T.; Quach, T.H.; Truong, V.A. An improvement of maximum power point tracking algorithm based on particle swarm optimization method for photovoltaic system. In Proceedings of the 5th International Conference on Green Technology and Sustainable Development (GTSD), Ho Chi Minh City, Vietnam, 27–28 November 2020; pp. 53–58.
12. Pragallapati, N.; Sen, T.; Agarwal, V. Adaptive Velocity PSO for Global Maximum Power Control of a PV Array under Nonuniform Irradiation Conditions. *IEEE J. Photovolt.* **2017**, *7*, 624–639. [CrossRef]
13. Xu, L.; Cheng, R.; Xia, Z.; Shen, Z. Improved particle swarm optimization (PSO)-based MPPT method for PV string under partially shading and uniform irradiance condition. In Proceedings of the Asia Energy and Electrical Engineering Symposium (AEEES), Chengdu, China, 29–31 May 2020; pp. 771–775.
14. Dorigo, M.; Birattari, M.; Stutzle, T. Ant Colony Optimization. *IEEE Comput. Intell. Mag.* **2006**, *1*, 28–39. [CrossRef]
15. Jiang, L.L.; Maskell, D. A uniform implementation scheme for evolutionary optimization algorithms and the experimental implementation of an ACO based MPPT for PV systems under partial shading. In Proceedings of the IEEE Symposium on Computational Intelligence Applications in Smart Grid (CIASG), Orlando, FL, USA, 9–12 December 2014; pp. 1–8.
16. Luthfansyah, M.; Eka, P.; Farid, D.M. Performance evaluation of ACO-MPPT and constant voltage method for street lighting charging system. In Proceedings of the International Seminar on Application for Technology of Information and Communication (iSemantic), Semarang, Indonesia, 21–22 September 2019; pp. 411–416.
17. Hadji, S.; Gaubert, J.P.; Krim, F. Real-time Genetic Algorithms-based MPPT: Study and Comparison (theoretical and experimental) with Conventional Methods. *Energies* **2018**, *11*, 459. [CrossRef]
18. Daraban, S.; Petreus, D.; Morel, C. A Novel MPPT (Maximum Power Point Tracking) Algorithm Based on a Modified Genetic Algorithm Specialized on Tracking the Global Maximum Power Point in Photovoltaic Systems Affected by Partial Shading. *Energy* **2014**, *74*, 374–388. [CrossRef]
19. Megantoro, P.; Nugroho, Y.D.; Anggara, F.; Rusadi, E.Y. Simulation and characterization of genetic algorithm implemented on MPPT for PV system under partial shading condition. In Proceedings of the 3rd International Conference on Information Technology, Information System and Electrical Engineering (ICITISEE), Yogyakarta, Indonesia, 13–14 November 2018; pp. 74–78.
20. Smida, M.; Sakly, A. Genetic based algorithm for maximum power point tracking (MPPT) for grid connected PV systems operating under partial shaded conditions. In Proceedings of the 7th International Conference on Modelling, Identification and Control (ICMIC), Sousse, Tunisia, 18–20 December 2015; pp. 1–6.

21. Halim, A.A.E.; Saad, N.H.; Sattar, A.A.E. A comparative study between perturb and observe and cuckoo search algorithm for maximum power point tracking. In Proceedings of the 21st International Middle East Power Systems Conference (MEPCON), Cairo, Egypt, 17–29 December 2019; pp. 716–723.
22. Anand, R.; Swaroop, D.; Kumar, B. Global maximum power point tracking for PV array under partial shading using cuckoo search. In Proceedings of the 9th IEEE Power India International Conference (PIICON), Sonepat, India, 28 February–1 March 2020; pp. 1–6.
23. Chao, K.H.; Chang, L.Y.; Wang, K.W. Global Maximum Power Point Tracking of Photovoltaic Module Arrays Based on Improved Cuckoo Search Algorithm. *Electronics* **2022**, *11*, 1247. [CrossRef]
24. Chao, K.H.; Li, J.Y. Global Maximum Power Point Tracking of Photovoltaic Module Arrays Based on Improved Artificial Bee Colony Algorithm. *Electronics* **2022**, *11*, 1572. [CrossRef]
25. Chao, K.H.; Rizal, M.N. A Hybrid MPPT Controller Based on the Genetic Algorithm and Ant Colony Optimization for Photovoltaic Systems under Partially Shaded Conditions. *Energies* **2021**, *14*, 2902. [CrossRef]
26. Teo, J.C.; Tan, R.; Mok, V.H.; Ramachandaramurthy, V.K.; Tan, C. Impact of Partial Shading on the P-V Characteristics and the Maximum power of a Photovoltaic String. *Energies* **2018**, *11*, 1860. [CrossRef]
27. Zafran, M.; Khan, L.; Khan, Q.; Ullah, S.; Sami, I.; Ro, J.S. Finite-Time Fast Dynamic Terminal Sliding Mode Maximum Power Point Tracking Control Paradigm for Permanent Magnet Synchronous Generator-Based Wind Energy Conversion System. *Appl. Sci.* **2020**, *10*, 6361. [CrossRef]
28. Ali, K.; Khan, Q.; Ullah, S.; Khan, I.; Khan, L. Nonlinear Robust Integral Backstepping Based MPPT Control for Stand-alone Photovoltaic System. *PLoS ONE* **2020**, *15*, e0231749. [CrossRef] [PubMed]
29. Khan, I.U.; Khan, L.; Khan, Q.; Ullah, S.; Khan, U.; Anmad, S. Neuro-adaptive Backstepping Integral Sliding Mode Control Design for Nonlinear Wind Energy Conversion System. *Turk. J. Elec. Eng. Comp. Sci.* **2021**, *29*, 531–547. [CrossRef]
30. Khan, R.; Khan, L.; Ullah, S.; Sami, I.; Ro, J.S. Backstepping Based Super-Twisting Sliding Mode MPPT Control with Differential Flatness Oriented Observer Design for Photovoltaic System. *Electronics* **2020**, *9*, 1543. [CrossRef]
31. Ali, K.; Khan, L.; Khan, Q.; Ullah, S.; Ahmad, S.; Mumtaz, S.; Karam, F.W. Naghmash, Robust Integral Backstepping Based Nonlinear MPPT Control for a PV System. *Energies* **2019**, *12*, 3180. [CrossRef]

energies

MDPI

Article

The Study of a Magnetostrictive-Based Shading Detection Method and Device for the Photovoltaic System

Xiaolei Fu and Yizhi Tian *

School of Electrical Engineering, Xinjiang University, Urumqi 830017, China
* Correspondence: torsionscale@163.com; Tel.: +86-186-9915-6957

Abstract: When the photovoltaic (PV) system suffers shading problems caused by different degrees and areas, the shaded PV cells will consume electricity and generate heat, the corresponding bypass diode operating at a certain current will conduct, and a special magnetic field will be generated in space. In this study, a magnetostrictive-based shading detection method and device for the PV system are developed from theoretical, simulation, and physical experimental aspects. This study aims to detect the special magnetic field using magnetostrictive material with a certain response pattern under the magnetic field to detect and locate the shading problem of each module in the PV system. Theoretically, the analysis is carried out from the on–off situation of the bypass diodes of PV modules under different shading conditions and the response mechanism of magnetostrictive materials under the action of the magnetic field. During simulation, the finite element magnetic field simulations are performed for the diode and the series magnetic field coil, and the structural parameters of the magnetic field coil are designed based on the simulation results. After establishing the validation idea of the detection method in this study, the experimental platform is built and the experimental steps are designed. Finally, the feasibility of the method proposed in this study is verified, the detection range of the method is calculated, and the minimum spacing of adjacent magnetic field coils is determined by experimental validation. This study provides a novel magnetostrictive-based detection method, as well as a theoretical and experimental basis, for identifying and localizing PV system shading problems, and discusses the feasibility of shading detection at the system level.

Keywords: PV system shading detection; PV module bypass diode; magnetostrictive sensor; finite element magnetic field simulation

Citation: Fu, X.; Tian, Y. The Study of a Magnetostrictive-Based Shading Detection Method and Device for the Photovoltaic System. *Energies* **2023**, *16*, 2906. https://doi.org/10.3390/en16062906

Academic Editors: Annamaria Buonomano, Jayanta Deb Mondol and Biplab Das

Received: 11 February 2023
Revised: 13 March 2023
Accepted: 17 March 2023
Published: 21 March 2023

1. Introduction

In compliance with immense modern energy demands, the need for a cheaper and reliable energy supply is globally evolving. In global energy resource utilization, renewable energy, such as biomass, hydrogen, solar radiation, and wind speed, is considered to be a promising means of solving problems associated with the rise in alternatives to fossil fuels, environment pollution, and global temperature, [1]. Photovoltaic (PV) technologies have been widely and maturely used in the market and will play a leading role in the current energy transition in order to address the disadvantages of environmental issues posed by fossil fuels. The highest contribution of currently installed PV systems is identified in Asia, including China (175 GW), Japan (55.5 GW), and India (26.8 GW). Europe ranks second in terms of PV-installed capacity, with considerable shares in Germany (45.9 GW), Italy (20.12 GW), and the UK (13.4 GW) as of 2020 [2]. The currently available PV technologies possess less than 23% conversion efficiencies, which underlines the need for further improvements to ensure better technological competitiveness [3].

1.1. Definition of the Shading Problem

In the PV system, every PV array consists of PV strings connected in parallel, every PV string consists of PV modules connected in series, and every PV module consists of PV cells connected in series. Over the last few years, traditional crystalline PV technologies have mostly been applied to new PV power plants [4]. When the surface of a single PV cell or different PV cells receives uneven illumination, this situation is referred to as a shading problem [5]. In actual operation, PV cells are often affected by clouds, dust, trees, buildings, etc., and suffer from shading problems [6].

When a PV cell has a certain degree and area of shading problem, its output power decreases, and the output current shows a downward trend [7]. Due to the special structure of the PV string, the rest of the normal PV module outputs a current of the original size to the shaded PV module, resulting in a large amount of power consumption on the shaded PV modules, an abnormal temperature rise, and a further "hot spot effect" [8]. The frequent occurrence of the "hot spot effect" will cause irreversible damage to the internal physical structure and external packaging materials of PV modules, which will seriously affect the reliability and service life of PV modules and even endanger the normal operation of the whole PV system [9].

The addition of bypass diodes connected in reverse parallel to the output of each PV module is proposed to protect PV modules from operating at negative voltages [10]. When a PV module has shading problems, its bypass diodes provide a shunt path for the rest of the normal PV modules in the string, which reduces the negative impact of the shading problems on the PV system to a certain extent, but the shading problems still exist and cannot be located [11]. Due to the non-linear output characteristics of PV cells, the PV system needs to monitor the output power of PV arrays and provide maximum power point tracking (MPPT) [12]. In the uniform irradiance condition, the PV strings only have a single maximum power point (MPP). Due to the use of the bypass diodes, a portion of the PV modules located in shadow or at lower irradiance will cause multiple peaks in the output power of the PV string, which is not conducive to achieving MPPT and causes energy loss [13]. Therefore, successfully detecting and locating shading problems can help to improve the safety and output power of the PV system.

1.2. The Prior Technical Ideas for PV Shading Detection

Among the prior technical ideas for PV shading detection, these studies focus on the output characteristics of PV strings. The authors of [14] proposed a detection method to classify the PV strings into four working conditions based on the real electrical measurements. The authors of [15], based on the instantaneous current reduction between PV strings and MPPT sampling instants, proposed a sensor-less detection technique to monitor the output characteristics of the PV strings. The authors of [16] proposed an improved sine–cosine algorithm to find a periodic partial shading condition and a MPP, avoiding the local power point obstacle. The authors of [17] considered the meteorological data as input variables and the coordinates of the MPP as targets, and also used artificial neural networks to detect the presence of partial shading and infer the number of shaded PV modules. The above studies successfully identified shading problems in PV strings in PV arrays, but could not pinpoint the shading problem to each PV module.

These studies focus on the shading conditions of individual PV modules. The authors of [18] proposed a method and sensor to detect partial shading on a module during its MPPT operation. The authors of [19] proposed using dynamic reconfiguration to rearrange the connection structure of PV modules by controlling electrical switches to maximize the output power and realize shading detection. The above studies successfully identified the shading problems in individual PV modules, but the signal lines of the sensors or switches for each module could not pinpoint the shading problem on each PV module, which complicated the whole PV system.

These studies revolve around the image recognition of photovoltaic arrays. The literature [20] proposes the use of webcams to identify the shadow irradiance and provide the reference voltage of the MPP using image analysis techniques. In the literature [21], an artificial neural network tool for image recognition is proposed to quantify the power loss caused by shading and the hot spot effects of PV modules. The studies mentioned above successfully identified the shading problems of individual PV modules by uncomplicated equipment composition, but there are limitations due to weather, distance, and image capture angles. The literature [22] proposes the use of unmanned aerial vehicles to capture the images of PV arrays for judging the shading condition of each PV module. The literature [23] proposes the use of fully convolutional networks with depth training for improving the recognition accuracy of aerial images of PV arrays. The above literature solves some of the limitations compared to the solution using fixed camera devices, but does not allow for the real-time detection of shading conditions.

1.3. The Study Ideas in This Paper

Based on the structure of the connection between the PV module and the bypass diode, as well as the shunting effect of the bypass diode when the PV module has shading problems, after finding the correspondence between the shading problems and the on–off state of bypass diodes, if we identify and locate the bypass diodes under on-state conditions, we can then successfully detect shading problems. The literature [24] verified that the short circuit of the bypass diode would cause a power loss in systems due to the mismatch phenomenon, and the reverse current would create electro-thermal problems in the PV array. Due to the electro-magnetic effect, the magnetic field is instantaneous and homogeneous compared to the heat when the diode is conducting. Therefore, this study aims to successfully recognize PV shading problems via the real-time detection of the magnetic field generated when the bypass diode is under on-state conditions.

In the process of magnetization, the size of the magnetic material elongates or shortens in the direction of magnetization, i.e., the magnetostrictive effect [25]. The magnetostrictive displacement sensor made by this effect enables the non-contact long-distance detection of multiple magnetic fields [26]. In the literature [27], a magnetostrictive sensor with a 50 m range is proposed for detecting the liquid level.

In summary, the specific study idea of this paper is as follows. The special distribution of magnetic fields is considered in space when the bypass diodes of PV modules are turned on under different degrees and areas of shading conditions. The response pattern of magnetostrictive materials are studied under the action of this magnetic field, the detection device is developed, and the shading problems of PV modules are identified and localized.

2. Materials and Methods

2.1. Analysis of the Diode On–Off Situation

The 300 W PV module is taken as an example from the current model. After 20 PV cells are connected in series to form a substring, 3 substrings are connected in series to form a PV module. One bypass diode is connected in reverse parallel to the output of each substring, and, finally, the bypass diode and its connecting wires are concentrated in the junction box. When a substring appears with shaded cells, this substring is referred to as the shaded substring. The internal wiring structure of the 300 W PV module is shown in Figure 1 [28].

As shown in Figure 2, within the same substring, the output current I_n and voltage U_r of the PV cell in normal conditions operate in the first quadrant. When a PV cell has a short-circuit current drop due to shading, the output current I_n of the normal PV cells in series with it is higher, resulting in the higher reverse bias voltage U_r of the shaded cell, whose current–voltage curve works in the second quadrant, consuming power and generating a certain amount of heat.

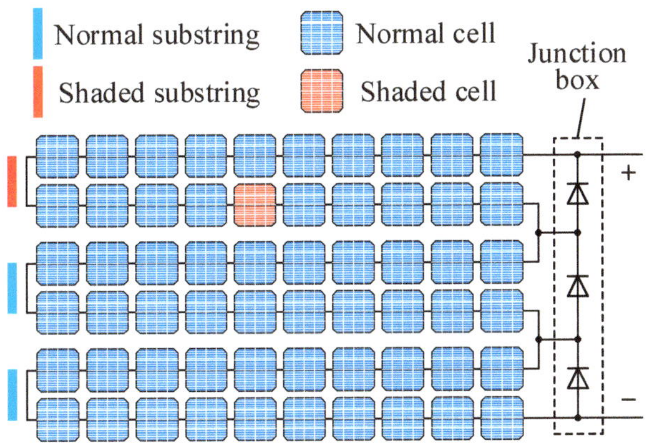

Figure 1. The internal wiring structure of the 300 W PV module.

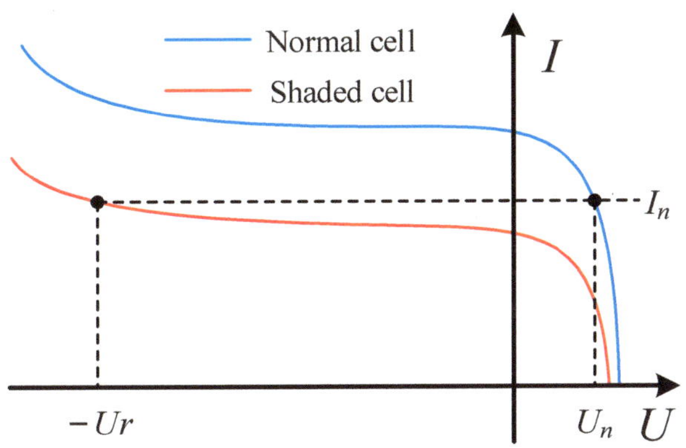

Figure 2. Current–voltage curves of normal and shaded cells in the same substring.

The on–off state of the bypass diode is analyzed. Assuming that there is one shaded cell in the substring and the rest are normal cells, the bypass diode will turn on when the output voltage of the substring is reverse-biased and higher than the conduction voltage of the bypass diode U_d. Thus, the following is satisfied [29]:

$$U_r - 19U_n \geq U_d \tag{1}$$

As shown in Figure 3, I_s is the short-circuit current of the normal substring and cell, I_{rc} is the short-circuit current of the shaded substring, I_g is the short-circuit current of the shaded cell, and U_{rc} is the critical reverse bias voltage of the shaded cell when the diode is switched on.

When the operating current of the substring is in interval 1, the output voltage of all cells within the substring operates is found in the first quadrant. Each cell outputs power, and each diode is switched off.

When the operating current of the substring is in interval 2, the shaded cell consumes power because its output voltage is reverse-biased. However, the output voltage of the substring is still in the first quadrant, making Equation (1) not valid and the diode is still switched off. Therefore, the current range in this interval is referred to as the bypass diode operating blind zone.

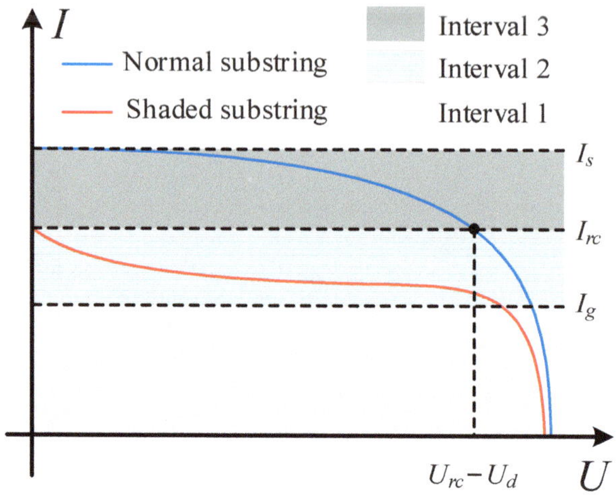

Figure 3. Current–voltage curves of normal and shaded substrings in the same PV module.

When the operating current of the substring is in interval 3, the output voltage of the shaded cell is reverse-biased and exceeds the critical value U_{rc}, so that Equation (1) becomes valid, and thus the diode conducts. Therefore, an appropriate increase in the operating current can turn on the bypass diode that works in the operating blind zone.

According to the results of the above analysis, due to the special wiring structure inside the PV module, the on–off state of the bypass diode is related to the interval where the operating current of the PV module is located under shading conditions.

By adjusting the operating current of the PV string, the bypass diode of the corresponding PV substring turns on and generates a specific magnetic field at a spatial location when there is a shading problem in the PV string. Detecting and locating this magnetic field helps to identify and localize the shading problems.

2.2. Analysis of the Response Mechanism of Magnetostrictive Materials under the Specific Magnetic Field Environment

Magnetostrictive materials are characterized by high permeability, low coercivity, easy magnetization, easy demagnetization, and small hysteresis line envelope area, which can be used to detect the location of specific magnetic fields in space over long distances. Under the action of the magnetic field, the length of magnetostrictive material in the direction of magnetic inductance changes from l_o to l and satisfies [30]:

$$\lambda = \frac{l - l_o}{l_o} \tag{2}$$

In Equation (2), λ is the line magnetostriction coefficient, which can be classified as a positive or negative magnetostriction coefficient according to its plus or minus value. The magnetostriction coefficient of a magnetostrictive material at saturation magnetic induction intensity is called the saturation magnetostriction coefficient λ_s.

The magnetostrictive waveguide wire with radius r has a uniform cross-sectional current distribution when a pulse current with amplitude I_c and pulse width τ passes, and the surface current density J can be expressed as:

$$J = \frac{I_c}{\pi r^2} \tag{3}$$

According to Ampere's law of circulation, the current I passing through a circular section of radius a inside the waveguide wire, with the axis as the center, is [31]:

$$I = J\pi a^2 = \frac{I_c a^2}{r^2} = \oint_C H \cdot dl = 2H\pi a \tag{4}$$

Furthermore, the intensity of the pulsed magnetic field H inside the waveguide wire at the position of distance a from the axis is obtained as:

$$H = \frac{I_c a}{2\pi r^2} \tag{5}$$

Equation (5) shows that the pulsed magnetic field is highest at the surface position of the waveguide wire with the value H_c:

$$H_c = \frac{I_c}{2\pi r} \tag{6}$$

The relative magnetic permeability of the waveguide wire surface when only subjected to the pulsed magnetic field is μ_{r1}. Based on the value of absolute magnetic permeability μ_0 of vacuum as $4\pi \times 10^{-7}$ Wb/(A·m), the pulsed magnetic induction intensity B_{c1} can be calculated as [32]:

$$B_{c1} = \frac{H_c}{\mu_0 \mu_{r1}} = \frac{I_c}{2\pi r \mu_0 \mu_{r1}} \tag{7}$$

Under the action of this magnetic induction intensity B_{c1}, the magnetostriction coefficient of the waveguide wire is λ_1, the circumferential magnetostriction deformation produced on the surface of the waveguide wire is Δl_1, and the original circumference of the circular section of the waveguide wire is l_0. Thus, the following is satisfied:

$$\lambda_1 = \frac{\Delta l_1}{l_0} = \frac{\Delta l_1}{2\pi r} \tag{8}$$

The magnetic induction intensity distribution on the surface of the waveguide wire can be analyzed when a permanent magnet generating axial magnetic field exists at the local position of the waveguide wire. With the center of the waveguide wire cross-section circle as the origin, the axial direction of the waveguide wire as the y-axis direction, and the waveguide wire cross-section as the plane where the xz-axis is located, a right-angle coordinate system is established, as shown in Figure 4.

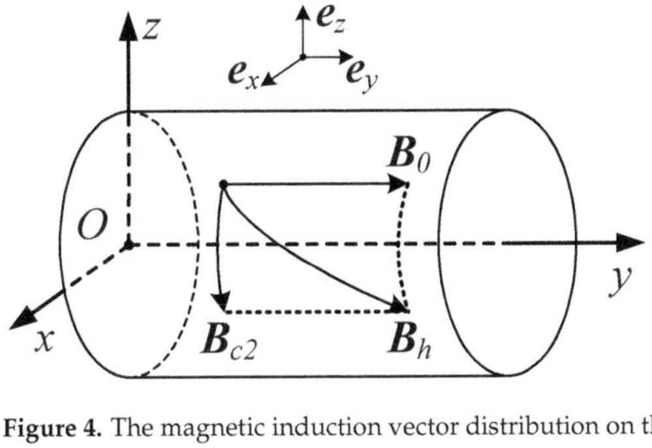

Figure 4. The magnetic induction vector distribution on the surface of the waveguide wire.

Taking the first quadrant of the xyz axis as an example, the coordinates of a point on the surface of the waveguide wire can be expressed as:

$$\left(x, y, \sqrt{r^2 - x^2}\right) \tag{9}$$

The magnetic induction intensity vector generated within the waveguide wire when subjected only to the constant axial magnetic field of the permanent magnet is $\mathbf{B}_0$, which is expressed as:

$$\mathbf{B}_0 = \mathbf{B}_0\mathbf{e}_y \tag{10}$$

In Equation (10), $\mathbf{e}_y$ denotes the unit magnetic induction intensity vector in the positive direction of the y-axis. Under the action of magnetic induction intensity $\mathbf{B}_0$, the relative magnetic permeability of the waveguide wire is μ_{r2}. By substituting the coordinate point (9) to Equation (7), the pulse magnetic induction intensity vector $\mathbf{B}_{c2}$ at this point can be expressed as:

$$\mathbf{B}_{c2} = \frac{I_c}{2\pi r \mu_0 \mu_{r2}} \cdot \left(\frac{\sqrt{r^2 - x^2}}{r}\mathbf{e}_x - \frac{x}{r}\mathbf{e}_z \right) \tag{11}$$

By superimposing the axial magnetic induction intensity vector $\mathbf{B}_0$ with the circumferential pulse magnetic induction intensity vector $\mathbf{B}_{c2}$, the torsional magnetic induction intensity vector $\mathbf{B}_h$ is calculated as follows:

$$\mathbf{B}_h = \frac{I_c\sqrt{r^2 - x^2}}{2\pi r^2 \mu_0 \mu_{r2}}\mathbf{e}_x + \mathbf{B}_0\mathbf{e}_y - \frac{I_c x}{2\pi r^2 \mu_0 \mu_{r2}}\mathbf{e}_z \tag{12}$$

In Equation (12), $\mathbf{e}_x$, $\mathbf{e}_y$, and $\mathbf{e}_z$ denote the unit magnetic induction intensity vectors in the positive direction of the x, y, and z axes, respectively. Under the action of torsional magnetic induction intensity B_h, the magnetostriction coefficient of the waveguide wire is λ_2.

During the action of the circumferential pulsed magnetic induction intensity vector $\mathbf{B}_{c2}$, the axial magnetic induction intensity vector $\mathbf{B}_0$ does not change, so the direction of magnetostrictive deformation on the surface of the waveguide wire remains circumferential. The deformation variable Δl_2 generated by the torsional magnetic field acting on the surface of the waveguide wire can be expressed as:

$$\lambda_2 = \frac{\Delta l_2}{l_0} = \frac{\Delta l_2}{2\pi r} \tag{13}$$

When the waveguide wire is mechanically stretched and fixed at both ends without allowing its deformation, a shear stress σ will be generated on the surface of the waveguide wire and transmitted along the waveguide wire in the form of a torque wave.

The shear modulus of the waveguide wire is G. According to the shear Hooke law, the relationship between the shear stress σ and the relative deformation variable Δl can be expressed as [33]:

$$\sigma = G\Delta l = G(\Delta l_2 - \Delta l_1) \tag{14}$$

Substituting Equations (8) and (13) into Equation (14) yields:

$$\sigma = 2\pi r G(\lambda_2 - \lambda_1) \tag{15}$$

The initial permeability of the waveguide wire in the absence of magnetic field influence is μ_{r0}, the permeability of its surface under the action of the shear stress σ is μ_σ, the saturation magnetic induction intensity of the waveguide wire material is B_s, and the corresponding saturation magnetostriction coefficient is λ_s. Then, according to the inverse magnetostriction effect, the following relationship exists between the change in the permeability of the waveguide wire and the stress applied to the waveguide wire [34]:

$$\frac{\mu_{r0} - \mu_\sigma}{\mu_\sigma} = \frac{2\lambda_s}{B_s}\sigma\mu_{r0} \tag{16}$$

From Equation (16), it can be deduced that:

$$d\mu = \mu_{r0} - \mu_{\sigma} = \frac{2\lambda_s \mu_{r0} \mu_{\sigma}}{B_s} d\sigma \tag{17}$$

Substituting Equation (15) into Equation (17) yields:

$$d\mu = \frac{4\pi r G \lambda_s \mu_{r0} \mu_{\sigma}}{B_s} d\lambda \tag{18}$$

A detection coil is installed along the axial direction of the waveguide wire, at a position far from the permanent magnet L, to detect the torque wave generated by the shear stress σ. By combining a permanent magnet with magnetic field intensity H_L with the detection coil, the magnetic induction intensity B_L induced by the waveguide wire with absolute permeability μ at the position of the detection coil satisfies:

$$B_L = \mu H_L \tag{19}$$

Substituting Equation (19) into Equation (18) yields:

$$dB_L = \frac{4\pi r G \lambda_s \mu_{r0} \mu_{\sigma} H_L}{B_s} d\lambda \tag{20}$$

According to Faraday's law of electromagnetic induction, the number of turns of the detection coil is N, and the area of the coil facing the magnetic field is S. Therefore, the induced electromotive force E caused by the transmission of the torque wave to the detection coil is:

$$E = NS \cdot \frac{dB_L}{dt} \tag{21}$$

Substituting Equation (20) into Equation (21) yields:

$$E = \frac{4\pi r G N S \lambda_s \mu_{r0} \mu_{\sigma} H_L}{B_s} \cdot \frac{d\lambda}{dt} \tag{22}$$

The magnetostriction coefficients of the waveguide wire are λ_1 and λ_2 when the pulsed magnetic induction intensity B_{c1} and the torsional magnetic induction intensity B_h are acting on the waveguide wire, respectively. Substituting them with the pulse width τ of the pulsed current into Equation (22) yields:

$$E = \frac{4\pi r G N S \lambda_s \mu_{r0} \mu_{\sigma} H_L (\lambda_2 - \lambda_1)}{\tau B_s} \tag{23}$$

The transmission speed of the torque wave is v. Based on the transmission time t of the torque wave from the position of the permanent magnet to the detection coil, the position information of the permanent magnet L can be further calculated as [35]:

$$L = vt \tag{24}$$

Since the transmission speed of the torque wave is a constant on the waveguide wire, the torque wave generated from multiple permanent magnet positions at different distances can be transmitted sequentially to the detection coil. Wave-absorbing rubbers are installed at both ends of the waveguide wire to reduce signal interference, thus enabling the non-contact long-distance detection of multiple magnetic fields.

According to Equation (23), it can be seen that the induced electromotive force E of the detection coil is positively related to the amount of variation in the magnetostriction coefficient of the waveguide wire at the position of the permanent magnet, so the axial magnetic field generated by the permanent magnet is the key influencing factor.

According to the simulation results in Figures 5 and 6, the size and direction of this magnetic field are not uniformly distributed, and are vulnerable to interference from the magnetic field generated by the internal energized conductor of the junction box and the external PV cells at work. This is not conducive to the direct detection of this magnetic field by magnetostrictive material.

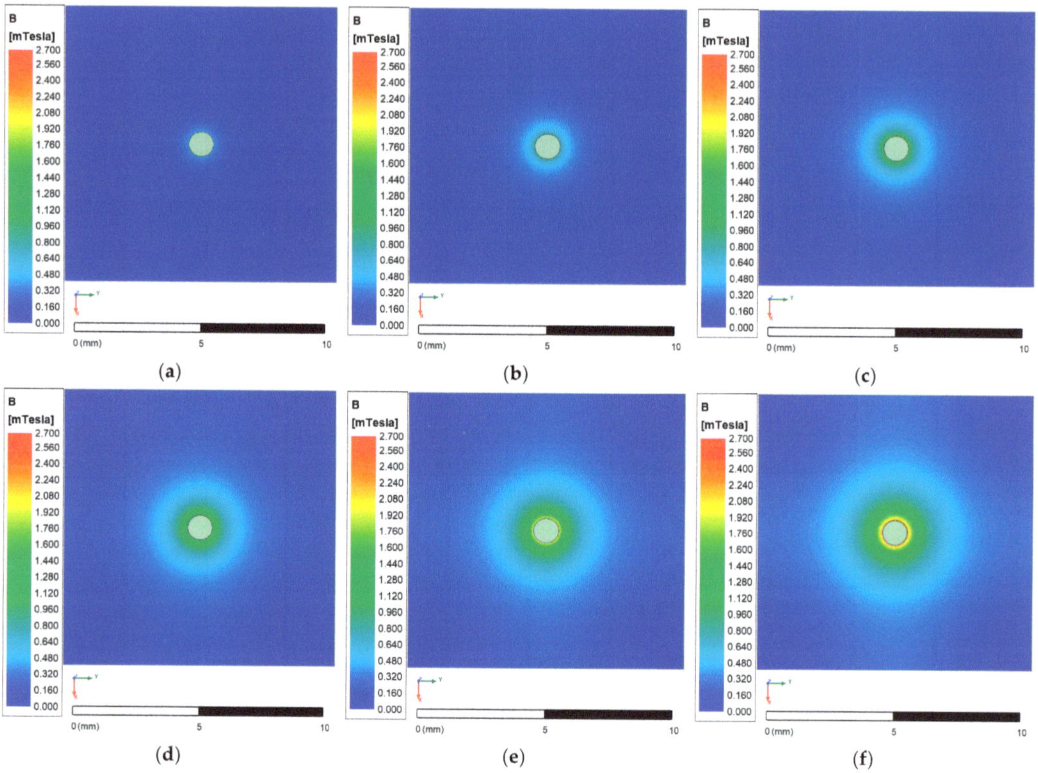

Figure 5. The results of the magnetic field simulation for the diode pin line when the diode is conducting at (**a**) 1A, (**b**) 2A, (**c**) 3A, (**d**) 4A, (**e**) 5A, and (**f**) 6A.

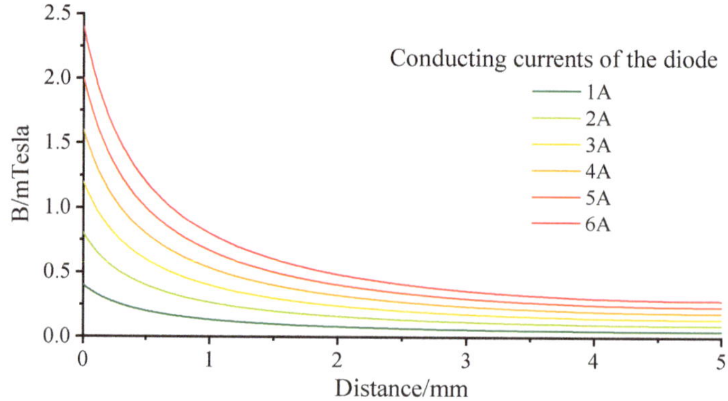

Figure 6. The curves indicating the relationship between the magnetic induction intensity and the distance in space to the surface position of the diode pin line.

2.3. Finite Element Simulation Analysis of the Magnetic Field Distribution When the Diode Is under On-State Conditions

Before the magnetic field simulation of the conduction diode, it is necessary to select a suitable type of PV module as the experimental object and then select the type of bypass diode. To facilitate the study of the protection role of bypass diodes in multi-substring PV modules, the single-substring PV module consisting of 12 PV cells connected in series is selected as the experimental object. Its type and parameters are selected, as shown in Table 1.

Table 1. The type and parameters of the experimental PV module.

Parameters	Type Specification or Numerical Value
Type specification	Monocrystalline PV module
Number of cells	12
Number of substrings	1
Number of bypass diodes	1
Maximum power	100 W
Open-circuit voltage	21.5 V
Short-circuit current	5.85 A
Maximum power operating voltage	18 V
Maximum power operating current	5.41 A
Dimension	920×670 mm^2

The type and parameters of the corresponding bypass diode, are selected as shown in Table 2.

Table 2. The type and parameters of the bypass diode.

Parameters	Type Specification or Numerical Value
Type specification	10A10MIC rectifier diode
Material of the pin line	Tinned copper
Diameter of the pin line	1.0 mm
Maximum forward current	10 A
Maximum forward voltage	1.1 V
Peak reverse voltage	700 V

According to the parameters listed in Table 2 and the actual installation position of the diode in the junction box on the backside of the PV module, the physical model of the diode pin line is established in the ANSYS 2020 R2 software (Canonsburg, PA, USA). Using the axial direction of the pin line as the z-axis direction and the cross-section of the pin line as the plane where the xy-axis is located, the maximum power operating currents of the PV modules (1, 2, 3, 4, 5, and 6 A) are selected as the on-state currents of the diode for magnetic field simulation, and the simulation results are shown in Figure 5.

The curve corresponding to the values of different distances in space to the surface position of the diode pin line and the simulation result of the magnetic induction intensity B is shown in Figure 6.

The simulation results in Figures 5 and 6 show that the spatial magnetic induction intensity at 1 mm from the diode pin line is less than 1.0 mT when the diode is used with 6 A as the on-state current. According to the above magnetic field simulation results, considering that the on–off state of each bypass diode in the whole PV string is affected by different shading conditions and conduction currents, magnetostrictive material with a certain response pattern under the action of the magnetic field is chosen to detect the special magnetic field.

2.4. Structure Design of Coil Connected in Series with the Diode

To increase the magnetic field strength generated by the current in the branch where the bypass diode is conducting and to distribute it more uniformly in a fixed direction, part of the conductor inside the branch is designed as a coil structure. The coil is densely winded so that the adjacent wire turns are closely spaced to reduce magnetization leakage after conduction. The coil of this structure is called the magnetic field coil.

Enameled copper wire is selected as the winding material. The winding length of the magnetic field coil is l_x, the inner radius is r_x, the diameter of the enameled wire is d_x, the diameter of the winding copper core is d_{Cu}, the cross-sectional area of the winding copper core is S_{Cu}, the thickness of the lacquer coating is d_s, and the maximum number of winding layers is m.

According to the maximum safe load current J_{Cu} of copper, the short-circuit current 5.85 A of the PV module is chosen as the maximum on-state current I_{dm} of the branch where the coil is located, and the maximum cross-sectional area of the copper core S_{Cumax} is calculated as:

$$S_{Cumax} = \frac{I_{dm}}{J_{Cu}} \tag{25}$$

The corresponding minimum diameter of copper core d_{Cumax} is:

$$d_{Cumax} = \sqrt{\frac{4S_{Cumax}}{\pi}} = \sqrt{\frac{4I_{dm}}{\pi J_{Cu}}} \tag{26}$$

When the bypass diode is conducted with the maximum value I_{dm}, the minimum equivalent resistance of the diode $R_{\min}$ is calculated based on the on-state voltage U_{dm}:

$$R_{\min} = \frac{U_{dm}}{I_{dm}} \tag{27}$$

The forward conduction current–voltage curve of the 10A10MIC rectifier diode is shown in Figure 7.

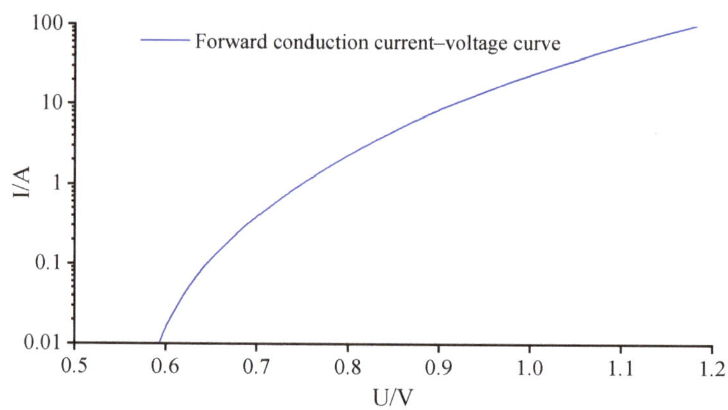

Figure 7. The forward conduction current–voltage curve of the 10A10MIC rectifier diode.

Based on the forward conduction current–voltage curve of the diode in Figure 7, the parameters in Equation (27) are selected and calculated, as shown in Table 3.

Table 3. The on-state voltage and equivalent resistance of the diode corresponding to the maximum on-state current.

Parameters	Numerical Value	Unit
Maximum on-state current I_{dm}	5.85	A
Corresponding on-state voltage U_{dm}	0.865	V
Minimum equivalent resistance $R_{\min}$	0.148	Ω

To reduce the impact of the magnetic field coil in the branch circuit due to voltage division on the protective effect of the diode, its maximum resistance value R_{xmax} should be much smaller than the minimum equivalent resistance value of the diode R_{min}. Thus, the following is satisfied:

$$100R_{xmax} \leq R_{min} \tag{28}$$

According to the resistivity ρ_{Cu} of copper, the maximum winding length of the coil wire l_{xmax} is satisfied:

$$l_{xmax} = \frac{R_{min}S_{Cumax}}{100\rho_{Cu}} = \frac{\pi R_{min}d_{Cumax}^2}{400\rho_{Cu}} \tag{29}$$

The number of each layer turns of the magnetic field coil winding can be calculated to be 1 and the perimeter of each layer can be calculated. The maximum value of the winding length l_{xmax} and the maximum number of winding layers m satisfy:

$$l_{xmax} \geq 2\pi m r_x + \pi m^2 d_x \tag{30}$$

According to the above calculation process, the winding length l_{xmax} limits the structure of the magnetic field coil, and the selection and calculation results of the parameters in relation to the magnetic field coil are shown in Table 4.

Table 4. The selection and calculation results of the parameters related to the magnetic field coil.

Parameters	Numerical Value	Unit
The maximum safe load current density J_{Cu} of copper	5.2	A/mm^2
The resistivity ρ_{Cu} of copper	0.0185	$\Omega\, mm^2/m$
The inner radius r_x of coil	0.75	mm
The diameter d_{Cu} of copper core	1.22	mm
The thickness d_s of lacquer coating	0.04	mm
The diameter d_x of enameled wire	1.28	mm
The maximum resistance value R_{xmax}	1.48	mΩ
The winding length l_{xmax}	90	mm
The maximum number m of winding layers	4	/

According to the parameters of Table 4, the maximum number of winding layers m is calculated in correspondence to the number of each layer turns. In the ANSYS finite element simulation software, the physical models are built sequentially, and 5.41 A is selected as the magnetic field coil conduction current I_d to characterize the operating current.

The maximum value of magnetic induction intensity B_{xmax} at the position of the magnetic field coil central axis is recorded, according to the magnetic field simulation results. According to the calculation and simulation results, the corresponding curve between the maximum value of magnetic induction intensity B_{xmax} and the serial number of structures of magnetic field coil is shown in Figure 8.

The specific numerical correspondence between the maximum value of magnetic induction intensity B_{xmax}, the serial number of the structure, the total turns, and the number of each layer turns is shown in Table 5.

According to the simulation results, it can be seen that the maximum value of magnetic induction intensity B_{xmax} generated at the position of the magnetic field coil central axis is the largest when the on-state current is the same and the 5th winding structures of 3 turns in layer 1, 2 turns in layer 2, and 1 turn in layer 3 are used. Based on this winding structure, experiments are carried out to verify the method used in this paper.

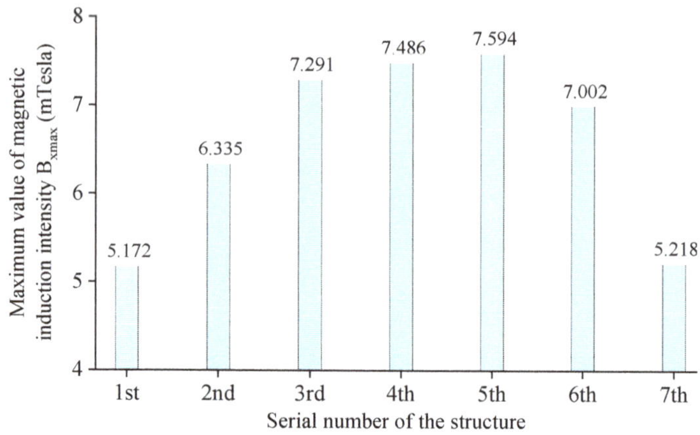

Figure 8. The correspondence curve between the winding structures of the magnetic field coil and the maximum value of magnetic induction.

Table 5. The numerical correspondence between the structure of each winding of the magnetic field coil and the maximum value of magnetic.

Serial Number of the Structure	Total Turns	Layer				B_{xmax} (mT)
		1st	2nd	3rd	4th	
1st	10	10	0	0	0	5.172
2nd	9	8	1	0	0	6.335
3rd	8	6	2	0	0	7.291
4th	6	3	3	0	0	7.486
5th	6	3	2	1	0	7.594
6th	7	5	1	1	0	7.002
7th	4	1	1	1	1	5.218

3. Results and Discussion

3.1. Validation Idea of the Experiment

As shown in Figure 9, to verify the magnetostrictive-based PV system shading detection method used in this paper, the experimental equipment to be used includes PV modules, bypass diodes, magnetic field coils, a voltage controller, a resistive load, a pulse power supply, pulse signal lines, waveguide wire, a detection coil, and an oscilloscope.

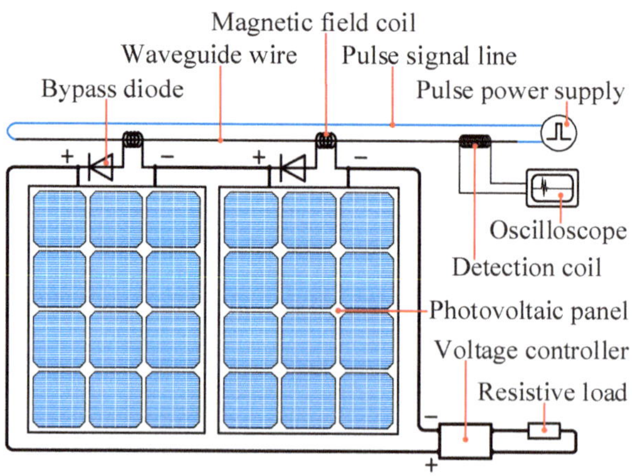

Figure 9. The relationship between each piece of experimental equipment.

The process of the whole detection method is shown in Figure 10.

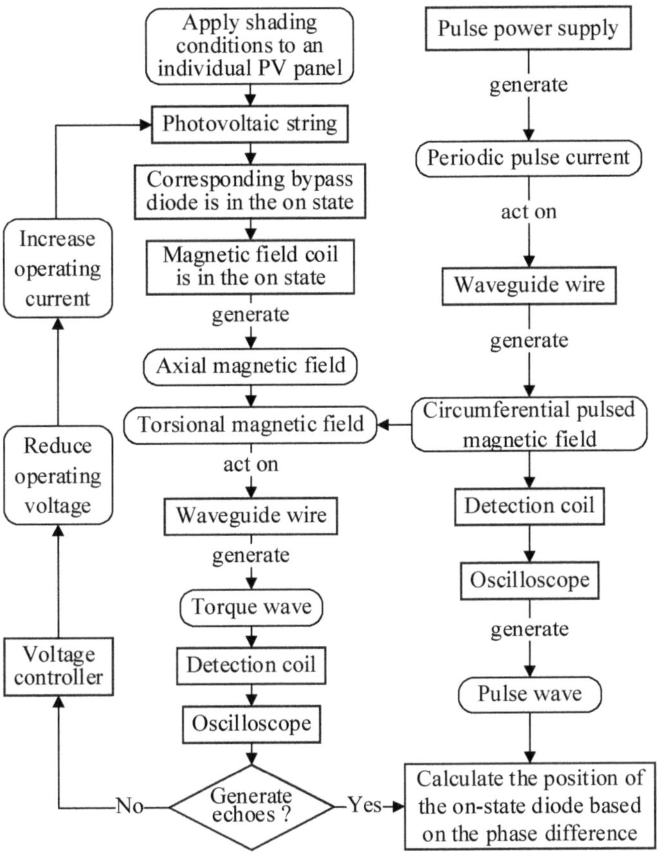

Figure 10. The process of the whole detection method.

Two PV modules are connected in series to form a PV string, and shading conditions are applied to single PV modules so that its corresponding bypass diode conducts and generates a local axial magnetic field from the magnetic field coil. This magnetic field is superimposed on the pulsed circumferential magnetic field generated when the waveguide wire is passed with a periodic pulsed current to form a torsional magnetic field.

Under the action of the torsional magnetic field, the waveguide wire produces a torque wave that propagates to both ends from the corresponding position. The magnetic field change caused by the end-position waveguide wire under the action of the pulse current and torque wave is detected by the detection coil, and the output waveform of the induced electromotive force generated by the detection coil is collected using an oscilloscope.

Based on the phase difference between the pulse wave and the corresponding echo, the specific position of the PV module corresponding to the magnetic field coil in the on-state current can be calculated.

3.2. Construction of the Experimental Platform

As shown in Figure 11, the experimental platform consists of two series-connected PV modules and their junction boxes, a DC voltage–current meter, a DC boost converter, a resistive load, a waveguide wire, signal lines, a pulse power supply, a detection coil, and an oscilloscope.

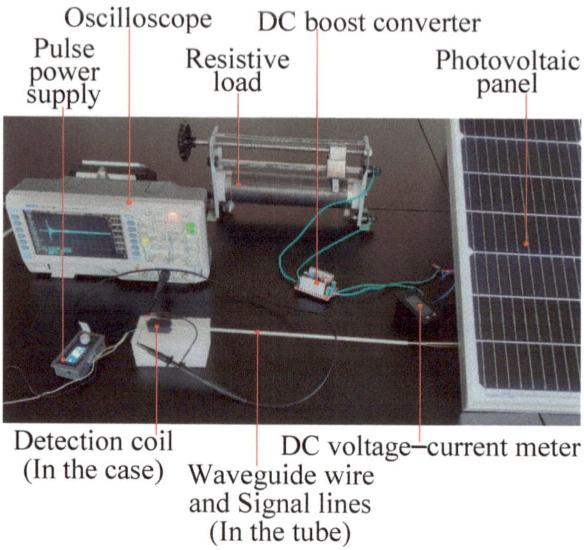

Figure 11. The experimental platform.

In the experimental platform, the magnetic field coil is connected in series with the branch where the bypass diode is in the junction box at the back of the PV module. The output of the PV module string is connected to the input of the DC boost converter and the resistive load is connected to the output of the DC boost converter. The waveguide wire passes through the detection coil and the magnetic field coils in turn and forms a circuit in series with the pulse power supply through signal lines. The output of the detection coil is connected to the probes of the oscilloscope. Wave-absorbing rubbers are installed at the ends of the waveguide wire to reduce the interference caused by the rebound torque waves transmitted from the end position of the waveguide wire.

The type and parameters of the PV module are shown in Table 1. The type and parameters of the bypass diode are shown in Table 2. The iron–nickel alloy is selected as the material for the waveguide wire with the specific dimensional parameters shown in Table 6.

Table 6. The dimensional parameters of the waveguide wire.

Parameters	Numerical Value	Unit
Length	3	m
Diameter	0.8	mm

Since the pulse signal is critical to the generation of the torque waves, the difference in the amplitude, width, duration, and period of the pulse affects the amplitude and stability of the output voltage waveform from the detection coil; thus, it is necessary to experimentally determine the appropriate pulse signal.

The PWM pulse generator is selected as the signal source of the pulse power supply and a direct current of 4A is selected on the magnetic field coil for conduction purposes. The maximum peak echo is induced by the detection coil when the pulse amplitude is selected as 12 V.

According to the output characteristics of the PWM pulse generator and the interference between different echoes, the echo duration is the shortest when the pulse width is chosen as 5 μs and the pulse duration is chosen as 20 μs.

It takes time for the torque wave to propagate to the detection coil from the time it is generated. The torque wave needs to be received before the next pulse signal is generated.

so the pulse period needs to be greater than the maximum time for the torque wave to be received. The period of the pulse signal is experimentally chosen as 20 ms.

The finalized output waveform of pulse power is shown in Figure 12.

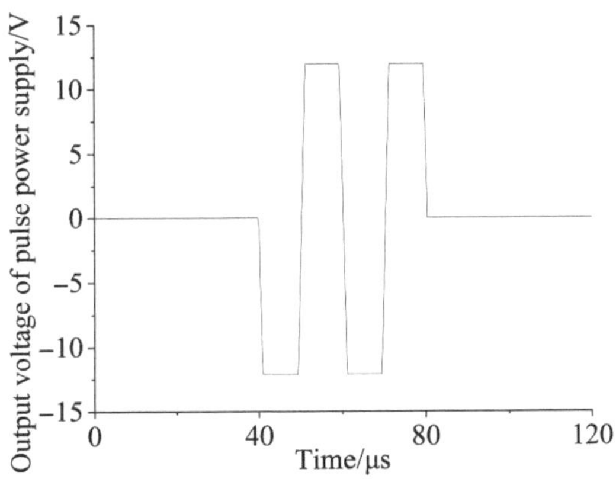

Figure 12. The output waveform of the pulse power.

3.3. Design of the Experimental Steps

The experiments based on the above experimental platform include PV shading experiment and magnetic field detection experiment.

3.3.1. PV Shading Experiment

The experimental platform is placed in a light environment and shading conditions are imposed on the individual PV module. The output voltage of the PV module string is changed by the DC boost converter, the reading of the DC voltage–current meter is recorded, and the output current–voltage and output power–voltage curves of the PV module string are plotted.

The testing device needs to run throughout the entire experiment. During the one-way recording of the PV module string output voltage growth from zero to open-circuit voltage, when the acquired waveform graph of the oscilloscope no longer shows any echoes, the output current of the PV module string is recorded at this time as the critical on-current value of the bypass diode under this masking condition.

3.3.2. Magnetic Field Detection Experiment

The minimum value of the operating current corresponding to the maximum power point of the PV module string under each shading condition is used as the on-state current of the magnetic field coil.

The distance between the magnetic field coil and the detection coil is changed, the echo acquisition is observed in the waveform graph collected by the oscilloscope, and the speed of torque wave propagation along the waveguide wire and the detection range of the method used are calculated in this paper. The separation distance of two magnetic field coils is changed and the minimum separation distance of adjacent magnetic field coils is calculated by observing the interference situation of echoes in the waveform graph collected by the oscilloscope.

3.4. Results and Analysis of the PV Shading Experiment

A sunny day at noon is selected as the experimental time, the experimental platform is placed in the outdoor light environment, and the PV module string is determined with the ground at an angle of 30° inclination. White paper is selected as a slight shade with low light transmittance and thick cardboard is selected as a heavy shade with no light transmittance. One PV module is kept in normal operation and different areas of lateral or longitudinal shading are applied to an individual cell in the other PV module while changing the output voltage of the PV module string and recording the readings of the DC voltage–current meter. To minimize the influence of the error caused by the change in light intensity and temperature on the experiment, each set of experimental data is measured within 15 min.

The PV module string output current–voltage curves for different degrees and areas of shading conditions when shading is applied laterally are shown in Figure 13a, and the output power–voltage curves are shown in Figure 13b. When the shading conditions are applied longitudinally, the corresponding output current–voltage curve is shown in Figure 13c and the output power–voltage curve is shown in Figure 13d.

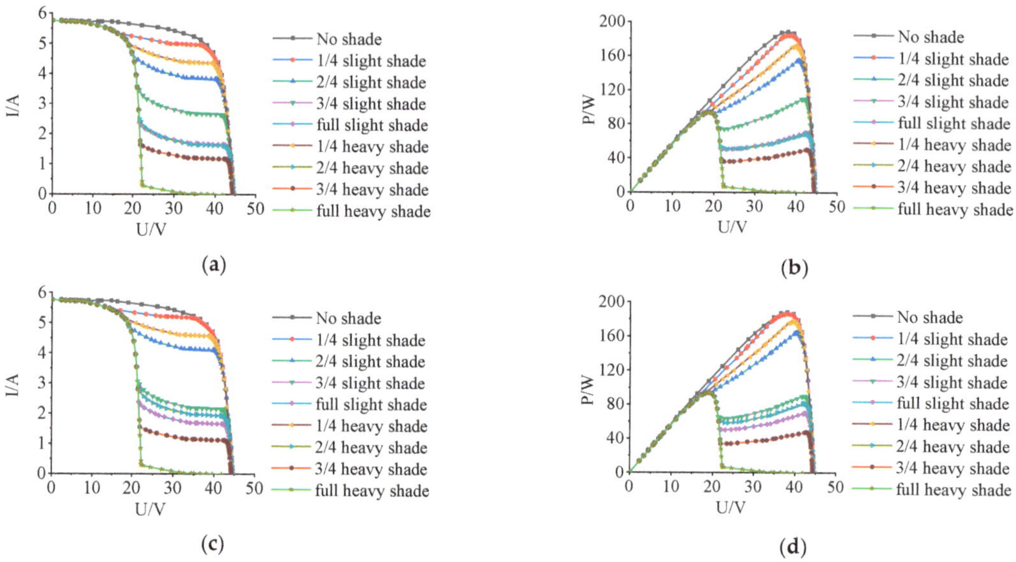

Figure 13. The output current–voltage and power–voltage curves of the PV module string in (**a**,**b**) lateral and (**c**,**d**) longitudinal shading conditions.

The maximum power point current of the PV module string is used as the operating current of the diode.

When the critical conduction current of the diode is greater than the operating current, the oscilloscope acquires no echo in the waveform graph, i.e., when the diode operates under off-state conditions. When the critical conduction current of the diode is less than the operating current, the oscilloscope acquires echoes in the waveform graph, i.e., when the diode operates under on-state conditions.

The corresponding relationships between the maximum power point current of the PV module string and the bypass diode on–off state and critical conduction currents under each shading condition are shown in Table 7.

Table 7. The bypass diode on–off state in each shading condition.

Shading Conditions	Maximum Power Point Current	Numerical Value	The On–Off State of Diode during Maximum Power Operation
No shade	4.948	/	off
Lateral 1/4 slight	4.726	5.342	off
Longitudinal 1/4 slight	4.931	5.479	off
Lateral 2/4 slight	3.814	4.650	off
Longitudinal 2/4 slight	4.089	4.896	off
Lateral 3/4 slight	2.634	3.546	off

Table 7. *Cont.*

Shading Conditions	Maximum Power Point Current	Numerical Value	The On–Off State of Diode during Maximum Power Operation
Longitudinal 3/4 slight	4.849	3.105	on
Full slight	4.831	2.360	on
Lateral 1/4 heavy	4.338	4.995	off
Longitudinal 1/4 heavy	4.553	5.131	off
Lateral 2/4 heavy	4.822	2.549	on
Longitudinal 2/4 heavy	4.852	2.796	on
Lateral 3/4 heavy	4.811	1.641	on
Longitudinal 3/4 heavy	4.834	1.538	on
Full heavy	4.859	0.302	on

According to the analysis results from Figure 3 and the information in Table 7, when the shading degree and area of an individual PV cell are small and the PV module string is operating at the maximum power point, the operating current of the PV module string is higher than the critical on-state current of the bypass diode, which restrains the bypass diode in the operating blind zone.

Therefore, appropriately increasing the operating current in a short period can make the bypass diode, originally found in the blind area, conduct, and further analyzing the echoes that appear in the waveform graph collected by the oscilloscope can help to realize the troubleshooting and positioning of potential shading problems.

3.5. Results and Analysis of the Magnetic Field Detection Experiment

The pulse current is conducted to the waveguide wire. After amplifying and limiting the output voltage waveform signal of the detection coil, the oscilloscope is used for acquisition and analysis.

When the magnetic field coil does not conduct, the waveform acquired by the oscilloscope is shown in Figure 14.

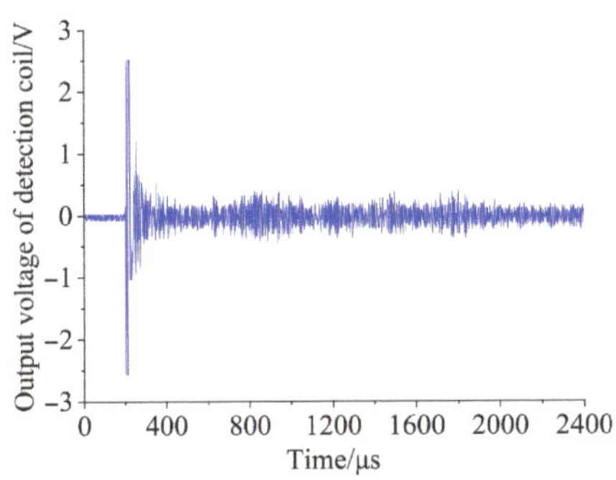

Figure 14. The waveform acquired by the oscilloscope when the magnetic field coil is not conducted.

The signal waves with peaks above 1 V in the figure show the pulse wave and the initial echo. The pulsed wave is caused by changes in the magnetic field when the waveguide wire conducts a pulsed current. The initial echo within 100 μs after the pulse wave is caused by the initial torque wave generated by the waveguide wire at the signal line connection position after the pulse current is applied.

The minimum value of the current corresponding to the maximum power point of the PV module string in Table 7 (2.634 A) is selected as the on-state current of the magnetic field coil. By adjusting the distance between the magnetic field coil and the detection coil, the effective waveforms acquired by the oscilloscope are shown in Figure 15a–c, respectively, when the magnetic field coil is placed at the head, middle, and end positions of the waveguide wire in turn after conducting.

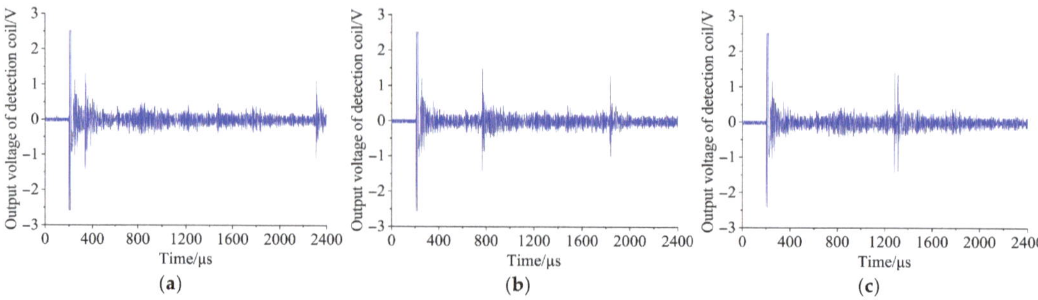

Figure 15. The waveform acquired by the oscilloscope when the magnetic field coil is conducting at the (**a**) head, (**b**) middle, and (**c**) end positions of the waveguide wire.

The signal waves with peaks above 1 V in the figures show the pulse wave, the initial echo, the target echo, and the elastic wave. The target echo is caused by the torque wave generated at the position corresponding to the waveguide wire after the magnetic field coil is turned on, with a minimum signal peak value of 1.28 V. The elastic wave is caused by the rebound torque wave generated after the torque wave is transmitted to the end of the waveguide wire, and its signal peak is smaller than that of the target echo.

Figure 15 shows that when the magnetic field coil is located at the position of the waveguide wire head, the time interval between the peak of the target echo and the pulse wave should be greater than 100 μs to reduce the interference of the initial echo. The pulse wave is the starting point, the midpoint of the horizontal line connecting the target echo peak and the elastic wave peak is the endpoint, and the interval time between the two points is 1080 μs. Therefore, waveform signals with peaks greater than 1.28 V and located within 100–1080 μs after the pulse wave are considered valid target echoes.

Based on the length of the waveguide wire of 3 m, the actual speed of the torque wave propagation along the waveguide wire is calculated as 2778 m/s. It is further calculated that the detection range of the method used in this paper is 0.28–3.00 m, and its maximum detection range is the same as the length of the waveguide wire.

After conducting the two magnetic field coils simultaneously with 2.634 A of direct current, the spacing distance of the magnetic field coils is adjusted so that there is no superposition effect between the target echo signals collected by the oscilloscope. The effective waveform graph acquired by the oscilloscope at the minimum interval distance is shown in Figure 16.

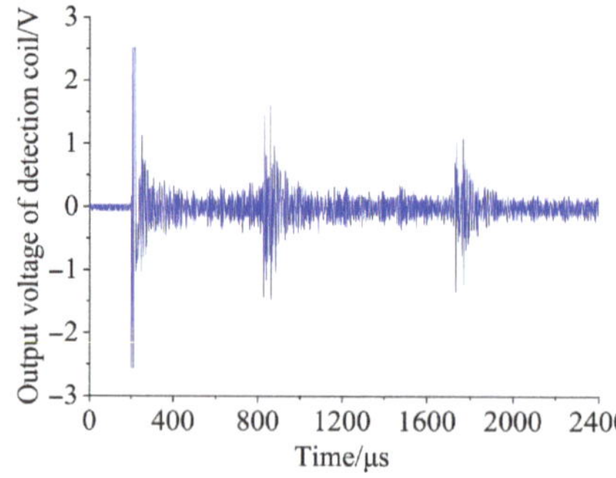

Figure 16. The effective waveform graph acquired by the oscilloscope when adjacent magnetic field coils reach the minimum interval distance.

In Figure 16, the torque waves generated by the conduction of the two magnetic field coils, after causing the two target echoes in turn, are rebounded by the end of the waveguide wire, causing an elastic wave of the same time interval. To observe the details of the time interval between the two target echoes in the figure, a local zoom is performed at the target echoes position, as shown in Figure 17.

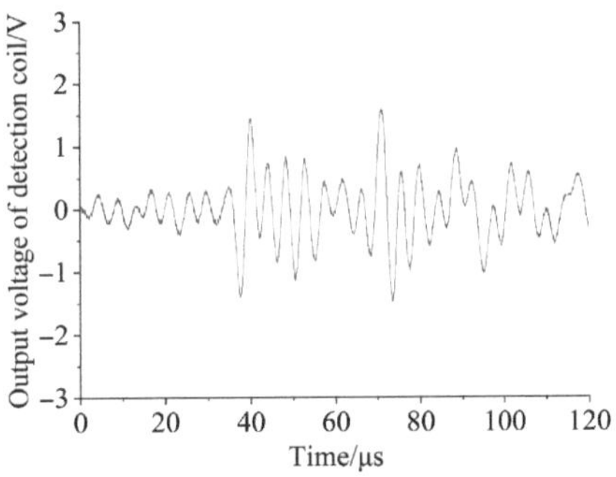

Figure 17. The details of the waveform acquired by the oscilloscope when adjacent magnetic field coils are at the minimum interval distance.

According to Figure 17, the minimum time interval between two adjacent target echoes is 31 μs. Based on the propagation speed of the torque wave along the waveguide wire, the minimum separation distance between adjacent magnetic field coils is calculated as 86 mm. After several groups of experiments are verified, excluding any invalid data caused by improper operation, the detection error of the experimental device is 4 mm, so the installation distance of the magnetic field coil is not affected by this error.

In summary, in the actual installation of the detection equipment designed in this paper, the distance between the first magnetic field coil and the detection coil should be greater than 0.28 m, and the interval distance between adjacent magnetic field coils should be greater than 86 mm.

4. Conclusions

The size and structural parameters of the magnetic field coil are determined based on the results of the magnetic field simulation. By building an experimental platform, the output waveform of the induced voltage of the detection coil is calculated and analyzed based on the conclusion of the analysis of the diode on–off situation in the PV shading experiment. The feasibility of the detection method is verified, the detection range of the method is calculated, and the minimum spacing of adjacent diodes corresponding to the magnetic field coils is determined. This study proposes a novel detection method based on magnetostriction and offers a theoretical and experimental basis for the shading problem of the PV system.

When the PV module is shaded by a smaller degree and area, if the operating current of the PV module is higher than the critical on-current of the bypass diode, the bypass diode will work in the blind area and remain off, which is not conducive to detection.

The PV string voltage controller is studied so that the bypass diode, originally found in the blind area of operation, can be conducted by appropriately increasing the operating current for a short period. Further magnetic field detection and localization are performed by magnetostrictive sensors to enable the identification and localization of potential shading problems. Specifically, it needs to be paired with a control algorithm that performs maximum power tracking while selecting moments of fixed period to increase the operating current of the PV string. During this period, the magnetostrictive sensors in this paper keep working and finally achieve MPPT, while monitoring the operating status of each PV module in the PV array in real time.

5. Patent

Tian, Y.; Wei, B.; Fu, X. A Shielded Condition Monitoring and Recognition Device for PV Modules Based on Magnetostriction. Xinjiang Uygur Autonomous Region CN114285376A, 5 April 2022.

Author Contributions: Conceptualization, Y.T.; methodology, X.F.; software, X.F.; validation, X.F.; formal analysis, X.F.; investigation, X.F.; data curation, X.F.; writing—original draft preparation, X.F.; writing—review and editing, Y.T. and X.F.; visualization, Y.T.; supervision, Y.T.; project administration, Y.T.; funding acquisition, Y.T. All authors have read and agreed to the published version of the manuscript.

Funding: This research was funded by the Natural Science Foundation of Xinjiang Uygur Autonomous Region (2022D01C364).

Data Availability Statement: The data are available from the corresponding author upon reasonable request.

Acknowledgments: The authors acknowledge the technical support of the School of Electrical Engineering, Xinjiang University.

Conflicts of Interest: The authors declare no conflict of interest.

Nomenclature

PV	photovoltaic
MPP	the maximum power point
MPPT	the maximum power point tracking
I_n	the output current of the PV cell in normal conditions
U_r	the output voltage of the PV cell in normal conditions
U_r	the reverse bias voltage of the shaded PV cell
U_d	the conduction voltage of the bypass diode
I_s	the short-circuit current of the normal PV cell
I_{rc}	the short-circuit current of the shaded PV cell
I_g	the short-circuit current of the shaded PV cell
U_{rc}	the critical reverse bias voltage of the shaded PV cell
l_o	the original length of magnetostrictive material
l	the present length of magnetostrictive material
λ	the line magnetostriction coefficient
λ_s	the saturation magnetostriction coefficient
r	the radius of the magnetostrictive waveguide wire
I_c	the amplitude of the pulse current
τ	the pulse width
J	the density of the surface current
a	the radius of the waveguide wire
H	the intensity of the pulsed magnetic field
H_c	the intensity of the pulsed magnetic field at the position of the waveguide wire surface
μ_{r1}	the relative magnetic permeability of waveguide wire surface when subjected only to the pulsed magnetic field
μ_0	the absolute magnetic permeability of the vacuum
B_{c1}	the pulsed magnetic induction intensity
λ_1	the magnetostriction coefficient of the waveguide wire
l_0	the original circumference of the circular section of the waveguide wire
$\mathbf{B}_0$	the axial magnetic induction intensity vector
$\mathbf{B}_{c2}$	the circumferential pulse magnetic induction intensity vector
$\mathbf{B}_h$	the torsional magnetic induction intensity vector
$\mathbf{e}_x$	unit magnetic induction intensity vectors in the positive direction of the x axis
$\mathbf{e}_y$	unit magnetic induction intensity vectors in the positive direction of the y axis
$\mathbf{e}_z$	unit magnetic induction intensity vectors in the positive direction of the z axis

λ_2	the magnetostriction coefficient of the waveguide wire under the action of torsional magnetic induction intensity
σ	the shear stress
G	the shear modulus
Δl	the deformation variable of the waveguide wire
μ_{r0}	the initial permeability of the waveguide wire in the absence of magnetic field influence
μ_σ	the permeability of the waveguide wire under the action of the shear stress
B_s	the saturation magnetic induction intensity of the waveguide wire
λ_s	the saturation magnetostriction coefficient
L	the position information of the permanent magnet
H_L	the magnetic field intensity of the permanent magnet
B_L	the magnetic induction intensity induced by the waveguide wire from the permanent magnet
μ	the absolute permeability
N	the number of turns of the detection coil
S	the area of coil facing the magnetic field
E	the induced electromotive force of the detection coil
v	the transmission speed of the torque wave
t	the transmission time of the torque wave
l_x	the winding length of the magnetic field coil
r_x	the inner radius of the magnetic field coil
d_x	the diameter of the enameled wire
d_{Cu}	the diameter of the winding copper core
S_{Cu}	the cross-sectional area of the winding copper core
d_s	the thickness of the lacquer coating
m	the maximum number of winding layers
J_{Cu}	the maximum safe load current of copper
I_{dm}	the maximum on-state current of the magnetic field coil
$S_{Cu\max}$	the maximum cross-sectional area of the copper core
$d_{Cu\max}$	the minimum diameter of the copper core
$R_{\min}$	the minimum equivalent resistance of the diode
U_{dm}	the maximum on-state voltage of the magnetic field coil
$R_{x\max}$	the maximum resistance of the magnetic field coil
ρ_{Cu}	the resistivity of copper
$l_{x\max}$	the maximum winding length of the coil wire
I_d	the conduction current of the magnetic field coil
$B_{x\max}$	the maximum magnetic induction intensity at the position of the magnetic field coil central axis

References

1. Saha, S.; Saini, G.; Mishra, S.; Chauhan, A.; Upadhyay, S. A comprehensive review of techno-socio-enviro-economic parameters, storage technologies, sizing methods and control management for integrated renewable energy system. *Sustain. Energy Technol. Assess.* **2022**, *54*, 102849. [CrossRef]
2. Allouhi, A.; Rehman, S.; Buker, M.S.; Said, Z. Up-to-date literature review on Solar PV systems: Technology progress, market status and R&D. *J. Clean. Prod.* **2022**, *362*, 132339. [CrossRef]
3. Alami, A.H.; Rabaia, M.K.H.; Sayed, E.T.; Ramadan, M.; Abdelkareem, M.A.; Alasad, S.; Olabi, A.G. Management of potential challenges of PV technology proliferation. *Sustain. Energy Technol. Assess.* **2022**, *51*, 101942. [CrossRef]
4. Zsiboracs, H.; Zentko, L.; Pinter, G.; Vincze, A.; Baranyai, N.H. Assessing shading losses of photovoltaic power plants based on string data. *Energy Rep.* **2021**, *7*, 3400–3409. [CrossRef]
5. Vunnam, S.; VanithaSri, M.; RamaKoteswaraRao, A. Performance analysis of mono crystalline, poly crystalline and thin film material based 6 × 6 T-C-T PV array under different partial shading situations. *Optik* **2021**, *248*, 168055. [CrossRef]
6. Saiprakash, C.; Mohapatra, A.; Nayak, B.; Ghatak, S.R. Analysis of partial shading effect on energy output of different solar PV array configurations. *Mater. Today Proc.* **2021**, *39*, 1905–1909. [CrossRef]
7. Bressan, M.; Gutierrez, A.; Gutierrez, L.G.; Alonso, C. Development of a real-time hot-spot prevention using an emulator of partially shaded PV systems. *Renew. Energy* **2018**, *127*, 334–343. [CrossRef]
8. Tang, S.; Xing, Y.; Chen, L.; Song, X.; Yao, F. Review and a novel strategy for mitigating hot spot of PV panels. *Sol. Energy* **2021**, *214*, 51–61. [CrossRef]

9. Skomedal, Å.F.; Aarseth, B.L.; Haug, H.; Selj, J.; Marstein, E.S. How much power is lost in a hot-spot? A case study quantifying the effect of thermal anomalies in two utility scale PV power plants. *Sol. Energy* **2020**, *211*, 1255–1262. [CrossRef]
10. Bastidas-Rodríguez, J.D.; Ramos-Paja, C.A.; Serna-Garcés, S.I. Improved modelling of bypass diodes for photovoltaic applications. *Alex. Eng. J.* **2022**, *61*, 6261–6273. [CrossRef]
11. Kreft, W.; Przenzak, E.; Filipowicz, M. Photovoltaic chain operation analysis in condition of partial shading for systems with and without bypass diodes. *Optik* **2021**, *247*, 167840. [CrossRef]
12. Jha, V. Generalized modelling of PV module and different PV array configurations under partial shading condition. *Sustain. Energy Technol. Assess.* **2023**, *56*, 103021. [CrossRef]
13. Ragb, O.; Bakr, H. A new technique for estimation of photovoltaic system and tracking power peaks of PV array under partial shading. *Energy* **2023**, *268*, 126680. [CrossRef]
14. Fadhel, S.; Delpha, C.; Diallo, D.; Bahri, I.; Migan, A.; Trabelsi, M.; Mimouni, M.F. PV shading fault detection and classification based on I-V curve using principal component analysis: Application to isolated PV system. *Sol. Energy* **2019**, *179*, 1–10. [CrossRef
15. Li, C.; Yang, Y.; Zhang, K.; Zhu, C.; Wei, H. A fast MPPT-based anomaly detection and accurate fault diagnosis technique for PV arrays. *Energy Convers. Manag.* **2021**, *234*, 113950. [CrossRef]
16. Chandrasekaran, K.; Sankar, S.; Banumalar, K. Partial shading detection for PV arrays in a maximum power tracking system using the sine-cosine algorithm. *Energy Sustain. Dev.* **2020**, *55*, 105–121. [CrossRef]
17. Salem, F.; Awadallah, M.A. Detection and assessment of partial shading in photovoltaic arrays. *J. Electr. Syst. Inf. Technol.* **2016** *3*, 23–32. [CrossRef]
18. Seapan, M.; Hishikawa, Y.; Yoshita, M.; Okajima, K. Detection of shading effect by using the current and voltage at maximum power point of crystalline silicon PV modules. *Sol. Energy* **2020**, *211*, 1365–1372. [CrossRef]
19. Sugumar, S.; Winston, D.P.; Pravin, M. A novel on-time partial shading detection technique for electrical reconfiguration in solar PV system. *Sol. Energy* **2021**, *225*, 1009–1025. [CrossRef]
20. Martin, A.D.; Vazquez, J.R.; Cano, J.M. MPPT in PV systems under partial shading conditions using artificial vision. *Electr. Power Syst. Res.* **2018**, *162*, 89–98. [CrossRef]
21. Cavieres, R.; Barraza, R.; Estay, D.; Bilbao, J.; Valdivia-Lefort, P. Automatic soiling and partial shading assessment on PV modules through RGB images analysis. *Appl. Energy* **2022**, *306*, 117964. [CrossRef]
22. Di Tommaso, A.; Betti, A.; Fontanelli, G.; Michelozzi, B. A multi-stage model based on YOLOv3 for defect detection in PV panels based on IR and visible imaging by unmanned aerial vehicle. *Renew. Energy* **2022**, *193*, 941–962. [CrossRef]
23. Sizkouhi, A.M.; Aghaei, M.; Esmailifar, S.M. A deep convolutional encoder-decoder architecture for autonomous fault detection of PV plants using multi-copters. *Sol. Energy* **2021**, *223*, 217–228. [CrossRef]
24. Lee, C.G.; Shin, W.G.; Lim, J.R.; Kang, G.H.; Ju, Y.C.; Hwang, H.M.; Chang HSKo, S.W. Analysis of electrical and thermal characteristics of PV array under mismatching conditions caused by partial shading and short circuit failure of bypass diodes. *Energy* **2021**, *218*, 119480. [CrossRef]
25. Ahmed, U.; Jeronen, J.; Zucca, M.; Palumbo, S.; Rasilo, P. Finite element analysis of magnetostrictive energy harvesting concept device utilizing thermodynamic magneto-mechanical model. *J. Magn. Magn. Mater.* **2019**, *486*, 165275. [CrossRef]
26. Seco, F.; Martin, J.M.; Jimenez, A.R.; Calderon, L. A high accuracy magnetostrictive linear position sensor. *Sens. Actuators A Phys.* **2005**, *123*, 216–223. [CrossRef]
27. Magnetostrictive liquid-level sensors. *Ultrasonics* **1967**, *5*, 196. [CrossRef]
28. Ko, S.W.; Ju, Y.C.; Hwang, H.M.; So, J.H.; Jung, Y.S.; Song, H.J.; Song, H.E.; Kim SHKang, G.H. Electric and thermal characteristics of photovoltaic modules under partial shading and with a damaged bypass diode. *Energy* **2017**, *128*, 232–243. [CrossRef]
29. Teo, J.C.; Tan, R.H.; Mok, V.H.; Ramachandaramurthy, V.K.; Tan, C. Impact of bypass diode forward voltage on maximum power of a photovoltaic system under partial shading conditions. *Energy* **2020**, *191*, 116491. [CrossRef]
30. Deng, C.; Kang, Y.; Li, E.; Zhang, Y.; Cheng, J.; Ge, T. A new model of the signal generation mechanism on magnetostrictive position sensor. *Measurement* **2014**, *47*, 591–597. [CrossRef]
31. Tarasov, V.E. General non-local electrodynamics: Equations and non-local effects. *Ann. Phys.* **2022**, *445*, 169082. [CrossRef]
32. Wang, H.; Yang, W.; Huang, Y. Adaptive finite element method for two-dimensional time-harmonic magnetic induction intensity equations. *J. Comput. Appl. Math.* **2022**, *412*, 114319. [CrossRef]
33. Shrikanth, S.; Neelakantan, S.; Prasad, R. Planes of Isotropic shear moduli in anisotropic materials. *Mech. Mater.* **2023**, 104619 [CrossRef]
34. Bechtold, C.; Teliban, I.; Thede, C.; Chemnitz, S.; Quandt, E. Non-contact strain measurements based on inverse magnetostriction. *Sens. Actuators A Phys.* **2010**, *158*, 224–230. [CrossRef]
35. Seco, F.; Martín, J.M.; Pons, J.L.; Jiménez, A.R. Hysteresis compensation in a magnetostrictive linear position sensor. *Sens. Actuators A Phys.* **2004**, *110*, 247–253. [CrossRef]

energies

MDPI

Article

Simulation and Experimental Evaluation of a Refractive-Reflective Static Solar Concentrator

Guillermo Luque-Zuñiga [1], Rubén Vázquez-Medina [1], G. Ramos-López [1,*], David Alejandro Pérez-Márquez [2] and H. Yee-Madeira [3,*]

1 Instituto Politécnico Nacional, CICATA Unidad Querétaro, Querétaro 76090, Mexico; memo_luque88@hotmail.com (G.L.-Z.); ruvazquez@ipn.mx (R.V.-M.)
2 Departamento de Ingeniería en Diseño Industrial, Universidad Politécnica del Bicentenario, Guanajuato 36283, Mexico; dapm99@hotmail.com
3 Instituto Politécnico Nacional, Escuela Superior de Física y Matemáticas, Unidad Profesional Adolfo López Mateos Zacatenco, Ciudad de Mexico 07320, Mexico
* Correspondence: gramos@ipn.mx (G.R.-L.); hernaniyee@hotmail.com (H.Y.-M.); Tel.: +52-442-1587466 (G.R.-L.)

Abstract: Static solar devices have advantages over solar tracking systems. In pure reflective systems, solar reception is limited by the entry angle of the reflector. Many reflective systems are based on mirror Compound Parabolic Concentrators. The solar collection can be improved by placing a lens on top of the reflector. In this work, a static system is proposed, consisting of a mirror funnel concentrator with a prism on top. The system is designed using ray-tracing software and is subsequently built and experimentally evaluated. The system designed for an effective concentration factor of 4× reaches an effective concentration of 3.2× at 11:30 a.m. and has an acceptance angle of 60°. Considering the time interval from 8 a.m. to 4 p.m., the system harvests 30.7% more energy than the flat surface. If the time interval considered is from 9:30 a.m. to 2:30 p.m., the increase in harvest is ∼77%. The incorporation of the prism represents an increase of ∼6% compared to the bare reflective system.

Keywords: solar concentrator; refractive-reflective system; collected energy; solar energy

Citation: Luque-Zuñiga, G.; Vázquez-Medina, R.; Ramos-López, G.; Pérez-Márquez, D.A.; Yee-Madeira, H. Simulation and Experimental Evaluation of a Refractive-Reflective Static Solar Concentrator. *Energies* **2023**, *16*, 1071. https://doi.org/10.3390/en16031071

Academic Editor: Manolis Souliotis

Received: 19 December 2022
Revised: 6 January 2023
Accepted: 13 January 2023
Published: 18 January 2023

1. Introduction

Solar static systems are cheaper than solar tracker systems. Even designing solar collection devices with larger acceptance angles for less accurate solar trackers is seen as a cost reduction [1]. Static concentration systems, having a low concentration ratio, can be used with common commercial silicon cells, so-called "one sun" cells [2,3]. Coello et al. observed an increase in PV generation with common solar cells up to a concentration ratio of 15×. Another study, choosing PV cells randomly in the market, shows that in some cases this benefit is limited to 3× [4]. The combination of static concentrators and common PV cells can reduce costs in photovoltaic applications.

Many of the solar concentrators presented in the literature use mirrors or reflecting surfaces [5–13]. Others make use of total internal reflection in the walls of solid dielectric materials, or a combination of both [1,6,9]. Many of the systems use Compound Parabolic Concentrators (CPC), either as mirrors or as solid dielectric concentrators. CPCs only collect energy at angles of incidence smaller than the half acceptance angle [11]. In a recent study, it was observed that other geometries, like the funnel, can have a better energy collection [14]. There is another group of systems, that combine a lens (or prism) on top of the reflective concentrating device [1,10,12,13,15]. The use of combined systems allows the improvement of certain aspects, such as increasing the optical efficiency, increasing the time of collection of solar energy and maintaining the temperature above certain values. Su et al. remarks on the importance of reducing the amount of dielectric material in the concentrating devices [6].

A direct comparison between systems is difficult since there are linear [1,12] and axial systems [6,9,13,15]. Some systems include only numerical simulations [1], whereas other systems include also experimental evaluations of certain aspects [5–7,9,12,13,15].

In this work, the design, construction and experimental evaluation of a funnel refractive static solar concentration system with a prismatic lens on top is carried out. The investigation is divided into 3 sections, in the first, the concentration system is presented, in the second the details of both theoretical and experimental evaluation are described and finally the comparison of results is done.

2. System Details

The concentration system is composed of a reflective Funnel and a prismatic refractive lens on top of it, as shown schematically in the cross-section of Figure 1. Only the inner wall of the Funnel is reflective with a reflectivity of 96%, and a concentration factor of 4× This concentration factor is the Geometrical Concentration Ratio (GCR) calculated using Equation (1). The receptor is placed at the bottom of the Funnel.

The funnel has a height of 75 mm, an inlet diameter of 75 mm and an outlet diameter of 37.5 mm. So, the inlet opening area is 4417.86 mm^2 while the outlet opening is 1104.47 mm^2.

The prism is a single piece of dielectric material (Polymethylmethacrylate PMMA) The base is a cylinder 5 mm high and the top is a cone with an angle of 12°. The outer diameter is 75 mm (see Figure 1). The dielectric material has a nominal refractive index of n = 1.49 [16] .

$$GCR = \frac{\text{aperture area}}{\text{receiver area}} \tag{1}$$

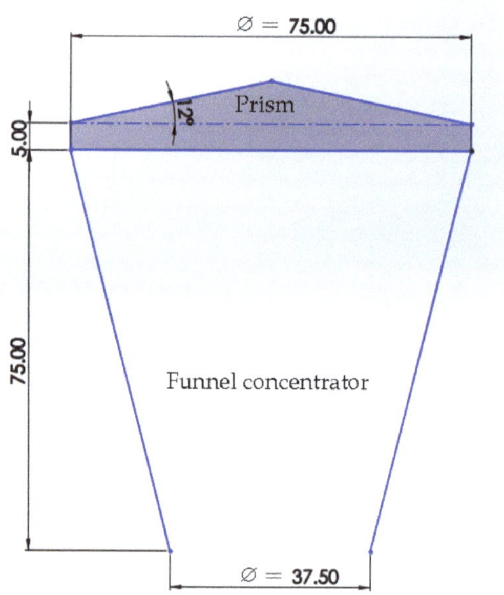

Figure 1. Cross section of the concentration system composed by the reflective Funnel and the refractive Prism (Dimensions in mm).

3. Evaluation Details

3.1. Ray-Tracing Evaluations

For the ray-tracing evaluation, OptiCAD [17] was used. For the sun, an irradiance of 1000 W/m^2 was considered for normal incidence (zenith). The simulations were carried out under the following considerations: The system is placed in the Equator, with the receptor horizontally, in the Equinox; the sun varies its apparent position from 8:00 a.m. to 4:00 p.m.; the simulations are made at 30 min intervals; the sun rays have a divergence of 0.53° and all have the same energy.

The prism is simulated with a refractive material with a refractive index of 1.49, but without surface reflectivity. The receiver is a circular absorbing plate with 0% reflectivity and 100% absorbance, placed 3 mm below the system, as shown in Figure 2. This position of the absorber was chosen to adequately represent the experimental conditions that are used in the laboratory prototype for evaluation.

Figure 2 shows the typical path of rays falling on the system for an angle of incidence of 22.5°. In the absence of the prism, the rays that enter through the upper opening can reach the receiver either directly (ray R1), or after one or several reflections (rays R2 and R3). They can also be reflected out of the system, as is the case of ray R4. Ray R4 undergoes several reflections but finally leaves the system through the upper opening (Figure 2a). In the presence of the prism, the trajectory of the rays is more complex. In Figure 2b the rays that fall on the central part of the prism will suffer a slight deviation and most of them will reach the receiver either directly (ray R1′) or after one or more reflections (rays R2′, R3′ and R4′). The R4′ ray corresponds to the R4 ray of Figure 2a, which was reflected out the system. As can be seen, at this angle of incidence, the prism helps rays such as R4′ reach the receiver, thus increasing the solar harvest.

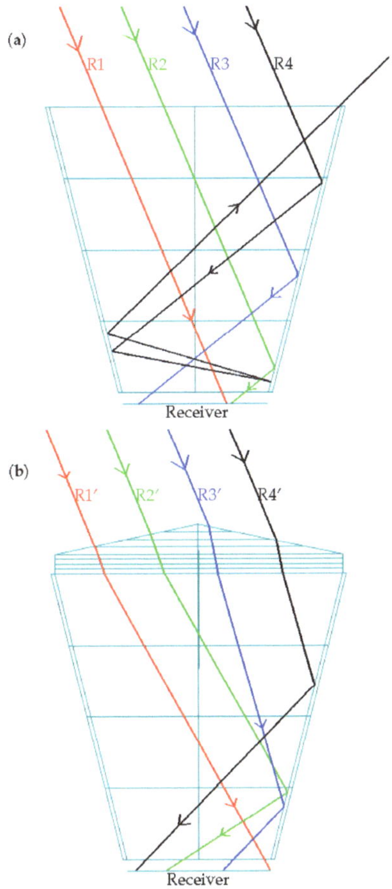

Figure 2. Path of typical rays falling on the system: (**a**) bare funnel concentrator, (**b**) refractive-reflective system, with an incidence angle of $\theta = 22.5°$.

The beneficial effect of the prism is only for angles of incidence smaller than ±37.5°. For larger angles, the majority of the rays undergo multiple reflections and leave the concentrator.

Figure 3 shows the results of the simulations made for the funnel concentrator with prism (Funnel-Prism SIM) and without prism (Funnel SIM), respectively. The collected energy is the total energy flux that strikes the receiver at a given time of day, in units of W/m^2. For the bare funnel, the solar harvest before 9:00 a.m. is negligible. At 9:30 a.m. 217.45 W/m^2 are harvested in the receiver. Later the harvest increases, until it reaches a maximum of 3203.65 W/m^2 at solar noon, to decrease afterwards. The addition of the prism to the funnel has very similar behaviour, however, there is a solar harvest already at 8:30 a.m. (11.08 W/m^2). While the harvest without the prism is higher in the interval between 9:05 a.m. and 10:15 a.m., the prism allows a better solar harvest between 10:15 a.m. and 11:30 a.m. The effect is more visible at 10:30 a.m. when the prism allows a 23% higher harvest. However, at solar noon the prism produces a slight decrease in the harvest, of less than 2%. In this way the total improvement of the addition of the prism is approximately 1% compared to the bare funnel, giving a total harvest 54% larger than for a flat plate.

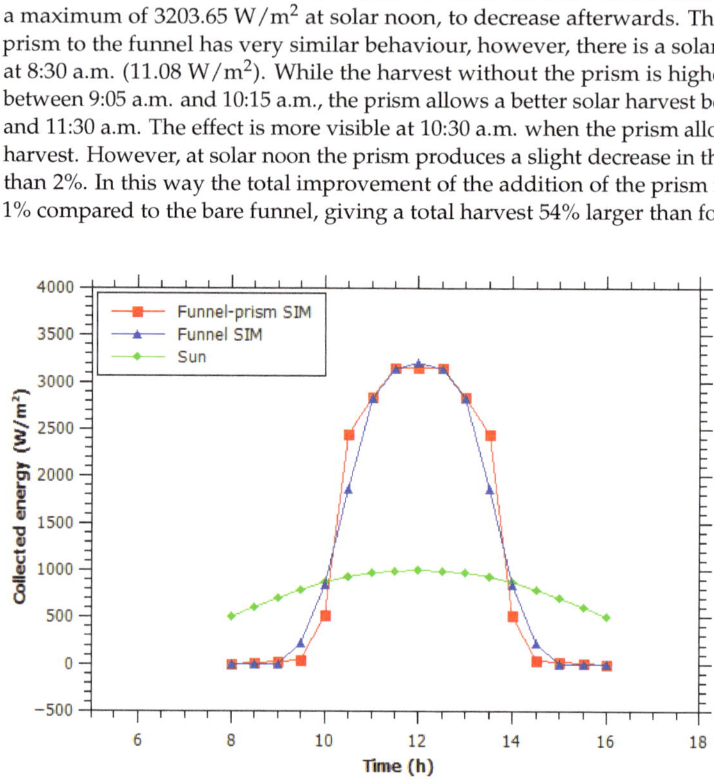

Figure 3. Collected energy as obtained with ray-tracing simulations.

3.2. *Experimental Evaluation*

While optical ray-tracing evaluations are very reliable, they consider idealized surfaces and optical properties. Sometimes these idealized conditions are difficult or very expensive to reproduce in the lab or in the prototypes. The experimental evaluation of a prototype developed without highly sophisticated tools can give us an idea of what, in general, can be expected from the design.

The Funnel was manufactured with a reflective aluminium sheet of about 1 mm thickness. A truncated cone-shaped wooden core was manufactured with the appropriate diameters (75 mm and 37.5 mm, respectively). On this core, the aluminium sheet was moulded. To keep the shape it was wound with wire and finally glued with epoxy glue. Figure 4a shows an upper image of the funnel and the concentration pattern at normal incidence. The manufactured shape is not perfect, particularly in the position of the joints of the sheets.

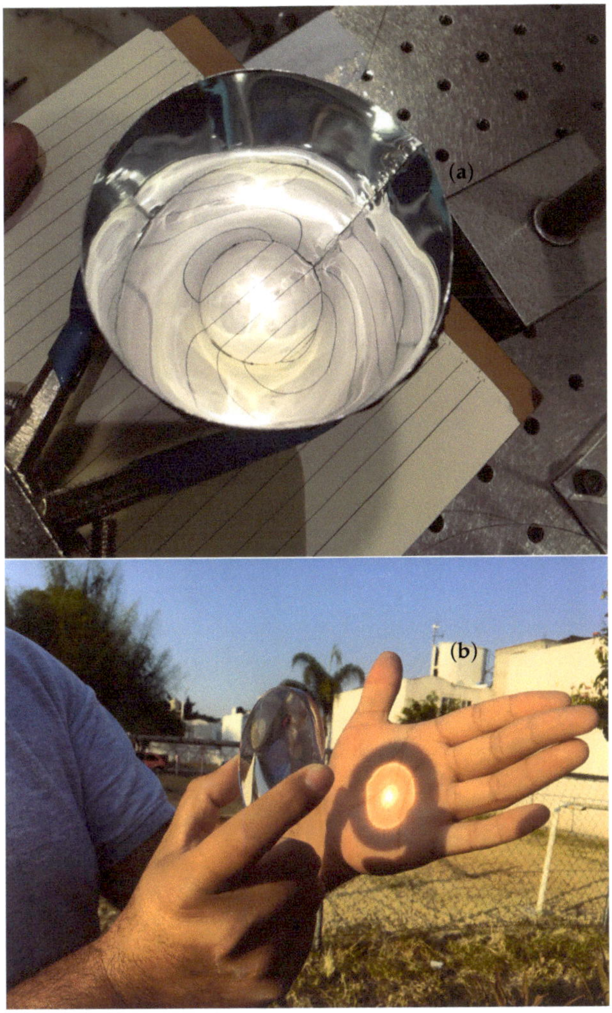

Figure 4. (**a**) Upper view of the funnel and concentration pattern, (**b**) Image of the prism and concentration pattern.

The prism was cut from a 3″ diameter PMMA cylinder. Subsequently, it was polished using different grades of silicon carbide sandpaper, until a surface of sufficient optical quality was obtained. The final polish was done with car headlight repair fluid. Figure 4b shows the polished prism and the typical concentration pattern in the sun.

A lateral image of the system is shown in Figure 5. To minimize the deformation that could be caused by the clamping of the system, a cylindrical support was developed, on which the system was seated for evaluation (black in Figure 5).

For the measurements, a 100 W tungsten halogen lamp powered by a regulated voltage supply of radiometric quality was used as the light source. The lamp bulb was placed at the distance at which the dimensions of the filament produced a divergence of 0.53°. This was done to emulate the ray divergence of the solar disk. To do the measurements for the different hours of the day, the light source was placed at the corresponding height and distance, as shown in Figure 6. The marks corresponding to each measurement can be seen in the table.

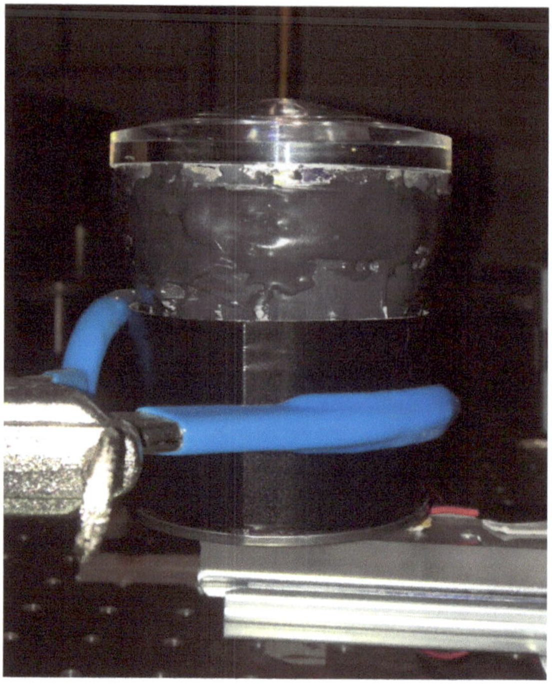

Figure 5. Lateral view of the system ready to be measured.

Figure 6. Experimental setup showing the lamp, the marks on the table, the system and XY positioning system together with the electronics.

An XY displacement system was developed to obtain the intensity measurements and the concentration patterns. The signal of the silicon photodiode is amplified using an operational amplifier, and the position is controlled by stepper motors. A microcontroller moves the diode to the desired position and stores the data in a non-volatile memory. The XY displacement system is programmed to obtain the measurements in a matrix of 20×20 bins, which covers the entire area of the receiver.

First, the measurements in the absence of the concentration system were made, adjusting the intensity of the light source so that it gave the equivalent of 1000 W/m^2 at solar noon. Moving the light source to the different positions corresponding to the different hours of the day, the data of the "Sun" curve of Figure 7 was obtained. Although it cannot be seen very well in the graph, the data agree good with the cosine curve.

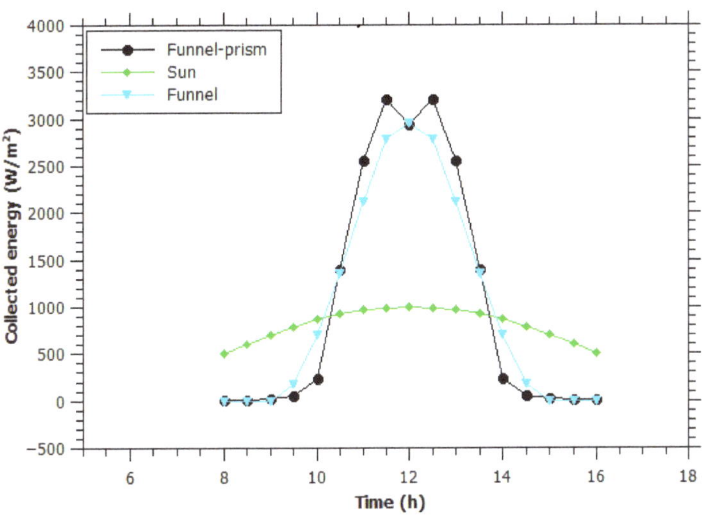

Figure 7. Experimental data of the collected energy.

Next, the curves with the bare funnel (Funnel curve in Figure 7) and the complete system (Funnel-Prism curve in Figure 7) were measured. For the Funnel curve (cyan curve), the energy harvest before 9:00 a.m. is negligible. Energy harvest starts at 9:30 a.m. (180 W/m^2), and increases rapidly between 9:30 and 11:30 a.m. Harvest reaches a maximum of 2953.27 W/m^2 at solar noon. The Funnel-prism curve (black curve), although hardly seen in Figure 7, does not start at zero. Already at 8:00 a.m. 11.97 W/m^2 are harvested. As in the previous case, the harvest increases rapidly between 9:30 and 11:30 a.m., reaching a maximum of 3203.36 W/m^2 at 11:30 a.m. For this time of the day, the angle of incidence is 7.5° and the harvest factor is 3.2 times the harvest without the device. In contrast to the bare funnel, the energy harvest is slightly less at solar noon (2940.15 W/m^2). This implies a decrease of 8%, which contrasts with the value of less than 2% from the simulations.

As can be seen, the addition of the prism has the most noticeable effect increasing the solar harvest between 10:30 a.m. and 1:30 p.m. The solar harvest between 8 a.m. and 4 p.m. without the system is 6.85 kW/m^2. The harvest in the same period with the bare funnel concentrator is 8.6 kW/m^2. This implies a 25.47% increase in harvest. The complete system has a harvest of 8.96 kW/m^2 in the same period (30.7% larger than without the system). If the time interval considered is only from 9:30 a.m. to 2:30 p.m., the improvement in the harvest is 76.78%.

4. Results

Figure 8 shows the comparison between the measurements and the simulations for the complete system. The general trend of the two curves is similar. However, the experimental curve is narrower than the simulation and the depression at solar noon is larger in the measurement than in the simulation. Another difference is that the harvest at 10:30 a.m. in the measurement is approximately 60% of the expected value according to the simulations. This difference is attributable to manufacturing inaccuracies, mainly in the funnel.

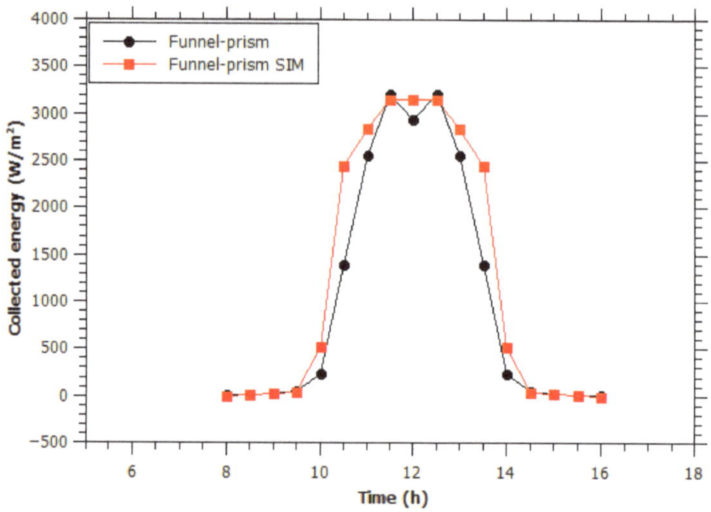

Figure 8. Comparison of experimental(Funnel-prism) and simulated data (Funnel-prism SIM) for the complete system.

In compound parabolic concentrators (CPCs) the acceptance angle and the concentration ratio are linked. Su et al. [6] obtained a concentration ratio of 4× with an acceptance angle of 29°. To have an acceptance angle comparative to ours of 47° they had to reduce the concentration ratio to 2.5×. In our system, using the Funnel, a geometric concentration ratio of 4× can be achieved with an acceptance angle of about 52°. Baig et al. [9] using a refractive tridimensional solid dielectric CPC together with a reflective casing designed the system for a concentration ratio of 3.6×. This system obtained under the best experimental conditions a power factor of 2.76. In contrast, our system reaches an experimental concentration factor of 3.2× using much less dielectric material.

As mentioned before, the comparison with linear systems is difficult. Vu et al. [17] designed a linear system with some elements similar to ours: a concentrating CPC with a prism on top of it. The system is intended to reduce costs using less accurate solar trackers instead of highly accurate ones. Different to ours, the CPC is a solid dielectric and concentrates due to total internal reflection. The system is designed for a high concentration ratio of 50× with an acceptance angle of 6°, thus making a direct comparison almost impossible. Li et al. presented a linear concentrator with a curved PMMA Fresnel lens on top [18] intended for air heating. The linear concentrator was an aluminium V-channel with 0.92 reflectivity. A concentration factor of 2.5× was achieved, but with an acceptance angle of only 19°. Our system has a higher concentration factor with more than twice the acceptance angle. It also has the advantage of being a simple system, since it uses a prism instead of a complicated curved Fresnel lens.

Coello et al. [3] describes only small variations of efficiency of a one-sun solar cell when used with concentrations less than 15×. A solar cell with an efficiency of 14% at one sun shows the efficiency of 15% at 3×. Taking this into account if used for photovoltaic applications the electrical harvest at the best performance of our system would imply more than three times the electrical production of the bare cell. Integrated over the whole day, the benefit would be about 1.6 times the energy production of the bare cell, as our system only collects sunlight for a little more than 4 h. All these considerations hold only if the solar cell is kept at 25° and the illumination pattern is reasonably homogeneous, a topic that requires further evaluation.

Figure 9 shows the optical efficiency of the system measured experimentally, at different times. For this evaluation, the light intensity at the entrance of the system and the integral measurement of light at the receiver was measured. The maximum optical efficiency is reached at 11:30 with a value of 69.36%. This value is slightly lower than that reported by other authors for static concentration systems (Baig et al. [9]).

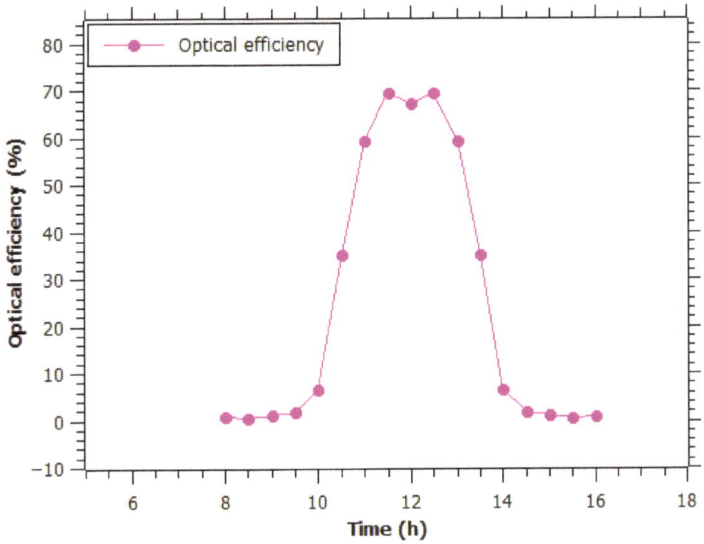

Figure 9. Experimental optical efficiency.

5. Conclusions

This work describes a double static concentrator, based on a refractive prism on top of a funnel with internally reflective walls. The geometry of the pieces is relatively simple and easy to manufacture. The best performance is achieved for acceptance angles between −22.5° and +22.5°, for which the concentration factor is larger than 1×. An effective concentration factor of 3.2× is reached at 11:30 h.

With the incorporation of the prism, the solar harvest of the funnel increases. The funnel has an acceptance angle of ∼ 52.5°, and with the addition of the prism, it increases to ∼ 60°. The double system has a maximum experimental optical efficiency of 69.36%, which could be increased by improving the manufacturing quality of the prism and the funnel.

Considering the time interval from 8 a.m. to 4 p.m., the system harvests 30.7% more energy than the flat surface. If the time interval considered is from 9:30 a.m. to 2:30 p.m., the increase in harvest is ∼77%. The incorporation of the prism represents an increase of ∼6% compared to the bare reflective system.

As the receiver geometry is flat, the concentration system can be used for photovoltaic or photo-thermal applications. Besides its simplicity, the system has the additional benefit of using a little amount of refractive material.

Author Contributions: Conceptualization, G.L.-Z. and G.R.-L.; methodology, G.L.-Z., G.R.-L.and H.Y.-M.; validation,R.V.-M. and D.A.P.-M.; investigation, G.L.-Z., G.R.-L.; resources, G.R.-L.; data curation, H.Y.-M. and D.A.P.-M.; writing—original draft preparation, G.L.-Z.; writing—review and editing, G.R.-L., H.Y.-M., R.V.-M. and D.A.P.-M.; All authors have read and agreed to the published version of the manuscript.

Funding: This research was funded by Instituto Politécnico Nacional, México, with grant SIP20220933.

Data Availability Statement: All data supporting the experiments and the conclusions of the study are included within the article, and additional data can be obtained from the corresponding author upon request.

Conflicts of Interest: The authors declare no conflict of interest.

References

1. Vu, N.H.; Shin, S. A Concentrator Photovoltaic System Based on a Combination of Prism-Compound Parabolic Concentrators *Energies* **2016**, *9*, 645. [CrossRef]
2. Amanlou, Y.; Tavakoli, H.T.; Ghobadian, B.; Najafi, G.; Mamat, R. A comprehensive review of Uniform Solar Illumination at Low Concentration Photovoltaic (LCPV) Systems. *Renew. Sustain. Energy Rev.* **2016**, *60*, 1430–1441. [CrossRef]
3. Coello, J.; Castro, M.; Anton, I.; Sala, G.; Vazquez, M.A. Conversion of Commercial Si Solar Cells to Keep Their Efficient Performance at 15 Suns. *Prog. Photovoltaics Res. Appl.* **2004**, *12*, 323–331. [CrossRef]
4. Pérez-Márquez, D.A. Sistema de Concentración Solar Semiestático para Celdas Fotovoltaicas. Ph.D. Thesis, Instituto Politecnico Nacional, Ciudad de Mexico, Mexico, 2015. (In Spanish)
5. Kaiyan, H.; Hongfei, Z.; Tao, T.; Xiaodi, X. Experimental investigation of high temperature congregating energy solar stove with sun light funnel. *Energy Convers. Manag.* **2009**, *50*, 3051–3055. [CrossRef]
6. Su, Y.; Pei, G.; Riffat, S.B.; Huang, H. A Novel Lens-Walled Compound Parabolic Concentrator for Photovoltaic Applications *ASME J. Sol. Energy Eng.* **2012**, *134*, 021010. [CrossRef]
7. Zheng, H.; Wu, G.; Tao, T.; Su, Y.; Dai, J. Combination of a light funnel concentrator with a deflector for orientated sunlight transmission. *Energy Convers. Manag.* **2014**, *88*, 785–793. [CrossRef]
8. Shanks, K.; Baig, H.; Senthilarasu1, S.; Reddy, K.S.; Mallick, T.K. Conjugate refractive–reflective homogeniser in a 500× Cassegrain concentrator: Design and limits. *IET Renew. Power Gener.* **2016**, *10*, 440–447. [CrossRef]
9. Baig, H.; Chemisana, D.; Sundaram, S.; Mallick, T. Conjugate refractive–reflective based building integrated photovoltaic system *Mater. Lett.* **2018**, *228*, 25–28. [CrossRef]
10. Gupta, M.; Kumar, D.A.; Kumar, V.; Singh, M.D. Solar concentrator based multipurpose sunlight harvesting system without tracking. *OSA Continuum.* **2019**, *2*, 667. [CrossRef]
11. Guiqiang, L. Design and Development of a Lens-walled Compound Parabolic Concentrator—A Review. *J. Therm. Sci.* **2019**, *28* 17–29.
12. Jin, R.; Zheng, H.; Ma, X.; Zhao, Y. Performance investigation of integrated concentrating solar air heater with curved Fresnel lens as the cover. *Energy* **2020**, *194*, 116808. [CrossRef]
13. Kumar K.B.; Gupta, M.; Singh, M.D. Efficient sunlight harvesting with combined system of large Fresnel lens segmented mirror reflectors and compound parabolic concentrator without tracking sun for indoor daylight illumination. *Renew. Energy* **2023**, *202*, 1198–1214. [CrossRef]
14. Luque-Zuniga, G.; Ramos, G.; Yee-Madeira, H.T.; Perez-Marquez, D.A.; Vázquez-Medina, R. Numerical investigation of three solar static concentrators with the same geometrical concentration rate. *Energy Sci. Technol. Manag.* **2021**, *1*, 16–20.
15. Wang, D.; Lan, T. Design of a gradient-index lens with a compound parabolic concentrator shape as a visible light communication receiving antenna. *Appl. Opt.* **2018**, *57*, 1510–1517. [PubMed]
16. Baker, A.K.; Dyer P.E. Refractive-Index Modification of PolyMethylMethAcrylate (PMMA) Thin Films by KrF-Laser Irradiation *Appl. Phys.* **1993**, *A57*, 543–544. [CrossRef]
17. Optical Analysis Program OptiCAD™, Opticad Corporation, 511 Juniper Drive, Santa Fe (NM), 87501 USA. Available online https://www.phenix.bnl.gov/phenix/WWW/publish/barish/publish/wasiko/Copy/Solar/OP_MANUAL.pdf (accessed on 7 December 2022).
18. Li, G.; Pei, G.; Su, Y.; Wang, Y.; Yu, X.; Ji, J.; Zheng, H. Improving angular acceptance of stationary low-concentration photovoltaic compound parabolic concentrators using acrylic lens-walled structure. *J. Renew. Sustain. Energy* **2014**, *6*, 013122. [CrossRef]

MDPI AG
Grosspeteranlage 5
4052 Basel
Switzerland
Tel.: +41 61 683 77 34

MDPI Books Editorial Office
E-mail: books@mdpi.com
www.mdpi.com/books

www.ingramcontent.com/pod-product-compliance
Lightning Source LLC
Chambersburg PA
CBHW041419210926
51850CB00013B/3311
9783725834655